Organic Polymers for Encapsulation of Drugs, Food Ingredients and Agrochemicals

Organic Polymers for Encapsulation of Drugs, Food Ingredients and Agrochemicals

Editors

Lorenzo Antonio Picos Corrales
Angel Licea-Claverie
Grégorio Crini

MDPI • Basel • Beijing • Wuhan • Barcelona • Belgrade • Manchester • Tokyo • Cluj • Tianjin

Editors

Lorenzo Antonio Picos Corrales
Facultad de Ingeniería Culiacán
Universidad Autónoma de Sinaloa
Culiacan
Mexico

Angel Licea-Claverie
Centro de Graduados e Investigación en Química
Instituto Tecnológico de Tijuana
Tijuana
Mexico

Grégorio Crini
Chrono-Environnement
Université de Franche-Comté
Besançon
France

Editorial Office
MDPI
St. Alban-Anlage 66
4052 Basel, Switzerland

This is a reprint of articles from the Special Issue published online in the open access journal *Polymers* (ISSN 2073-4360) (available at: www.mdpi.com/journal/polymers/special_issues/organic_poly_encapsulation).

For citation purposes, cite each article independently as indicated on the article page online and as indicated below:

LastName, A.A.; LastName, B.B.; LastName, C.C. Article Title. *Journal Name* **Year**, *Volume Number*, Page Range.

ISBN 978-3-0365-6783-9 (Hbk)
ISBN 978-3-0365-6782-2 (PDF)

Cover image courtesy of Lorenzo Antonio Picos Corrales

© 2023 by the authors. Articles in this book are Open Access and distributed under the Creative Commons Attribution (CC BY) license, which allows users to download, copy and build upon published articles, as long as the author and publisher are properly credited, which ensures maximum dissemination and a wider impact of our publications.
The book as a whole is distributed by MDPI under the terms and conditions of the Creative Commons license CC BY-NC-ND.

Contents

About the Editors vii

Preface to "Organic Polymers for Encapsulation of Drugs, Food Ingredients and Agrochemicals" ix

Melissa Garcia-Carrasco, Lorenzo A. Picos-Corrales, Erick P. Gutiérrez-Grijalva, Miguel A. Angulo-Escalante, Angel Licea-Claverie and J. Basilio Heredia
Loading and Release of Phenolic Compounds Present in Mexican Oregano (*Lippia graveolens*) in Different Chitosan Bio-Polymeric Cationic Matrixes
Reprinted from: *Polymers* **2022**, *14*, 3609, doi:10.3390/polym14173609 1

Ana Teixeira, Marisa P. Sárria, Inês Pinto, Begoña Espiña, Andreia C. Gomes and Alberto C. P. Dias
Protection against Paraquat-Induced Oxidative Stress by *Curcuma longa* Extract-Loaded Polymeric Nanoparticles in Zebrafish Embryos
Reprinted from: *Polymers* **2022**, *14*, 3773, doi:10.3390/polym14183773 21

María Carolina Otálora, Andrea Wilches-Torres and Jovanny A. Gómez Castaño
Spray-Drying Microencapsulation of Pink Guava (*Psidium guajava*) Carotenoids Using Mucilage from *Opuntia ficus-indica* Cladodes and Aloe Vera Leaves as Encapsulating Materials
Reprinted from: *Polymers* **2022**, *14*, 310, doi:10.3390/polym14020310 37

Mirela Kopjar, Ivana Buljeta, Mario Nosić, Ivana Ivić, Josip Šimunović and Anita Pichler
Encapsulation of Blackberry Phenolics and Volatiles Using Apple Fibers and Disaccharides
Reprinted from: *Polymers* **2022**, *14*, 2179, doi:10.3390/polym14112179 53

Hend Mohamed Hasanin Abou El-Naga, Samah A. El-Hashash, Ensaf Mokhtar Yasen, Stefano Leporatti and Nemany A. N. Hanafy
Starch-Based Hydrogel Nanoparticles Loaded with Polyphenolic Compounds of Moringa Oleifera Leaf Extract Have Hepatoprotective Activity in Bisphenol A-Induced Animal Models
Reprinted from: *Polymers* **2022**, *14*, 2846, doi:10.3390/polym14142846 71

Htet Htet Moe San, Khent Primo Alcantara, Bryan Paul I. Bulatao, Waraluck Chaichompoo, Nonthaneth Nalinratana and Apichart Suksamrarn et al.
Development of Turmeric Oil—Loaded Chitosan/Alginate Nanocapsules for Cytotoxicity Enhancement against Breast Cancer
Reprinted from: *Polymers* **2022**, *14*, 1835, doi:10.3390/polym14091835 91

Doha Berraaouan, Kamal Essifi, Mohamed Addi, Christophe Hano, Marie-Laure Fauconnier and Abdesselam Tahani
Hybrid Microcapsules for Encapsulation and Controlled Release of Rosemary Essential Oil
Reprinted from: *Polymers* **2023**, *15*, 823, doi:10.3390/polym15040823 109

Aurica Ionela Gugoasa, Stefania Racovita, Silvia Vasiliu and Marcel Popa
Grafted Microparticles Based on Glycidyl Methacrylate, Hydroxyethyl Methacrylate and Sodium Hyaluronate: Synthesis, Characterization, Adsorption and Release Studies of Metronidazole
Reprinted from: *Polymers* **2022**, *14*, 4151, doi:10.3390/polym14194151 125

Carlos Enrique Osorio-Alvarado, Jose Luis Ropero-Vega, Ana Elvira Farfán-García and Johanna Marcela Flórez-Castillo
Immobilization Systems of Antimicrobial Peptide Ib−M1 in Polymeric Nanoparticles Based on Alginate and Chitosan
Reprinted from: *Polymers* **2022**, *14*, 3149, doi:10.3390/polym14153149 . . . . **155**

Nareekan Chaiwong, Yuthana Phimolsiripol, Pimporn Leelapornpisid, Warintorn Ruksiriwanich, Kittisak Jantanasakulwong and Pornchai Rachtanapun et al.
Synergistics of Carboxymethyl Chitosan and Mangosteen Extract as Enhancing Moisturizing, Antioxidant, Antibacterial, and Deodorizing Properties in Emulsion Cream
Reprinted from: *Polymers* **2022**, *14*, 178, doi:10.3390/polym14010178 . . . . **169**

Wesam A. Hatem and Yakov Lapitsky
Accelerating Payload Release from Complex Coacervates through Mechanical Stimulation
Reprinted from: *Polymers* **2023**, *15*, 586, doi:10.3390/polym15030586 . . . . **189**

About the Editors

Lorenzo Antonio Picos Corrales

Lorenzo Antonio Picos Corrales is a Research Professor at the Autonomous University of Sinaloa (UAS, since 2012), Mexico, with scientific interests focused on synthesis and characterization of polymers for drug delivery systems and water treatment. He studied Chemical Engineering at the Faculty of Chemical-Biological Sciences-UAS. Master in Chemistry by the Technological Institute of Tijuana (IT-Tijuana), Mexico; his work obtained the 2008 Best Master's Thesis Award from the National Technological System. In 2012, he got his Ph.D. in Chemistry (Polymer Science) at the IT-Tijuana, Mexico, obtaining a DAAD (German Academic Exchange Service, 2010) scholarship for the "Sandwich PhD Model" at the Technische Universität Dresden, Germany. He has also received research training at the Leibniz-Institut für Polymerforschung (IPF) Dresden, Germany, and at the Autonomous University of Baja California, Tijuana, Mexico. At present, Lorenzo Picos is a full-time Professor at the Faculty of Engineering Culiacán-UAS in the Environmental Laboratory, and Lecturer in the Department of Civil Engineering; member of the Basic Academic Nucleus of the Postgraduate Program in Biological Sciences at the Faculty of Biology-UAS. Additionally, he is member of the National System of Researchers from Mexico. He is a Special Guest Editor of the journal *Polymers* and Assistant Editor of the journal *Hybrid Advances*, and has also contributed as a reviewer to more than 15 different journals.

Angel Licea-Claverie

Angel Licea-Claverie is a full-time Professor at the Instituto Tecnológico de Tijuana, Mexico, which is part of the Tecnológico Nacional de México. He has 20 years of experience in the study of temperature and pH sensitive polymers and 14 years of experience on controlled radical polymerization techniques. His research group have studied the controlled release of anti-inflammatory and anticancer drugs from sensitive micelles, polymersomes and nanogels developed. He is member of the American Chemical Society, Division of Polymer Chemistry, and of the Polymer Society of Mexico, for which he served as National President 2015–2017.

Grégorio Crini

Grégorio Crini is an environmental polymer scientist and research director at the Université de Franche-Comté, Chrono-environnement, Besançon, France. He received his Ph.D. degree in organic and macromolecular chemistry from the Université de Lille in 1995 under supervision of Professor Michel Morcellet. He then spent 2 years as a postdoctoral fellow at the G. Ronzoni Institute for Chemical and Biochemical Research in Milan, Italy, working with Research Director Giangiacomo Torri and Professor Benito Casu on the characterization of polysaccharide-based hydrogels by NMR techniques. In 1997, he joined the Université de Franche-Comté where he set up a research group working on the use of oligosaccharides (dextrins, cyclodextrins) and polysaccharides (starch, chitin, chitosan, cellulose) in industrial applications. In 2000, he completed his authorization to supervise research activities. His current interests focus on the design of new functional macromolecular networks and the environmental aspects of oligosaccharide, polysaccharide, and natural fibre (hemp, flax) chemistry for applied research. Grégorio CRINI has published more than 215 papers, a patent, and 17 books. He is a highly cited researcher (h-index of 45, more than 17,100 citations) and since 2021, he has been listed as one of the world's most cited researchers by the American university Stanford. Grégorio CRINI has also conducted consulting projects for many companies.

Preface to "Organic Polymers for Encapsulation of Drugs, Food Ingredients and Agrochemicals"

Active compounds (e.g., drugs and food ingredients) help to improve the quality of life around the world. In order to increase the effectiveness in each intended application, research groups have worked under the motivation of demonstrating that the encapsulation helps protect guest compounds against premature degradation and can enhance their transport in aqueous medium, increasing the percentage of substance that is available for a biological action in a controlled manner over time. For that purpose, the encapsulation of guest molecules has been accomplished using organic polymers (natural, synthetic, and semi-synthetic chains) as platforms. This is an interesting topic in which the scholars have designed systems that reach the desired efficiency reducing side effects and/or contamination derived from the overuse of active compounds. Hence, this reprint aims to provide an update regarding this topic. The content may be helpful for the colleagues and students working on encapsulation of active compounds.

The contribution of all authors to this collection is deeply appreciated. It is important to highlight the active participation of the reviewers and staff of the section "Polymer Networks". They were key persons who helped complete this Reprint Book.

Lorenzo Antonio Picos Corrales, Angel Licea-Claverie, and Grégorio Crini
Editors

Article

Loading and Release of Phenolic Compounds Present in Mexican Oregano (*Lippia graveolens*) in Different Chitosan Bio-Polymeric Cationic Matrixes

Melissa Garcia-Carrasco [1], Lorenzo A. Picos-Corrales [2], Erick P. Gutiérrez-Grijalva [3], Miguel A. Angulo-Escalante [1], Angel Licea-Claverie [4,*] and J. Basilio Heredia [1,*]

1. Nutraceuticals and Functional Foods Laboratory, Centro de Investigación en Alimentación y Desarrollo, A.C., Carretera a Eldorado Km. 5.5, Col. Campo El Diez, Culiacán 80110, Sinaloa, Mexico
2. Facultad de Ingeniería Culiacán, Universidad Autónoma de Sinaloa, Ciudad Universitaria, Culiacán 80013, Sinaloa, Mexico
3. Cátedras CONACYT-Centro de Investigación en Alimentación y Desarrollo, A.C., Carretera a Eldorado Km. 5.5, Col. Campo El Diez, Culiacán 80110, Sinaloa, Mexico
4. Centro de Graduados e Investigación en Química, Tecnológico Nacional de Mexico/Instituto Tecnológico de Tijuana, A.P. 1166, Tijuana 22000, Baja California, Mexico
* Correspondence: aliceac@tectijuana.mx (A.L.-C.); jbheredia@ciad.mx (J.B.H.)

Abstract: Mexican oregano (*Lippia graveolens*) polyphenols have antioxidant and anti-inflammatory potential, but low bioaccessibility. Therefore, in the present work the micro/nano-encapsulation of these compounds in two different matrixes of chitosan (CS) and chitosan-*b*-poly(PEGMA$_{2000}$) (CS-*b*-PPEGMA) is described and assessed. The particle sizes of matrixes of CS (~955 nm) and CS-*b*-PPEGMA (~190 nm) increased by 10% and 50%, respectively, when the phenolic compounds were encapsulated, yielding loading efficiencies (LE) between 90–99% and 50–60%, correspondingly. The release profiles in simulated fluids revealed a better control of host–guest interactions by using the CS-*b*-PPEGMA matrix, reaching phenolic compounds release of 80% after 24 h, while single CS retained the guest compounds. The total reducing capacity (TRC) and Trolox equivalent antioxidant capacity (TEAC) of the phenolic compounds (PPHs) are protected and increased (more than five times) when they are encapsulated. Thus, this investigation provides a standard encapsulation strategy and relevant results regarding nutraceuticals stabilization and their improved bioaccessibility.

Keywords: chitosan; polyethylene glycol methacrylate; oregano; phenolic compounds; nanoencapsulation; bioaccessibility

1. Introduction

The secondary metabolites present in plants, such as phenolic compounds, better known as phytochemicals, have gained great interest in recent years since they play an important role in the prevention of different diseases such as cancer, diabetes, and obesity, which are related to oxidative stress [1]. Mexican oregano (*L. graveolens*) is an endemic species from northwestern Mexico mostly known for its culinary uses but is also a rich source of phenolic compounds that can bring great benefits to human health due to their pharmacological properties, which include the anti-inflammatory, antifungal, and antibacterial activities, among others [2–4]. In this subject, essential oils and polyphenols are the major secondary metabolites found in oregano that are responsible for its biological properties [2,5]. Previous works have shown that oregano polyphenols have low bioaccessibility, and in-vitro digestion assays have demonstrated that phenolic acids, flavones, and flavanones of oregano seem to be susceptible to the pH changes during each digestion stage [6–8]. Thus, despite the bioactive properties of oregano polyphenols, they may be affected during their journey through the gastrointestinal (GI) tract, mainly due to pH changes favoring the ionization of these compounds or they are degraded, and thus the active compounds could lose their properties [1,9,10].

The low bioaccessibility of polyphenols is a subject highly reported; hence, different alternatives have been proposed to protect these metabolites from degradation. Among the technologies useful for this purpose are: (1) spray drying (for microencapsulation), (2) biopolymeric-type systems such as pectin, alginate, gums, and chitosan used to encapsulate different phytochemical compounds such as curcumin, thymol, and carvacrol [11–17]. Although most of these loading and release studies have been carried out with isolated compounds, it has also been found that when there is a combination of phenolic compounds, they can create synergism between them and enhance their activity [18].

The biopolymer chitosan (CS), which is obtained from a natural polysaccharide called chitin, represents a biocompatible and biodegradable platform with outstanding performance in sorption-oriented processes. This is one of the main polymeric matrixes used for the encapsulation of synthetic or natural agents due to polyelectrolyte character, which means that the charge of its functional groups can be modified depending on the pH of the medium, where the amino and the hydroxyl groups give this character [19–22]. However, this biopolymer is insoluble in normal deionized water, limiting biological applications; regarding this, some studies have been focused on modifying these molecules with different polymers to improve their solubility, such as polyethylene glycol (PEG). PEG has been one of the polymers with better biocompatibility [23]. Modified biopolymers have shown the capacity to adhere to peptide sequences [24], growth factors [25], and the ability to control mechanical properties regardless of polymerization conditions [26].

As it is well known, PEG is one of the few polymers approved by the U.S. FDA; moreover, the polyethylene glycol methyl methacrylate (PEGMA) properties are attributed to PEG due to their similar structures [27]. In addition, in case a certain percentage passes into the blood system, these particles avoid a response by the immune system, thus expanding the areas of application [28,29]. In some approaches, the use of PEGMA is related to the increase in the lower critical solution temperature (LCST) of the synthesized copolymers, increasing the LCST at temperatures greater than 32 °C, which ensures that the matrix can remain stable at temperatures higher than the LCST in aqueous medium [30].

Based on the abovementioned, in the present work the encapsulation of phenolic compounds extracted from the aerial part of Mexican oregano (L. graveolens) was carried out in systems based on chitosan (CS) and chitosan modified with PEGMA (Chitosan-block-poly(PEGMA$_{2000}$) (CS-b-PPEGMA), their release profiles at different pH levels were studied and their antioxidant activity before and after encapsulation was assessed. For that, a simple encapsulation method involving mechanical stirring was used.

2. Materials and Methods

2.1. Reagents

Polyethylene glycol methyl methacrylate 2000 g mol^{-1} (PEGMA), ammonium persulfate (APS, 98%, Sigma Aldrich, Toluca, Mexico), sodium chloride (NaCl, Jalmek, San Nicolás de los Garza, Mexico), chitosan (Low weight, 98%, Sigma Aldrich), glacial acetic acid (99.7%, Fermont, Monterrey, Mexico), deuteride chloride/deuterium oxide (D$_2$O/DCl 35% by weight, 99.9% deuterium, Sigma Aldrich), Folin-Ciocalteu reagent, aluminum chloride, potassium acetate, quercetin, DPPH radical, ABTS radical, potassium persulfate, HPLC grade water, and formic acid were purchased from Sigma-Aldrich (Toluca, Mexico). Moreover, sodium hydroxide (97.8%, Chemical Products of Monterrey SA de CV, Monterrey, Mexico) was purchased through a local provider.

2.2. Chitosan Purification

CS was dissolved in an aqueous acetic acid solution at 1% in volume, up to a concentration of 10 mg mL^{-1}; afterward, the mixture was filtered under reduced pressure with Büchner. CS was precipitated from the acidic solution using 1 M sodium hydroxide solution. The alkaline CS suspension was filtered under reduced pressure with a 5 µm particle cutoff filter. The CS was washed with deionized water until neutralization, frozen, and finally lyophilized (Labconco, FreeZone® 1 L, Kansas City, MI, USA).

2.3. Plant Material and Extraction of Free Polyphenols

Oregano (*L. graveolens*) was collected in Santa Gertrudis, Durango, Mexico. Dried oregano leaves were ground using an Ika Werke M20 grinder (IKA, Staufen, Germany) until a fine powder consistency was obtained. Oregano powder was stored at −4 °C until use. The extract of phenolic compounds was obtained using 25 mL of absolute ethanol for 1 g of oregano powder, where the mixture was stirred and homogenized on a stirring plate (Thermo Scientific Cimarec, Waltham, MA, USA) at room temperature for 18 h. Subsequently, the extract obtained was centrifuged at 10,000 rpm for 15 min, and the supernatant was collected and stored at 4 °C until use. This technique was performed repeatedly to obtain approximately 1 L of extract. Figure 1 is a schematic representation for the process involving the extraction of phenolic compounds.

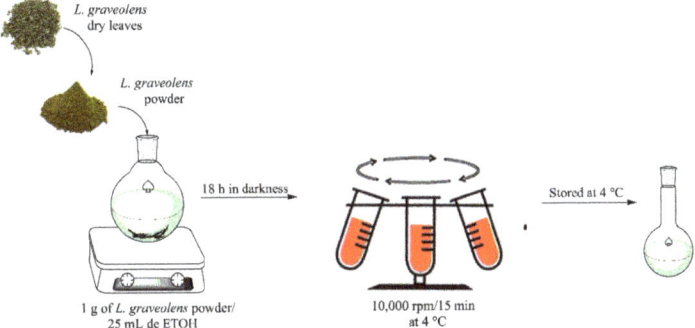

Figure 1. Schematic representation for free phenolic compounds extraction.

2.4. Characterization of the Extract of L. graveolens

2.4.1. Total Reducing Capacity

The total reducing capacity was evaluated through phenolic content analysis using the Folin–Ciocalteu (FC) method proposed by Swain and Hillis [31], with some modifications. The procedure consisted of mixing 10 µL of the samples, 230 µL of distilled water, and 10 µL of FC reagent in a 96-well microplate. The mixture was incubated for 3 min, and then 25 µL of 4N Na_2CO_3 were added, incubating again at room temperature for 2 h in the darkness. After incubation, absorbance at 725 nm was measured (Synergy HT microplate reader). Calculations were made using a gallic acid standard curve (from 0 to 0.4 mg mL^{-1}) and the results were expressed in milligrams of gallic acid equivalents per gram of powder obtained (mg AG g^{-1}). Each sample was measured in triplicate (n = 3).

2.4.2. Total Flavonoids Content (TFC)

The total flavonoid content was performed according to the methodology described by Ghasemi, et al. [32], with slight modifications. The process consists of taking 30 µL of the extract, then 250 µL of distilled water are added, and then 10 µL of aluminum chloride and 10 µL of 1 M potassium acetate, and it is left incubating in the darkness for 30 min; after incubation, absorbance is read at 415 nm in a Synergy HT microplate reader (Synergy HT, Bio-Tek Instruments, Inc., Winooski, VT, USA). The content of total flavonoids is determined from a quercetin standard curve (from 0 to 0.4 mg mL^{-1}); the results are expressed in equivalent mg of quercetin per gram of dry extract (mg QE g^{-1} of dry sample). Each sample was measured in triplicate (n = 3).

2.4.3. Antioxidant Capacity Methods

Inhibition of the 2,2-Diphenyl-1-Picrylhydrazyl Radical (DPPH)

This method uses the DPPH radical, which reduces its purple chromogen by the action of an antioxidant compound to hydrazine, a compound that colors a pale-yellow tone. This DPPH radical scavenging assay was carried out according to Karadag, et al. [33], for which 20 µL of the sample was placed in a 96-well flat-bottomed transparent microplate. Then, 280 µL of the DPPH radical were added and incubated for 30 min in the absence of white light. Finally, the absorbance at 515 nm was measured (Synergy HT microplate reader). A Trolox curve from 0.1 to 1 mmol TE g^{-1} was used to calculate the results, which are expressed as mmol Trolox equivalent per gram of powder (mmol TE g^{-1}). Each sample was measured in triplicate (n = 3).

Trolox Equivalent Antioxidant Capacity (TEAC)

The antioxidant capacity by the TEAC assay of the encapsulated sample was determined as described by Thaipong, et al. [34]. ABTS was dissolved in distilled water at a concentration of 7.4 mM (stock solution). The ABTS•+ radical was produced by mixing the ABTS stock solution with 2.6 mM potassium persulfate (1:1 v/v) and incubating the mixture in the dark at 25 °C for 12–16 h before use. Subsequently, the reaction solution was prepared by taking 100 µL of the radical and dissolving in 2900 µL of solvent to adjust the absorbance. For the assay, aliquots of 15 µL of extract and 285 µL of the reaction solution were added and homogenized using a vortex. Subsequently, it was incubated in the darkness for 2 h. After the time elapsed, the absorbance at 734 nm was read in a Synergy HT microplate using transparent 96-well flat-bottom plates. The reaction solution was taken as a blank. A Trolox curve from 0.1 to 1 mmol TE g^{-1} was used to calculate the results, which are expressed as mmol Trolox equivalent per gram of powder (mmol TE g^{-1}). Each sample was measured in triplicate (n = 3).

2.5. Identification of Phenolic Compounds by Ultra High-Resolution Liquid Chromatography/Mass Spectrometry (UPLC/MS)

Mass-liquid chromatography was used to carry out the separation for the identification of individual phenolic compounds from unencapsulated oregano extract. The analysis was performed in a class H UPLC unit (Waters Corporation, Milford, MA, USA) coupled to a G2-XS QT of mass analyzer (quadrupole and time of flight). The separation of phenolic acids was performed with a UPLC BEH C18 column (1.7 µm × 2.1 mm × 100 mm) at 40 °C, with gradient elution solution A (water-0.1% formic acid) and solution B (methanol), which is supplied at a flow rate of 0.3 mL min^{-1}. On the other hand, the separation of flavonoids was performed with a different set of conditions, including a UPLC BEH C18 column (1.7 µm × 2.1 mm × 100 mm) at 30 °C, with gradient elution solution A (water-0.05% formic acid) and solution B (acetonitrile), which is supplied at a flow rate of 0.3 mL min^{-1}. The ionization of the compounds was performed by electrospray (ESI), and the parameters used consisted of a capillary voltage of 1.5 kV, sampling cone: 30 V, desolvation gas of 800 (L h^{-1}), and a temperature of 500 °C. A collision ramp of 0–30 V was used. The Massbank of North America (MoNA) database was used for compound identification. The identification of phenolic compounds by UPLC was performed in duplicate (n = 2).

2.6. Synthesis of Chitosan-Block-Poly(PEGMA)

The methodology for the synthesis of CS-b-PPEGMA blocks was carried out as published by Ganji and Abdekhodaie [35], with slight modifications. Briefly, the preparation was done using conventional free radical polymerization with a weight ratio of 50:50 CS:PEGMA and 0.01 M free radical initiator (KPS). CS (0.5 g), PEGMA (0.46 mL, 0.5 g), KPS (0.135 g, 0.01 M), an inert atmosphere (N_2) in a three-necked flask with a magnetic stir bar, and 50 mL of water containing 1% (v/v) acetic acid were used. First, CS and initiator (KPS) were added followed by stirring for 30 min at 60 °C using an oil bath, and then the PEGMA$_{2000}$ was added dropwise, and the reaction was stirred (350 rpm) for 6 h. After the reaction time, the flask was removed from the oil bath and placed in a cold-water bath. For the purification of the solution, NaOH 4M was first added to the solution to precipitate the CS; next, the product was filtered and subsequently washed with acetone to remove the residual PEGMA$_{2000}$. Finally, the sample was dialyzed for 48 h, with changes of water periodically; after that, the recovered sample was lyophilized and weighed to determine the yield of the reaction, resulting around 70%. Chitosan-b-PPEGMA; ^{1}H-NMR (400 MHz, CDCl$_3$, δ, ppm): 4.00–4.30 (HOCH$_2$CH$_2$OCHCH$_2$OH of the polysaccharide ring of Chitosan), 3.87 (CH$_2$CH$_2$O of the PEG chain), 5.13 (CHNH of acetylglucosamine ring), 3.50 (CHNH$_2$ of glucosamine ring), 2.29 (CH$_2$ aliphatic from the CS backbone).

2.7. Preparation of Nanometric Polymer Aggregates

The copolymer aggregates were prepared through a direct dissolution consisting of the solubilization of the bulk copolymer CS-b-PPEGMA (10 mg) in distilled water (10 mL) under magnetic stirring at room temperature for 24 h. For CS, a solution of 1 wt% was prepared in 15 mL of water with 1% (v/v) of acetic acid under magnetic stirring at room temperature for 24 h.

2.8. Loading of Phenolics Compounds

The loading was performed based on a solvent evaporation method [36] and adapted from the methodology reported by Picos-Corrales, et al. [37]. Briefly, 10 mg of block copolymers were dissolved in 10 mL of distilled water, and 1.5 mg of phenolic compounds were dissolved in 5 mL of ethanol. The phenolic compounds (PPHs) solution was added dropwise into the polymer solution and left under magnetic stirring for 24 h. The phenolic compounds that were not loaded were removed by centrifugation for 20 min. The purified material was filtered using a disc filter with a pore size of 1 μM and then frozen and freeze-dried. The mass of phenolic compounds loaded in the CS and block copolymers was determined by preparing a 0.3 mg mL^{-1} solution in PBS pH 2, measuring the absorbance by UV analysis at a wavelength (λ_{max}) of 280 nm, and then quantified by using a calibration curve of phenolic compounds in PBS pH 2. The loading efficiency of phenolic compounds (LE) and the loaded PPHs content (LC) were calculated using Equations (1) and (2).

LE(%) = (mass of PPHs in polymer/mass of PPHs in loading solution) × 100   (1)

LC(%) = (mass of PPHs in polymers/mass of dry polymer) × 100   (2)

2.9. Measurements

Hydrogen nuclear magnetic resonance (^{1}H-NMR) spectra were collected on a Bruker AMX-400 (Bruker Corporation, Billerica, MA, USA) (400 MHz) spectrometer and reported in ppm using tetramethylsilane (TMS) as the NMR reference standard. The solvent used was deuterium chloride (37%), D$_2$O + DCl, for all samples.

Thermogravimetric analysis (TGA) was performed on a TA-Instruments Discovery-TGA equipment (TA-Instruments, New Castle, DE, USA). Measurements were performed for cationic matrices and phenolic compounds loaded by heating under nitrogen flow from room temperature up to 600 °C using a heating rate of 10 °C min^{-1}.

Differential Scanning Calorimetry (DSC) was performed on a TA Instrument DSC2000, New Castle, DE, USA). Measurements were performed for cationic matrices and were used to determine the melting point (T_m), the glass transition temperature (T_g) and thermal decomposition temperatures (T_d). For measurements, samples were cooled to −10 °C; maintained isothermally for 5 min, and afterward heated with modulation (±0.5 °C every 60 s) at a rate of 5 °C min^{-1} to 375 °C in a nitrogen atmosphere using. Two cycles of measurement were run, and the results reported corresponded to the second cycle.

Dynamic light scattering (DLS) measurements were carried out on 1.0 mg mL^{-1} block copolymer and 1 wt% of CS solutions at 25 °C using a Malvern Instruments Nano-ZS Nanosizer (ZEN 3690) (Malvern, Worcestershire, UK) equipment. The instrument is equipped with a helium-neon laser (633 nm) with size detection between 0.6 nm and 5 μm. DLS experiments were performed at the scattering angle of 90°, and the distribution of sizes was calculated using Malvern Instruments dispersion technology software, based on CONTIN analysis and Stokes-Einstein equation for spheres as usual.

UV–Vis spectra of Phenolics Compounds (PPHs) dispersions for the measurement were acquired using a UV-Vis Varian Cary 100 spectrophotometer system (Agilent Technologies, Santa Clara, CA, USA) at room temperature from aqueous dispersion.

The Zeta potential (ζ) of PPHs and PPHs-loaded polymeric matrixes dispersions (at 1 mg mL^{-1}) was measured using a Malvern ZetaSizer Nano ZS instrument (Malvern, Worcestershire, UK). The measurements were the average of three runs performed at 25 °C and distilled water.

2.10. In Vitro Release Studies

For release profile studies, 0.3 mg mL^{-1} of PPHs-loaded polymeric matrixes were dispersed in 10 mL of distilled water (pH ~5 for CS-*b*-PPEGMA@PPHs) or aqueous acetic acid solution (1% v/v) (pH ~4.5 for CS@PPHs) and then added to a dialysis tube (Spectra/Por® MWCO: 12–14 KDa, diameter 10 mm, Spectrum, Los Angeles, CA, USA). The dialysis tube was introduced into a 100 mL release medium, with mixture of 30% ethanol and 70% PBS inside an Erlenmeyer flask. The flask was placed in a shaking bath (Shel Lab, model SWBR17, Sheldon Manufacturing, Inc., Cornelius, OR, USA), operating at 37 °C and a shaking speed of 100 rpm. Medium aliquots of 2 mL were taken out at different times and replaced by 2 mL of fresh PBS at every sampling point. The released fraction of phenolic compounds was calculated from UV measurements at λ_{max} = 280 nm and 320 nm depending on the pH of the release medium and was then quantified using a calibration curve of phenolics compound in PBS.

2.11. In Vitro Gastrointestinal Digestion

A simulation of gastrointestinal digestion was performed according to the static in-vitro digestion method reported by Brodkorb, et al. [38]. This standardized procedure simulates the physiological conditions in the mouth, stomach, and small intestine, mimicking the chemical and pH conditions. Briefly, 1 mg of sample was mixed with simulated salivary fluids (SSF), then pH was adjusted to pH 7 with 6M NaOH; after that, the mixture was incubated for 5 min at 37 °C. Then, 1 mL of simulated gastric fluids (SGF), pH was adjusted to pH 3 and incubated for 2 h at 37 °C. Finally, 2 mL of simulated intestinal fluids (SIF) were added, pH was adjusted to pH 7, and the mixture was incubated for 2 h at 37 °C. In the final digestion step, the samples were centrifuged (9390× g at 4 °C for min), and the supernatant was collected and freeze-dried. After that, samples were resuspended in ethanol for antioxidant capacity assays.

3. Results and Discussion

3.1. Characterization of the Phenolic Compounds Present in L. graveolens

The results obtained from the nutraceutical characterization of the phenolic compounds in *L. graveolens* obtained by maceration in ethanol are found in Table 1, where the extracted compounds show antioxidant capacity. It can be seen that the extracted compounds have antioxidant activity against the different free-radical DPPH, AAPH, and ABTS. Comparing our results with the literature (TRC, 51.26 ± 2.36 mg EAG g^{-1}; TFC, 11.80 ± 0.12 mg QE g^{-1}; DPPH, 500.54 ± 9.63 µmol TE g^{-1}; ORAC, 812.31 ± 35.46 µmol TE g^{-1}; and ABTS, 350.07 ± 0.45 µmol TE g^{-1}) [9], it was observed that the TRC of both extracts is very similar, showing differences mainly in the TFC where a lower value was registered with the extraction in absolute ethanol. Cortes-Chitala, et al. [39], reported that the TRC of the *L. graveolens* was around 99.71 mg AGE g^{-1} dried weight; the higher reducing activity might be mainly because they used ethanol:water (58:42) as a solvent for extraction, plus the region where the sample was taken is different [39].

Table 1. Nutraceutical results of the ethanolic extract of oregano (*L. graveolens*).

TRC * (mg GAE g^{-1} Dried Oregano Leaf)	TFC ** (mg QE g^{-1} Dried Oregano Leaf)	DPPH ***	ORAC ***	ABTS ***
		(µmol TE g^{-1} Dried Oregano Leaf)		
50 ± 5.5	0.59 ± 0.019	339 ± 26.56	2639 ± 12.7	476 ± 1.27

* Total Reducing Capacity (TRC), ** Total Flavonoids Content (TFC) *** DPPH and ABTS are scavenging capacity, and ORAC is antioxidant capacity. Values represent the mean ± standard deviation (n = 3).

Characterization by UPLC-MS

The *L. graveolens* ethanolic extract was characterized by UPLC-MS, where it was possible to observe that the sample had a diversity of flavonoids and some phenolic acids (Table 2). Some of the identified compounds are flavones such as luteolin-glycoside, cosmoside, eriodictyol, and cirsimaritin; others are flavanones such as naringin, naringenin, and pinocembrin (Figure 2). The phenolic profile in this study is consistent with previous studies in *L. graveolens* extracts [9,39,40]. However, the distribution and content of phenolic content in our study could be different, owing to factors such as recollection time and the extraction method used for phenolic recovery. Moreover, in previous studies, the most abundant phenolics in methanolic extracts were naringenin, kaempferol-glucoside, kaempferide, caffeic acid, cirsimaritin, kaempferol, and taxifolin; most of these molecules were also found in our extracts. These metabolites have been associated with anti-inflammatory, antiapoptotic, and antimicrobial activity [4,39,41].

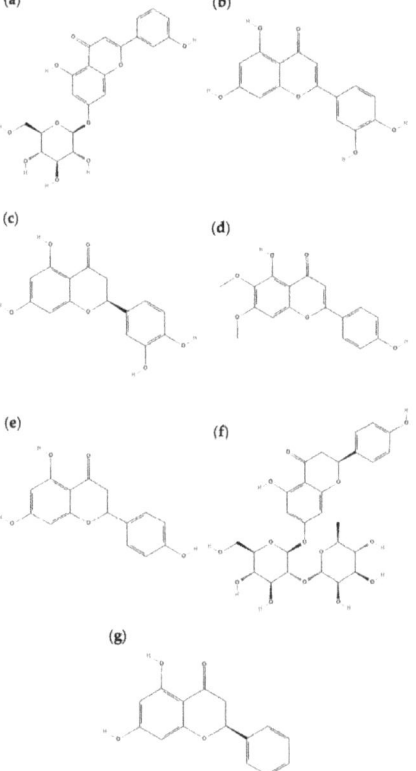

Figure 2. Chemical structures of some compounds contained in the phenolic extract characterized by UPLC-MS: (**a**) cosmoside, (**b**) luteolin, (**c**) eriodyctiol, (**d**) cirsimaritin, (**e**) naringenin, (**f**) naringin, (**g**) pinocembrin.

Table 2. Flavonoids from the ethanolic extract of *L. graveolens* characterized by UPLC-MS.

Compound	Compound Type	Molecular Mass [M-H]⁻
Luteolin-glucoside	Flavone	447.1
Cosmoside	Flavone	431.1
Naringin	Flavanone	579.17
Quercetin	Flavonol	301.04
Kaempferol	Flavonol	285.04
Eriodictyol	Flavone	287.06
Naringenin	Flavanone	271.06
Pinocembrin	Flavanone	255.07
Taxifolin	Flavanonol	305.05
Cirsimaritin	Flavone	313.07

3.2. Synthesis and Characterization of Cationic Matrixes
3.2.1. Chitosan

After performing the purification of CS by precipitation in aqueous medium, it was characterized by ^{1}H-NMR (Figure 3). The assignment of the hydrogens corresponding to CS, which is a polysaccharide containing units of glucosamine (GlcN) and acetyl-glucosamine (GlcNAc), is the following: at 2.36 ppm, the signal corresponds to the hydrogens of the methyl group of the acetyl-glucosamine units (H_7); a signal with an intensity of around 2.67 ppm can also be observed, which corresponds to the signal of residual acetic acid (HAc), which comes from the purification of CS and was difficult to remove completely. Around 3.5 ppm, a multiplet is observed, and the signal is associated to the alpha hydrogen signal to the amino group of the repeating unit of GlcN (H_2); the signals located between 3.55 and 4.5 ppm correspond to the hydrogens from the polysaccharide ring (H_3-H_6); and the signal at 5.14 ppm corresponds to the H_1 of the repetitive unit of glucosamine [42].

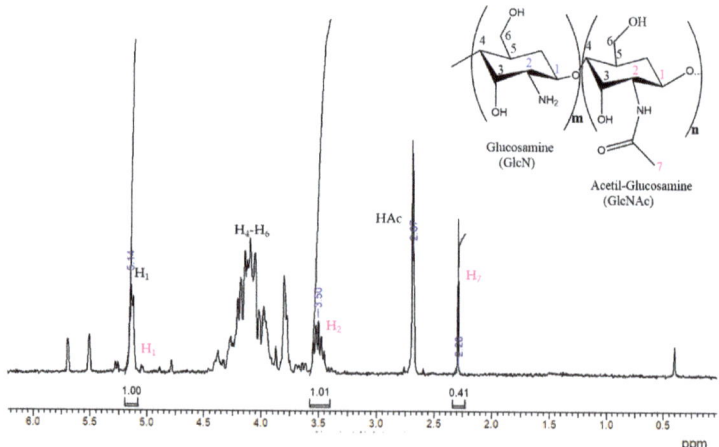

Figure 3. ^{1}H-NMR spectrum of chitosan in D_2O/DCl at 30 °C.

The degree of deacetylation (*DD*) of CS was calculated with Equation (3), using the integrations under the curve of the signal of H_1 (GlcN) and H_7 (GlcNAc), which gave us an 88% deacetylation degree of CS.

After obtaining the intrinsic viscosity, the viscosity-average molecular weight was calculated using the Mark–Houwink–Sakurada constants, with *K* and *a* being 2.14×10^{-3} mL g^{-1}

and 0.657, respectively [43]. As a result, the viscosimetric molecular weight obtained was 114 KDa, classified as relatively low molecular weight, which is suitable for using CS as a nanocarrier of different drugs.

$$DD = \left\{ \left(\frac{H_{1(GlcN)}}{H_{1(GlcN)} + \frac{H_{7(GlcNAc)}}{3}} \right) \times 100 \right\} \qquad (3)$$

3.2.2. Chitosan-b-PPEGMA

The block copolymer based on CS and PEGMA$_{2000}$, was prepared following the methodology reported by Ganji and Abdekhodaie [35] applying some modifications. The product was characterized by ^{1}H-NMR (Figure 4). In the spectrum, the intense signal can be highlighted at 3.87 ppm that corresponds to the hydrogens (H$_9$ and H$_{10}$) of the repetitive ethoxy units in PEGMA$_{2000}$; and the signal of lower intensity at 3.55 ppm (H$_{11}$) corresponds to the hydrogen of methylene bound to oxygen at the end of the PPEGMA chain. Besides, the signals previously described for the chitosan backbone are also observed. This analysis indicated that the synthesis of chitosan-b-poly(PEGMA$_{2000}$) was successfully accomplished.

The percentage of PEGMA$_{2000}$ present as polyPEGMA$_{2000}$ block was determined by ^{1}H-NMR. For that, the integration from 3.6 ppm to 4.4 ppm was related to the signal at 5.15 ppm, which turned out to be higher than 90%, which makes the copolymer very soluble in an aqueous medium. This is an important factor when using this type of biopolymer in different pharmacological applications because of its good reservoir capacity for loading different drugs. Furthermore, the average size of the CS backbone was shortened during the free-radical high-temperature synthesis; this was also determined by viscosimetry in an experiment without PEGMA$_{2000}$. The results showed that the size of the polysaccharide chains decreased by more than 90%, therefore it is not surprising that the CS content in the block copolymer is much lower than expected from the synthetic recipe [44].

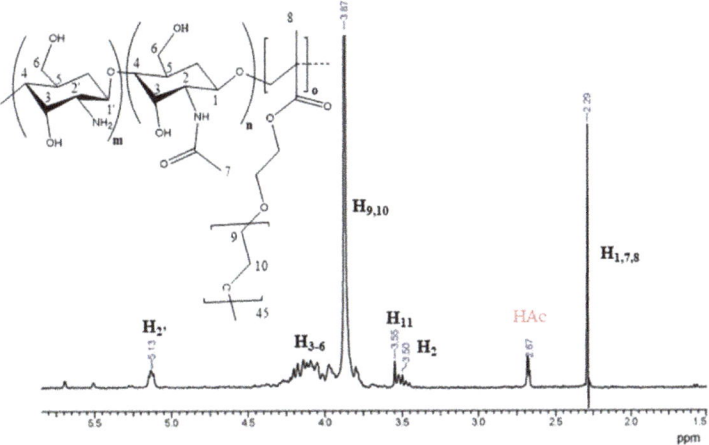

Figure 4. ^{1}H-NMR spectrum of chitosan-b-poly(PEGMA$_{2000}$) in D$_2$O/DCl at 30 °C.

3.2.3. Thermal Characterization

A change in the structure of CS can be observed in the differential scanning calorimetry analysis (Figure 5); in the respective thermogram, only one thermal event can be observed at 303 °C, which is characteristic of the decomposition of glucosamine groups [45,46]. On the other hand, in the thermogram of the CS-b-PPEGMA blocks, two transition temperatures can be observed: the first at −12.5 °C, representing the glass transition temperature (T$_g$) of PEGMA$_{2000}$ (an expanded view is presented), and the second at 47 °C representing the melting point (T$_m$) also of PEGMA$_{2000}$ [35,47]; a third thermal event observed at

242 °C is related to the decomposition temperature of PEGMA and CS. The decrease in the decomposition temperature of CS could be attributed to the smaller size of the CS chain after the synthetic procedure.

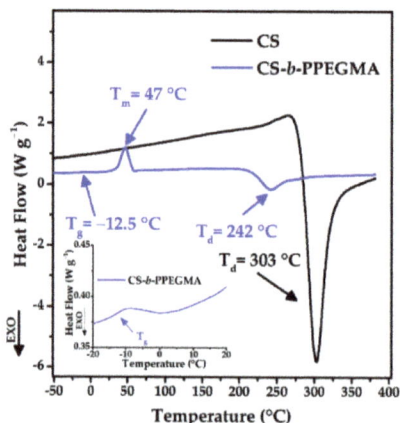

Figure 5. DSC thermograms of CS and CS-*b*-PPEGMA.

Thermogravimetric analysis shows differences between both compounds, highlighting that the decomposition temperatures of CS and the CS-*b*-PPEGMA block copolymer are similar (Figure 6a). Figure 6b shows the decomposition of these biopolymers; in the case of CS, a single decomposition around 300 °C is shown [48]. As for the CS-*b*-PPEGMA block copolymer, four decomposition steps can be observed: the first one at temperatures below 100 °C, which represents the loss of moisture from the sample; at 125 °C, this first decomposition step can be attributed to the low molecular weight chains of CS that formed after breaking the backbone, the next step at 227 °C corresponds to the first decomposition of PEGMA$_{2000}$, the third step at 334 °C corresponds to the decomposition of CS chains, and finally a weight loss at 456 °C, which would correspond to the second decomposition of PEGMA$_{2000}$ [49,50]. In both systems, a residue of around 40% is observed, being attributed to the formation of coke.

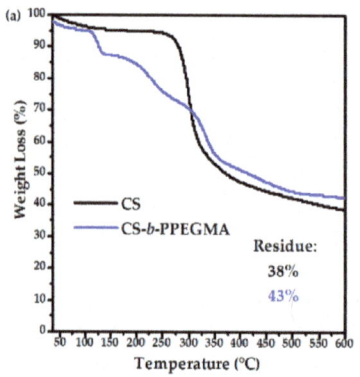 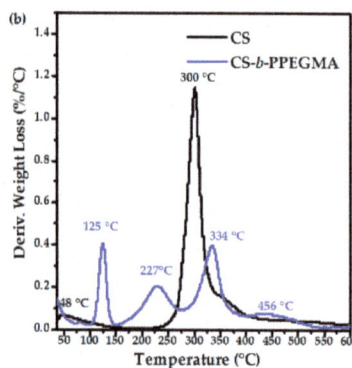

Figure 6. Thermograms of CS and CS-*b*-PPEGMA: (a) TG-thermogram and (b) DTG-thermogram.

3.3. Loading of Phenolic Compounds into Different Cationic Polymers

After loading the different matrixes, samples were analyzed by UV-vis spectroscopy to determine the content of phenolic compounds (Table 3). It was observed that CS had a loading efficiency between 90–99%, while for the blocks of CS-*b*-PPEGMA the efficiency was lower from 50–60%; on the other hand, the loading content was 65% and 99%, respectively [51,52]. The pK$_a$ of phenolic compounds varies around 6–9 [53–58], and the solubility of these is affected depending on the pH of the medium; considering that, their loading was carried out in an aqueous medium acidified with 1% v/v of acetic acid (pH = 4.5), it could be said that some of the compounds were partially soluble. An opposite case for CS with a pK$_a$ around 6.17–6.51 [59–61], is completely soluble at acidic pH due to its protonated amines. When those amines are in contact with the hydroxyl groups (-OH) of the phenolic compounds (Figure 2) they can form hydrogen bonds or present electrostatic interactions (neutralization of charges) (Figure 7). Due to this molecular interaction in the case of CS, which has a higher content of free amines, it can encapsulate a higher concentration of the phenolic compounds; on the other hand, in the CS-b-PPEGMA blocks, the shorter CS chains and the fact that the ends are blocked with PPEGMA leads to less free amines per chain that can be available for interaction with the -OH of the phenolic compounds (Table 3). Moreover, a significant difference was registered in ζ change when a simple mixture of matrix and PPHs was prepared as compared with PPHs-loaded polymeric matrixes, so the stabilization and loading with the different matrixes can be inferred.

Table 3. Loading Efficiency (LE) and Loading Content (LC) of PPHs in polymer matrixes; zeta potential (ζ) and particle sizes of loaded (@PPHs) and unloaded (Matrix) polymer matrixes. Values represent the mean ± standard deviation (n = 3).

Polymer	LE (%)	LC (%)	ζ (mV)			D$_h$ (nm)	
			Mixture	@PPHs	Matrix	@PPHs	Matrix
CS	90–99	65	10.2 ± 4.32	50.4 ± 3.27	55.4 ± 5.07	1106 ± 87	955 ± 75
CS-b-PPEGMA	50–60	99	7.91 ± 4.37	−15.5 ± 4.57	−9.07 ± 4.86	458 ± 0.01	190 ± 17
PPHs			−8.79 ± 4.29	-	-	-	-

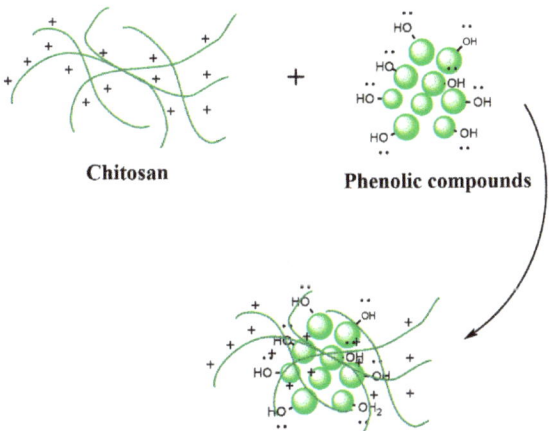

Figure 7. Scheme of the possible interaction between CS and the polyphenolic compounds.

To determine if these matrixes protected the phenolic compounds, the stability of these systems before and after loading was determined by zeta potential (Figure 8a) and particle size (Figure 8b) in measurements for about 15 days. From these experiments, the most stable system resulted to be the CS matrix, whilst the particle size of the CS-b-PPEGMA blocks increased with the time, although the latter presented good size stability for up to 5 days. On the other hand, the study of zeta potential for the two different systems and the unencapsulated polyphenols can be observed, and it was found that the zeta potential (positive or negative) of the samples increased with the time. This effect may be because most of these compounds are not soluble in aqueous medium, so they precipitate, and those partially soluble compounds can be ionized [9,12].

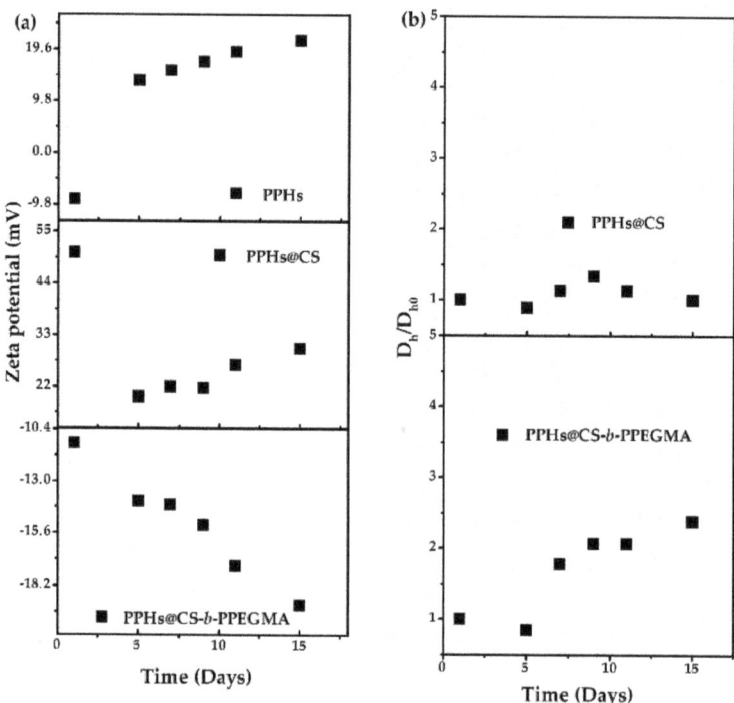

Figure 8. Stability study of systems loaded with phenolic compounds from *L. graveolens* zeta potential (a) and particle size (b).

On the other hand, in the case of CS systems, it can be observed how there was a drastic decrease in the surface charge after day 3 (compared with day 5), which could be caused by a decrease in the number of free amino groups improving the polymer-organic compounds interactions, maintaining similar particle size and colloidal stability along the days (Figure 8). In the case of the CS-PPHs system (PPHs@CS), the variations in the zeta potential did not significantly affect the size, and formation of more complex aggregates was not detected. In general, surface charge variations depend on free groups in their ionized form or forming hydrogen bonds, belonging to the polymer or the nutraceuticals. Figure 7 shows the possible interaction between CS and the polyphenolic compounds.

Thermal Characterization of Phenolic Compounds

The stability of the PPHs in the two different cationic matrixes was determined by thermogravimetric analysis, as is shown in Figure 9. It is shown that the decomposition signals of the single PPHs sample were not overlapped with the signals of the compounds loaded in the different cationic matrixes (CS@PPHs and CS-*b*-PPEGMA@PPHs). Also, a displacement of the decomposition temperatures of the PPHs can be observed, which would be associated with the cationic matrixes protecting these systems through different interactions, which could also impart on them higher thermal stability [62].

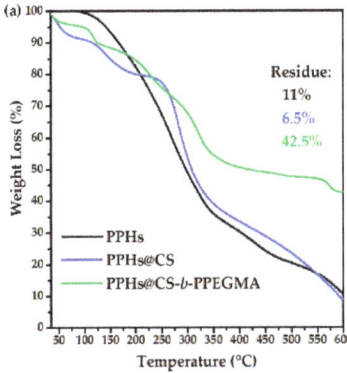 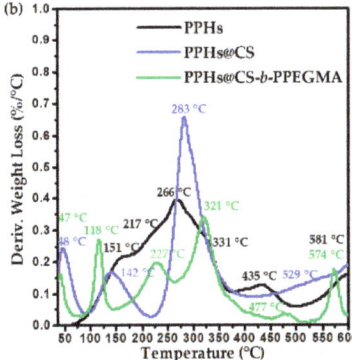

Figure 9. Thermal stability of loaded and unloaded PPHs: (a) TG-thermogram and (b) DTG-thermogram of Polyphenolics Compounds (PPHs), PPHs loaded in CS (CS@PPHs), and PPHs loaded in CS-*b*-PPEGMA (CS-*b*-PPEGMA@PPHs).

3.4. Release of Phenolic Compounds

The release profiles were performed by simulating pH conditions in the gastrointestinal tract using phosphate buffers with pH 7.4, 2, and 8 (Figure 10); each simulation was incubated for 24 h. Subsequently, a continuous release was also performed (pH 2, 6.8, 7.4 and 8); in the same way, the surface charge of both systems was determined at the different pH to study their behavior. In Figure 10a, it can be seen that the phenolic compounds show a higher percentage of release at pH 8, while at pH 7.4 a drop in their concentration is observed after 24 h of release; furthermore, the maximum release percentage of these compounds was 45%. It is relevant to specify that the rest of the PPHs precipitate as they are not completely soluble in an aqueous medium. According to the release profiles recorded from the two different matrixes, it can be seen that the CS-*b*-PPEGMA blocks provide a more controlled release of these compounds, releasing more than 80% of the PPHs after 24 h. On the other hand, CS only releases a maximum of around 40% of the guest compounds, and this is mainly due to the solubility of CS in the different pH values [63], observing that at pH 2, CS is completely soluble and releases a higher percentage of the phenolic compounds.

The release in all cases started quickly at the beginning and then a slower release of nutraceuticals compounds over time was recorded; this could result from a fraction of the molecules just being adsorbed onto the matrixes' surface, triggering an initial burst release [64].

In Figure 11, the continuous release profiles of the PPHs loaded in CS and CS-*b*-PPEGMA are shown—the systems presented different behavior when continuous release was performed, changing the pH and simulating the gastrointestinal tract passage as compared to individual pH values. It is observed that the maximum release for the CS-*b*-PPEGMA matrix was less than 20%, and in the case of CS a maximum of 8% was released; this shows a gap where more than 80% of these compounds are protected within the different matrixes [65]. The difference between the release of the different matrixes is mainly due to their solubility in the aqueous medium, as well as the solubility/stability of the phenolic compounds released in the medium; these tend to precipitate when they are in an aqueous medium. Nevertheless, there is still a possibility that the enzymes present in *in-vivo* systems can degrade these matrixes and release a higher percentage of these compounds [13].

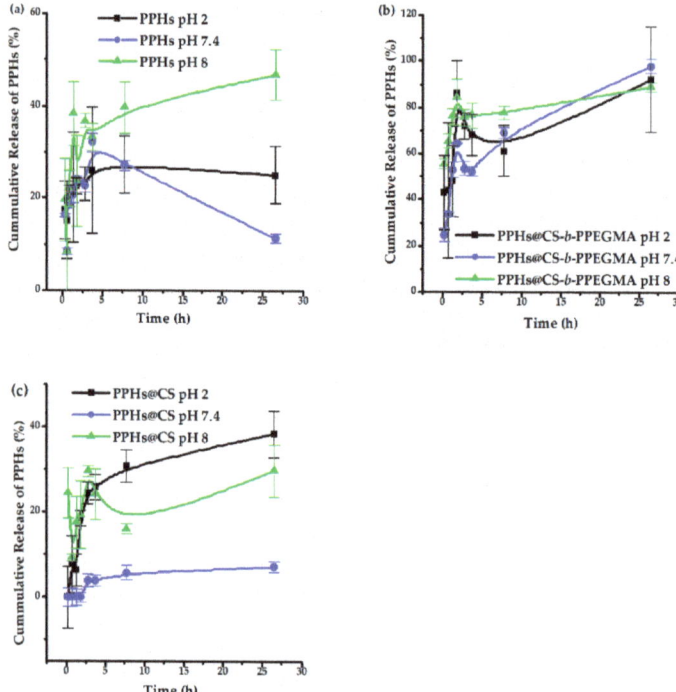

Figure 10. Release profiles of (**a**) PPHs, (**b**) PPHs@CS-*b*-PPEGMA, and (**c**) PPHs@CS. Values represent the mean ± standard deviation (n = 3).

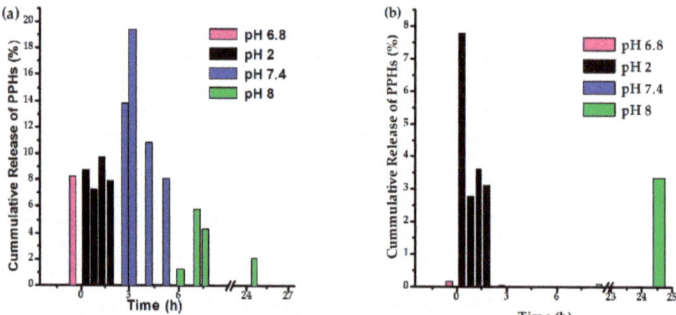

Figure 11. Cumulative release profiles of PPHs from CS-*b*-PPEGMA (**a**) and CS (**b**). Note: the cumulative release is only within a pH value; after every pH change, a new cumulative release is calculated.

3.5. In-Vitro Gastrointestinal Digestion

Encapsulation has been used both to enhance the bioaccessibility of phenolic compounds and to protect them from degradation [12]. In this sense, the properties of the charged and unloaded phenolic compounds were determined in both systems after the gastric and intestinal phase by the total reducing capacity and TEAC assays to evaluate if the compounds maintained their antioxidant properties.

Table 4 shows a summary of the results obtained, and it can be observed that both the total reducing capacity and the antioxidant capacity of the PPHs are diminished in the SGF phase. This may be mainly due to the ionization of the phenolic compounds present, which causes a decrease in their activity towards the free radical target; in the intestinal phase, a decrease is seen mainly by the ABTS assay, which may have been because a large proportion of the released phenolic compounds could have been degraded or ionized in the gastric phase [9]. During in-vitro gastrointestinal digestion, encapsulated polyphenols can be partially released from the matrix; in this case, the matrix material can be affected by the pH changes during each digestive stage, CS matrix is most affected by this change in pH since it changes its solubility as it passes through the GI tract. At low pH, it is completely soluble, while at neutral-basic pH it is the opposite. The released polyphenols from the CS matrix are exposed to the different pH in the simulated gastrointestinal solutions, undergoing deprotonation of their chemical structures and partial hydrolysis, which has also been reported. Deprotonated polyphenols can affect the way they interact with the targeted free radicals in each antioxidant assay. The rate at which polyphenols are affected by the gastrointestinal environment is dependent on many factors, and one of the most reported is the matrix in which they are contained. In this work, we used the TEAC assay, which is based on the transfer method and depends on the reducing capacity of the evaluated substance [66,67].

Table 4. Results of the total reducing capacity (TRC) and Trolox equivalent antioxidant capacity (TEAC) of phenolic compounds loaded in chitosan (PPHs@CS), phenolic compounds loaded in CS-block-poly(PEGMA) (PPHs@CS-b-PPEGMA), and non-encapsulated phenolics (PPHs) after gastric and intestinal in-vitro digestion.

	TRC * (mg QE g^{-1} Dried Oregano)		TEAC ** (μmol TE g^{-1} Dried Oregano)	
	SGF	SIF	SGS	SIF
PPHs@CS	81.19 ± 4.18 [b]	111.70 ± 9.90 [a]	163.12 ± 79.11 [b]	446.56 ± 9.01 [a]
PPHs@CS-b-PEGMA	89.03 ± 2.39 [b]	135.50 ± 3.15 [a]	79.15 ± 23.15 [b]	415.79 ± 7.07 [a]
PPHs	16.68 ± 1.41 [b]	50.54 ± 2.90 [a]	104.19 ± 0.22 [a]	100.82 ± 7.99 [a]

* Total Reducing Capacity (TRC), ** Trolox equivalent antioxidant capacity (TEAC) is a scavenging capacity. Values represent the mean ± standard deviation (n = 3). [a, b] means are significantly ($\alpha > 0.05$) different according to the Tukey test.

In this approach, the encapsulated phenolic compounds have a higher activity than when they are not protected by one of the different matrixes [68]. In the same way, little difference can be observed between both matrix systems, and this may be mainly due to the effect of the surface charge that each system presents, causing the final TCR to be higher in the CS-b-PPEGMA matrix, having a negative partial charge because the amines are not protonated, and a higher concentration of these compounds remains inside; nonetheless, CS has a positive surface charge due to the number of amines that can be protonated, showing lower activity. The opposite is seen in the results obtained by TEAC, but in this case, it could be because the CS has a greater amount of compounds within the matrix; namely, there is a greater PPHs fraction compared to the block copolymers [52].

Moreover, it has been reported that CS-based polymers improve the disturbance in glucose metabolism in diabetic mice, such as reducing blood glucose, reversing insulin resistance, enhancement of the colonic epithelial integrity, and a modulatory effect on the gut microbiota. In addition, CS-based polymers loaded with phenolic compounds have been proposed as promising nanochemopreventive agents against cancer. Also, CS has been reported as a permeability enhancer, which must be addressed to evaluate if phenolic-loaded CS can increase the cell permeability of phenolics. Thus, CS and CS-based polymers loaded with phenolic compounds can result in multiple beneficial effects in the development of functional foods [11,41,69,70].

4. Conclusions

From the characterization, it can be concluded that the synthesis of block copolymers based on CS and PEGMA was successfully accomplished. The phenolic compounds were efficiently encapsulated with the CS and CS-*b*-PPEGMA matrixes, which was verified by determining the surface charge (zeta potential) of the colloidal systems before and after the loading. This change is derived from the type of interaction taking place between the -OH groups of the phenolic compounds and the $-NH_2$ onto the CS backbone in the different aqueous medium. PEGMA$_{2000}$ led to the development of formulations having smaller particle size; however, CS plays a key role improving the matrix–guest compounds interactions and colloidal stability. The CS-*b*-PPEGMA matrix allowed a higher percentage of release, which can be attributed to the improved solubility of these platforms, as compared with single CS, and during cumulative release experiments only a maximum of 20% of active compounds were released from this platform; namely, 80% of the guest substances remained trapped in the polymeric matrix after the gastric and intestinal phases, indicating that the CS-*b*-PPEGMA may offer greater protection for the active compounds in the gastric phase. This could be related to the negative surface charge of this matrix undergoing delayed protonation, and an inverse effect was exhibited by CS. Thus, the results demonstrated that the encapsulation of nutraceuticals can improve their stability, solubility, and activity, and the CS-*b*-PPEGMA matrix can help improve the controlled release of compounds present in Mexican oregano (*Lippia graveolens*).

Author Contributions: Conceptualization, J.B.H. and A.L.-C.; Investigation, M.G.-C.; Resources, J.B.H. and A.L.-C.; Supervision, J.B.H. and A.L.-C.; Writing—original draft, M.G.-C.; Writing—review & editing, J.B.H., A.L.-C., E.P.G.-G., L.A.P.-C. and M.A.A.-E. All authors have read and agreed to the published version of the manuscript.

Funding: The APC was funded by CIAD, Mexico.

Institutional Review Board Statement: Not applicable.

Informed Consent Statement: Not applicable.

Data Availability Statement: The data that support the findings of this study are available from the corresponding author upon reasonable request.

Acknowledgments: We thank MC. L. Contreras (UPLC-MS measurements, CIAD-Culiacán) and V. Miranda (NMR measurements, ITT NMR facilities funded by CONACYT Grant: INFR-2011-3-173395). E.P.G.-G. thanks CONACYT for the Catedras Project #397.

Conflicts of Interest: The authors declare no conflict of interest.

References

1. Peanparkdee, M.; Iwamoto, S. Encapsulation for Improving In Vitro Gastrointestinal Digestion of Plant Polyphenols and Their Applications in Food Products. *Food Rev. Int.* **2020**, *38*, 335–353. [CrossRef]
2. Leyva-Lopez, N.; Gutierrez-Grijalva, E.P.; Vazquez-Olivo, G.; Heredia, J.B. Essential Oils of Oregano: Biological Activity beyond Their Antimicrobial Properties. *Molecules* **2017**, *22*, 989. [CrossRef] [PubMed]
3. Herrera-Rodriguez, S.E.; Lopez-Rivera, R.J.; Garcia-Marquez, E.; Estarron-Espinosa, M.; Espinosa-Andrews, H. Mexican oregano (*Lippia graveolens*) essential oil-in-water emulsions: Impact of emulsifier type on the antifungal activity of *Candida albicans*. *Food Sci. Biotechnol.* **2019**, *28*, 441–448. [CrossRef] [PubMed]
4. Subramanian, A.P.; Jaganathan, S.K.; Manikandan, A.; Pandiaraj, K.N.; Gomathi, N.; Supriyanto, E. Recent trends in nano-based drug delivery systems for efficient delivery of phytochemicals in chemotherapy. *RSC Adv.* **2016**, *6*, 48294–48314. [CrossRef]
5. Gutiérrez-Grijalva, E.P.; Picos-Salas, M.A.; Leyva-López, N.; Criollo-Mendoza, M.S.; Vazquez-Olivo, G.; Heredia, J.B. Flavonoids and Phenolic Acids from Oregano: Occurrence, Biological Activity and Health Benefits. *Plants* **2018**, *7*, 2. [CrossRef] [PubMed]
6. Gutiérrez-Grijalva, E.P.; Antunes-Ricardo, M.; Acosta-Estrada, B.A.; Gutiérrez-Uribe, J.A.; Basilio Heredia, J. Cellular antioxidant activity and in vitro inhibition of α-glucosidase, α-amylase and pancreatic lipase of oregano polyphenols under simulated gastrointestinal digestion. *Food Res. Int.* **2019**, *116*, 676–686. [CrossRef] [PubMed]
7. Gayoso, L.; Roxo, M.; Cavero, R.Y.; Calvo, M.I.; Ansorena, D.; Astiasarán, I.; Wink, M. Bioaccessibility and biological activity of Melissa officinalis, Lavandula latifolia and Origanum vulgare extracts: Influence of an in vitro gastrointestinal digestion. *J. Funct. Foods* **2018**, *44*, 146–154. [CrossRef]
8. Sęczyk, Ł.; Król, B.; Kołodziej, B. In vitro bioaccessibility and activity of Greek oregano (*Origanum vulgare* L. ssp. hirtum (link) Ietswaart) compounds as affected by nitrogen fertilization. *J. Sci. Food Agric.* **2020**, *100*, 2410–2417. [CrossRef]

9. Gutierrez-Grijalva, E.P.; Angulo-Escalante, M.A.; Leon-Felix, J.; Heredia, J.B. Effect of In Vitro Digestion on the Total Antioxidant Capacity and Phenolic Content of 3 Species of Oregano (*Hedeoma patens, Lippia graveolens, Lippia palmeri*). *J. Food Sci.* **2017**, *82*, 2832–2839. [CrossRef]
10. de Torre, M.P.; Vizmanos, J.L.; Cavero, R.Y.; Calvo, M.I. Improvement of antioxidant activity of oregano (*Origanum vulgare* L.) with an oral pharmaceutical form. *Biomed. Pharmacother.* **2020**, *129*, 110424. [CrossRef]
11. Martau, G.A.; Mihai, M.; Vodnar, D.C. The Use of Chitosan, Alginate, and Pectin in the Biomedical and Food Sector-Biocompatibility, Bioadhesiveness, and Biodegradability. *Polymers* **2019**, *11*, 1837. [CrossRef] [PubMed]
12. Grgic, J.; Selo, G.; Planinic, M.; Tisma, M.; Bucic-Kojic, A. Role of the Encapsulation in Bioavailability of Phenolic Compounds. *Antioxidants* **2020**, *9*, 923. [CrossRef] [PubMed]
13. Li, S.; Zhang, H.; Chen, K.; Jin, M.; Vu, S.H.; Jung, S.; He, N.; Zheng, Z.; Lee, M.S. Application of chitosan/alginate nanoparticle in oral drug delivery systems: Prospects and challenges. *Drug Deliv.* **2022**, *29*, 1142–1149. [CrossRef] [PubMed]
14. Rezagholizade-Shirvan, A.; Najafi, M.F.; Behmadi, H.; Masrournia, M. Design and Synthesis of Novel Curcumin/Chitosan-PVA-Alginate Nanocomposite to Improve Chemico-Biological and Pharmaceutical Curcumin Properties. *SSRN* **2022**, *24*, JDDST-D-22-00009. [CrossRef]
15. Sheorain, J.; Mehra, M.; Thakur, R.; Grewal, S.; Kumari, S. In vitro anti-inflammatory and antioxidant potential of thymol loaded bipolymeric (tragacanth gum/chitosan) nanocarrier. *Int. J. Biol. Macromol.* **2019**, *125*, 1069–1074. [CrossRef]
16. Niaz, T.; Imran, M.; Mackie, A. Improving carvacrol bioaccessibility using core-shell carrier-systems under simulated gastrointestinal digestion. *Food Chem.* **2021**, *353*, 129505. [CrossRef]
17. Bautista-Hernandez, I.; Aguilar, C.N.; Martinez-Avila, G.C.G.; Torres-Leon, C.; Ilina, A.; Flores-Gallegos, A.C.; Kumar Verma, D.; Chavez-Gonzalez, M.L. Mexican Oregano (*Lippia graveolens* Kunth) as Source of Bioactive Compounds: A Review. *Molecules* **2021**, *26*, 5156. [CrossRef] [PubMed]
18. Carrasco-Sandoval, J.; Aranda-Bustos, M.; Henríquez-Aedo, K.; López-Rubio, A.; Fabra, M.J. Bioaccessibility of different types of phenolic compounds co-encapsulated in alginate/chitosan-coated zein nanoparticles. *LWT* **2021**, *149*, 112024. [CrossRef]
19. Panda, P.K.; Yang, J.M.; Chang, Y.H. Preparation and characterization of ferulic acid-modified water soluble chitosan and poly (gamma-glutamic acid) polyelectrolyte films through layer-by-layer assembly towards protein adsorption. *Int. J. Biol. Macromol.* **2021**, *171*, 457–464. [CrossRef]
20. Panda, P.K.; Yang, J.M.; Chang, Y.H.; Su, W.W. Modification of different molecular weights of chitosan by p-Coumaric acid: Preparation, characterization and effect of molecular weight on its water solubility and antioxidant property. *Int. J. Biol. Macromol.* **2019**, *136*, 661–667. [CrossRef]
21. Silva, N.C.D.; Barros-Alexandrino, T.T.; Assis, O.B.G.; Martelli-Tosi, M. Extraction of phenolic compounds from acerola by-products using chitosan solution, encapsulation and application in extending the shelf-life of guava. *Food Chem.* **2021**, *354*, 129553. [CrossRef] [PubMed]
22. Hasan, K.M.F.; Wang, H.; Mahmud, S.; Jahid, M.A.; Islam, M.; Jin, W.; Genyang, C. Colorful and antibacterial nylon fabric via in-situ biosynthesis of chitosan mediated nanosilver. *J. Mater. Res. Technol.* **2020**, *9*, 16135–16145. [CrossRef]
23. Thomas, A.; Müller, S.S.; Frey, H. Beyond Poly(ethylene glycol): Linear Polyglycerol as a Multifunctional Polyether for Biomedical and Pharmaceutical Applications. *Biomacromolecules* **2014**, *15*, 1935–1954. [CrossRef] [PubMed]
24. Jun, H.W.; West, J.L. Endothelialization of microporous YIGSR/PEG-modified polyurethaneurea. *Tissue Eng.* **2005**, *11*, 8. [CrossRef]
25. Yu, H.; VandeVord, P.J.; Mao, L.; Matthew, H.; Wooley, P.H.; Yang, S.-Y. Improved tissue-engineered bone regeneration by endothelial cell mediated vascularization. *Biomaterials* **2009**, *30*, 508–517. [CrossRef]
26. Chiu, Y.-C.; Kocagöz, S.; Larson, J.C.; Brey, E.M. Evaluation of Physical and Mechanical Properties of Porous Poly (Ethylene Glycol)-co (L Lactic Acid) Hydrogels during Degradation. *PLoS ONE* **2013**, *8*, e60728. [CrossRef]
27. Martwong, E.; Tran, Y. Lower Critical Solution Temperature Phase Transition of Poly(PEGMA) Hydrogel Thin Films. *Langmuir* **2021**, *37*, 8585–8593. [CrossRef]
28. Zarei, B.; Tabrizi, M.H.; Rahmati, A. PEGylated Lecithin-Chitosan Nanoparticle-Encapsulated Alpha-Terpineol for In Vitro Anticancer Effects. *AAPS PharmSciTech* **2022**, *23*, 94. [CrossRef] [PubMed]
29. Cai, T.; Marquez, M.; Hu, Z. Monodisperse Thermoresponsive Microgels of Poly(ethylene glycol) Analogue-Based Biopolymers. *Langmuir* **2007**, *23*, 8663–8666. [CrossRef]
30. Matthes, R.; Frey, H. Polyethers Based on Short-Chain Alkyl Glycidyl Ethers: Thermoresponsive and Highly Biocompatible Materials. *Biomacromolecules* **2022**, *23*, 2219–2235. [CrossRef]
31. Swain, T.; Hillis, W.E. The phenolic constituents of Prunus domestica. I.—The quantitative analysis of phenolic constituents. *J. Sci. Food Agric.* **1959**, *10*, 63–68. [CrossRef]
32. Ghasemi, K.; Ghasemi, Y.; Ebrahimzadeh, M.A.; Ebrahimzadeh, M.A. Antioxidant activity, phenol and flavonoid contents of 13 citrus species peels and tissues. *Pak. J. Pharm Sci.* **2009**, *22*, 277–281. [PubMed]
33. Karadag, A.; Ozcelik, B.; Saner, S. Review of Methods to Determine Antioxidant Capacities. *Food Anal. Methods* **2009**, *2*, 41–60. [CrossRef]
34. Thaipong, K.; Boonprakob, U.; Crosby, K.; Cisneros-Zevallos, L.; Hawkins Byrne, D. Comparison of ABTS, DPPH, FRAP, and ORAC assays for estimating antioxidant activity from guava fruit extracts. *J. Food Compos. Anal.* **2006**, *19*, 669–675. [CrossRef]

35. Ganji, F.; Abdekhodaie, M.J. Synthesis and characterization of a new thermosensitive chitosan–PEG diblock copolymer. *Carbohydr. Polym.* **2008**, *74*, 435–441. [CrossRef]
36. Fang, Z.; Bhandari, B. Encapsulation of polyfenols-a review. *Trends Food Sci. Technol.* **2010**, *21*, 13. [CrossRef]
37. Picos-Corrales, L.A.; Garcia-Carrasco, M.; Licea-Claverie, A.; Chavez-Santoscoy, R.A.; Serna-Saldívar, S.O. NIPAAm-containing amphiphilic block copolymers with tailored LCST: Aggregation behavior, cytotoxicity and evaluation as carriers of indomethacin, tetracycline and doxorubicin. *J. Macromol. Sci.* **2019**, *56*, 759–772. [CrossRef]
38. Brodkorb, A.; Egger, L.; Alminger, M.; Alvito, P.; Assunção, R.; Ballance, S.; Bohn, T.; Bourlieu-Lacanal, C.; Boutrou, R.; Carrière, F.; et al. INFOGEST static in vitro simulation of gastrointestinal food digestion. *Nat. Protocols* **2019**, *14*, 991–1014. [CrossRef] [PubMed]
39. Cortes-Chitala, M.D.C.; Flores-Martinez, H.; Orozco-Avila, I.; Leon-Campos, C.; Suarez-Jacobo, A.; Estarron-Espinosa, M.; Lopez-Muraira, I. Identification and Quantification of Phenolic Compounds from Mexican Oregano (*Lippia graveolens* HBK) Hydroethanolic Extracts and Evaluation of Its Antioxidant Capacity. *Molecules* **2021**, *26*, 702. [CrossRef]
40. Lin, L.Z.; Mukhopadhyay, S.; Robbins, R.J.; Harnly, J.M. Identification and quantification of flavonoids of Mexican oregano (*Lippia graveolens*) by LC-DAD-ESI/MS analysis. *J. Food Compost Anal.* **2007**, *20*, 361–369. [CrossRef]
41. Paul, S.; Hmar, E.B.L.; Zothantluanga, J.H.; Sharma, H.K. Essential oils: A review on their salient biological activities and major delivery strategies. *Sci. Vis.* **2020**, *20*, 54–71. [CrossRef]
42. Hermosillo-Ochoa, E.; Picos-Corrales, L.A.; Licea-Claverie, A. Eco-friendly flocculants from chitosan grafted with PNVCL and PAAc: Hybrid materials with enhanced removal properties for water remediation. *Sep. Purif. Technol.* **2021**, *258*, 118052. [CrossRef]
43. Sabnis, S.; Block, L.H. Chitosan as an enabling excipient for drug delivery systems I. Molecular modifications. *Int. J. Biol. Macromol.* **2000**, *27*, 6. [CrossRef]
44. Kou, S.G.; Peters, L.; Mucalo, M. Chitosan: A review of molecular structure, bioactivities and interactions with the human body and micro-organisms. *Carbohydr. Polym.* **2022**, *282*, 119132. [CrossRef] [PubMed]
45. Nair, R.S.; Morris, A.; Billa, N.; Leong, C.O. An Evaluation of Curcumin-Encapsulated Chitosan Nanoparticles for Transdermal Delivery. *AAPS PharmSciTech* **2019**, *20*, 69. [CrossRef] [PubMed]
46. El-Sherbiny, I.M.; Smyth, H.D. Smart Magnetically Responsive Hydrogel Nanoparticles Prepared by a Novel Aerosol-Assisted Method for Biomedical and Drug Delivery Applications. *J. Nanomater* **2011**, *2011*, 910539. [CrossRef]
47. Paberit, R.; Rilby, E.; Göhl, J.; Swenson, J.; Refaa, Z.; Johansson, P.; Jansson, H. Cycling Stability of Poly(ethylene glycol) of Six Molecular Weights: Influence of Thermal Conditions for Energy Applications. *ACS Appl. Energy Mater.* **2020**, *3*, 10578–10589. [CrossRef]
48. Chuc-Gamboa, M.G.; Vargas-Coronado, R.F.; Cervantes-Uc, J.M.; Cauich-Rodriguez, J.V.; Escobar-Garcia, D.M.; Pozos-Guillen, A.; San Roman Del Barrio, J. The Effect of PEGDE Concentration and Temperature on Physicochemical and Biological Properties of Chitosan. *Polymers* **2019**, *11*, 1830. [CrossRef]
49. Hassani Najafabadi, A.; Abdouss, M.; Faghihi, S. Synthesis and evaluation of PEG-O-chitosan nanoparticles for delivery of poor water soluble drugs: Ibuprofen. *Mater. Sci. Eng. C Mater. Biol. Appl.* **2014**, *41*, 91–99. [CrossRef]
50. Li, R.; Wu, Y.; Bai, Z.; Guo, J.; Chen, X. Effect of molecular weight of polyethylene glycol on crystallization behaviors, thermal properties and tensile performance of polylactic acid stereocomplexes. *RSC Adv.* **2020**, *10*, 42120–42127. [CrossRef]
51. Kamel, K.M.; Khalil, I.A.; Rateb, M.E.; Elgendy, H.; Elhawary, S. Chitosan-Coated Cinnamon/Oregano-Loaded Solid Lipid Nanoparticles to Augment 5-Fluorouracil Cytotoxicity for Colorectal Cancer: Extract Standardization, Nanoparticle Optimization, and Cytotoxicity Evaluation. *J. Agric. Food Chem.* **2017**, *65*, 7966–7981. [CrossRef] [PubMed]
52. Espinosa-Sandoval, L.; Ochoa-Martinez, C.; Ayala-Aponte, A.; Pastrana, L.; Goncalves, C.; Cerqueira, M.A. Polysaccharide-Based Multilayer Nano-Emulsions Loaded with Oregano Oil: Production, Characterization, and In Vitro Digestion Assessment. *Nanomaterials* **2021**, *11*, 878. [CrossRef]
53. The Metabolomics Innovation Center. Quercetin. Available online: https://foodb.ca/compounds/FDB011904 (accessed on 15 June 2022).
54. The Metabolomics Innovation Center. Pinocembrin. Available online: https://foodb.ca/compounds/FDB002758 (accessed on 15 June 2022).
55. The Metabolomics Innovation Center. Naringin. Available online: https://foodb.ca/compounds/FDB011866 (accessed on 15 June 2022).
56. The Metabolomics Innovation Center. Cirsimaritin. Available online: https://foodb.ca/compounds/FDB001537 (accessed on 15 June 2022).
57. The Metabolomics Innovation Center. Luteolin. Available online: https://foodb.ca/compounds/FDB013255 (accessed on 15 June 2022).
58. The Metabolomics Innovation Center. Naringenin. Available online: https://hmdb.ca/metabolites/HMDB0002670 (accessed on 15 June 2022).
59. Wang, Q.Z.; Chen, X.G.; Liu, N.; Wang, S.X.; Liu, C.S.; Meng, X.H.; Liu, C.G. Protonation constants of chitosan with different molecular weight and degree of deacetylation. *Carbohydr. Polym.* **2006**, *65*, 194–201. [CrossRef]
60. Mazancová, P.; Némethová, V.; Treľová, D.; Kleščíková, L.; Lacík, I.; Rázga, F. Dissociation of chitosan/tripolyphosphate complexes into separate components upon pH elevation. *Carbohydr. Polym.* **2018**, *192*, 104–110. [CrossRef] [PubMed]

61. Ardean, C.; Davidescu, C.M.; Nemeş, N.S.; Negrea, A.; Ciopec, M.; Duteanu, N.; Negrea, P.; Duda-Seiman, D.; Musta, V. Factors Influencing the Antibacterial Activity of Chitosan and Chitosan Modified by Functionalization. *Int. J. Mol. Sci.* **2021**, *22*, 7449. [CrossRef] [PubMed]
62. Hussain, K.; Ali, I.; Ullah, S.; Imran, M.; Parveen, S.; Kanwal, T.; Shah, S.A.; Saifullah, S.; Shah, M.R. Enhanced Antibacterial Potential of Naringin Loaded β Cyclodextrin Nanoparticles. *J. Clust. Sci.* **2021**, *33*, 339–348. [CrossRef]
63. Hamdi, M.; Nasri, R.; Li, S.; Nasri, M. Design of blue crab chitosan responsive nanoparticles as controlled-release nanocarrier: Physicochemical features, thermal stability and in vitro pH-dependent delivery properties. *Int. J. Biol. Macromol.* **2020**, *145*, 1140–1154. [CrossRef]
64. Hosseini, S.F.; Zandi, M.; Rezaei, M.; Farahmandghavi, F. Two-step method for encapsulation of oregano essential oil in chitosan nanoparticles: Preparation, characterization and in vitro release study. *Carbohydr. Polym.* **2013**, *95*, 50–56. [CrossRef]
65. Maqsoudlou, A.; Assadpour, E.; Mohebodini, H.; Jafari, S.M. The influence of nanodelivery systems on the antioxidant activity of natural bioactive compounds. *Crit. Rev. Food Sci. Nutr.* **2022**, *62*, 24. [CrossRef]
66. Bermúdez-Soto, M.J.; Tomás-Barberán, F.A.; García-Conesa, M.T. Stability of polyphenols in chokeberry (*Aronia melanocarpa*) subjected to in vitro gastric and pancreatic digestion. *Food Chem.* **2007**, *102*, 865–874. [CrossRef]
67. Wootton-Beard, P.C.; Moran, A.; Ryan, L. Stability of the total antioxidant capacity and total polyphenol content of 23 commercially available vegetable juices before and after in vitro digestion measured by FRAP, DPPH, ABTS and Folin–Ciocalteu methods. *Food Res. Int.* **2011**, *44*, 217–224. [CrossRef]
68. Chew, S.-C.; Tan, C.-P.; Long, K.; Nyam, K.-L. In-vitro evaluation of kenaf seed oil in chitosan coated-high methoxyl pectin-alginate microcapsules. *Ind. Crops Prod.* **2015**, *76*, 230–236. [CrossRef]
69. Zheng, J.; Yuan, X.; Cheng, G.; Jiao, S.; Feng, C.; Zhao, X.; Yin, H.; Du, Y.; Liu, H. Chitosan oligosaccharides improve the disturbance in glucose metabolism and reverse the dysbiosis of gut microbiota in diabetic mice. *Carbohydr. Polym.* **2018**, *190*, 77–86. [CrossRef] [PubMed]
70. Terada, A.; Hara, H.; Sato, D.; Higashi, T.; Nakayama, S.; Tsuji, K.; Sakamoto, K.; Ishioka, E.; Maezaki, Y.; Tsugita, T.; et al. Effect of Dietary Chitosan on Faecal Microbiota and Faecal Metabolites of Humans. *Microb. Ecol. Health Dis.* **1995**, *8*, 15–21. [CrossRef]

Article

Protection against Paraquat-Induced Oxidative Stress by *Curcuma longa* Extract-Loaded Polymeric Nanoparticles in Zebrafish Embryos

Ana Teixeira [1,2], Marisa P. Sárria [3,†], Inês Pinto [1,2], Begoña Espiña [3], Andreia C. Gomes [2,4,*] and Alberto C. P. Dias [1,2,4]

1. Centre for the Research and Technology of Agro-Environment and Biological Sciences (CITAB), Department of Biology, University of Minho, Campus of Gualtar, 4710-057 Braga, Portugal
2. CBMA (Centre of Molecular and Environmental Biology), Department of Biology, University of Minho, Campus of Gualtar, 4710-057 Braga, Portugal
3. INL—International Iberian Nanotechnology Laboratory, Avenida Mestre José Veiga, 4715-330 Braga, Portugal
4. IB-S—Institute of Science and Innovation for Sustainability, University of Minho, Campus of Gualtar, 4710-057 Braga, Portugal
* Correspondence: agomes@bio.uminho.pt; Tel.: +351-253-601511; Fax: +351-253-604319
† Current affiliation: European Commission, Joint Research Centre (JRC), 21027 Ispra, Italy.

Abstract: The link between oxidative stress and environmental factors plays an important role in chronic degenerative diseases; therefore, exogenous antioxidants could be an effective alternative to combat disease progression and/or most significant symptoms. *Curcuma longa* L. (CL), commonly known as turmeric, is mostly composed of curcumin, a multivalent molecule described as having antioxidant, anti-inflammatory and neuroprotective properties. Poor chemical stability and low oral bioavailability and, consequently, poor absorption, rapid metabolism, and limited tissue distribution are major restrictions to its applicability. The advent of nanotechnology, by combining nanoscale with multi-functionality and bioavailability improvement, offers an opportunity to overcome these limitations. Therefore, in this work, poly-ε-caprolactone (PCL) nanoparticles were developed to incorporate the methanolic extract of CL, and their bioactivity was assessed in comparison to free or encapsulated curcumin. Their toxicity was evaluated using zebrafish embryos by applying the Fish Embryo Acute Toxicity test, following recommended OECD guidelines. The protective effect against paraquat-induced oxidative damage of CL extract, free or encapsulated in PCL nanoparticles, was evaluated. This herbicide is known to cause oxidative damage and greatly affect neuromotor functions. The overall results indicate that CL-loaded PCL nanoparticles have an interesting protective capacity against paraquat-induced damage, particularly in neuromuscular development that goes well beyond that of CL extract itself and other known antioxidants.

Keywords: curcumin; zebrafish embryogenesis; oxidative stress; neuroprotection; PCL nanoparticles; paraquat

1. Introduction

The world is experiencing a dramatic increase in age-related diseases like cancer, cardiovascular and neurodegenerative disorders. Besides genetic predisposition, there is significant evidence of the link between oxidative stress and environmental factors with incidence towards chronic pathologies [1–4]. In this regard, the development and progression of some of these pathologies like Parkinson and Alzheimer's disease have been linked to the exposure of environmental chemicals, such as herbicides [5–7]. These agrochemicals may induce overproduction of free radicals, mitochondrial dysfunction and damage of the antioxidant system, resulting in impaired cell functioning and production of more toxic species [6–8].

Regulating the production and elimination of these reactive species is the key for modulation of critical cellular functions [2,9,10]. Accordingly, it is suggested that the use of exogenous antioxidants could be an effective alternative to combat the progression of major human degenerative diseases [11–13]. Some studies have shown that medicinal plants are an excellent source of molecules with incomparable chemical diversity and high bioactivity, which can be effective preventives of various chronic diseases [12,14,15].

Curcuma longa L. (CL), commonly known as turmeric, is a popular food condiment that stands out due to a wide range of bioactivities. Turmeric contains a variety of interesting phytochemicals as curcumin (CC), which is assigned to be the most active agent [16,17]. This yellow-colored compound is a multivalent molecule described as having antioxidant activity [18,19], anti-inflammatory [20], anti-cancer [21], cardioprotective [22], neuroprotective [23–25] and hepatoprotective properties [26,27]. CC was also proven to be a safe agent for in vivo application [28]. However, the clinical use of this polyphenolic constituent is limited due to its poor chemical stability and low oral bioavailability and, consequently, poor absorption, rapid metabolism, and limited tissue distribution [29,30]. Due to its chemical instability, at physiological pH, CC rapidly degrades to bicyclopentadione by autoxidation. Furthermore, due to the presence of metabolizing enzymes, CC may undergo biotransformation to form reduction products and glucuronide in the liver and other organs [31]. As a result, low serum levels are quantified. The work [32] has shown that after CC oral administration to healthy volunteers, minimal quantities were quantified in serum level (in 10 g only 50.5 ng/mL has quantified). After absorption, CC is subjected to conjugation at various tissue sites, the liver being indicated as the major organ responsible for its metabolism [33]. Additionally, the systemic elimination from the organism is another contributing factor that affects CC bioavailability and consequently interferes with its biological characteristics [33]. Therefore, a formulation with higher solubility and controlled release property would be advantageous for the therapeutic application of CC.

Nanotechnological approaches, such as the development of an efficient drug delivery system [34–40], may help overcome these limitations of CC-based therapies by combining nanoscale with multi-functionality and bioavailability improvement of active molecules [24,30,41]. Polymeric nanoparticles, liposomes, microemulsions, and solid lipid nanoparticles have been designed and synthetized in order to do so [24,25,42]. These nanocarriers improve the solubility and stability of active compounds, prevent premature degradation, enhance uptake, control drug release, and target the formulation to specific tissues or organs [42]. Despite extensive research into drug delivery vehicles, new materials are being continuously developed without information about their long-term toxicological impact. Thus, prior to application as nanotherapeutics, their toxicity and bioavailability validation are required [43].

Animal models have recognized importance in the study of pathogenesis and therapeutic strategies of human diseases. Among small vertebrate species, zebrafish (*Danio rerio* Hamilton, 1822) has achieved high popularity in the last years for modeling human diseases and disorders [44–46]. This organism exhibits many organs and cell types similar to that of mammals, and most genes have significant homology with humans [44]. These characteristics, together with its low cost of maintenance, ease of husbandry, high fecundity, rapid development, external fertilization and development, and optical transparency of the embryos have made it a popular model in various fields of research, such as neuroscience, cardiovascular studies, genetics and in (eco)toxicology (e.g., [24,47–50]). The zebrafish embryotoxicity test (known as Fish Embryo Acute Toxicity test, FET) is a recommended OECD protocol for the assessment of the toxic effects from exposure to environmental chemicals to nano-/micro-sized particles [24,43,50–52]. This method offers the complexity of a whole vertebrate system and provides the possibility to study both acute and chronic toxicity effects at every moment of the early developmental window.

Taking into consideration the therapeutic potential assigned to CL and to CC, this work was focused on bioactivity assessment of the methanolic extract of this plant (i.e., CL), free or encapsulated in poly-ε-caprolactone (PCL) nanoparticles, against paraquat-induced oxidative damage. Herbicide paraquat is known to cause oxidative damage and greatly affect neuromotor functions [6,8,53]. The lower cost, high permeability, compatibility with a vast range of drugs, long-term degradation, non-toxic nature, and compatibility with several tissues are some advantages that make PCL, and the chosen polymer, suitable to compose an effective delivery system [25,54,55]. Toxicity assessment of the developed nanosystems was investigated using zebrafish embryos, following OECD FET test guideline 236 [56].

2. Materials and Methods

2.1. Chemicals and Reagents

Paraquat (PQ; CAS number 75365-73-0), gallic acid (GA), curcumin (CC) from *Curcuma longa* L. (turmeric), dimethyl sulfoxide (DMSO), methanol, and poly-ε-caprolactone (PCL) were purchased from Sigma-Aldrich, Lisbon, Portugal. Acetone was obtained from Merck, Lisbon, Portugal; Pluronic® F-68 from AppliChem, Lisbon, Portugal and formic acid from VWR, Lisbon, Portugal.

2.2. Compounds and Nanoparticles

PQ, GA, CL, and CL-PCL NPs were defined as experimental conditions. All tested solutions and suspensions were prepared in DMSO and then diluted using ultrapure water (18.2 MΩ cm at 25 °C). To guarantee that an increase in mortality or malformation in zebrafish embryos were not solvent-induced, the final concentration of DMSO considered was below 0.10% (v/v) [57,58].

2.3. Preparation and Characterization of CL Plant Extract

Rhizomes of CL collected from India were dried under shade and finely powdered. A volume of 50 g of biomass was extracted in 100% methanol solution with cycles of sonication at room temperature (RT), in the dark. Following filtration, the extract was concentrated using a rotary evaporator and stored at −80 °C. After lyophilization, CL was kept at RT in the dark and protected from moisture. For CL extract characterization, the liquid phase was filtered (Ø 20 μm), and the sample in the proportion 1:10 (extract: methanol) was analyzed by HPLC-DAD (HITACHI, LabChrom Elite, Tokyo, Japan) and monitored by the computer software EZChrome elite (Agilent Technologies, v 3.02, Santa Clara, CA, USA). The compounds separation was performed on a reversed phase LiChroCART 250-4 column (Phosursohere. RP-18e. 5 μm, Merck, Darmstadt, Germany) at RT, using acetonitrile containing 0.1% (v/v) formic acid (ACN–FA) and ultrapure water containing formic acid 0.1% (v/v) (upW-FA), as the mobile phases. The flow rate was 0.8 mL/min, and the elution gradient was 5% (v/v) of ACN-FA at time 0 min, 30% (v/v) of ACN-FA at time 30 min, 90% (v/v) of ACN-FA at time 40 min, and 95% (v/v) of upW-FA at time 60 min. Spectral data from all compounds were accumulated in the range of 230–550 nm, and chromatograms were recorded at 400 nm. CC (and other curcuminoids) were quantified at 400 nm by the external method, using a commercial curcumin standard (>65% purity).

2.4. Preparation and Characterization of Curcuma Longa-Polycaprolactone Nanoparticles

CL encapsulated PCL NPs were prepared by solvent displacement method, with a ratio of 1:10. The corresponding mass of PCL and CL were dissolved in 4 mL of acetone and transferred drop by drop into an aqueous solution containing 1% (v/v) of Pluronic® F-68, under continuous magnetic stirring (450 rpm) at RT, for 4 h. The NPs suspension was centrifuged (12,000 rpm for 30 min at 4 °C; Eppendorf centrifuge 5804 R, Hamburg, Germany), lyophilized, and stored in a moisture free environment for further use. Empty-PCL NPs were also prepared in similar conditions with an absence of the CL extract in its composition. The particle size and polydispersity index (PdI) were determined by dynamic light scattering (DLS), and the surface charge was determined by zeta potential using Zetasizer Nano ZS (Malvern Panalytical, Malvern, UK). Results are the mean of five test runs. All data were expressed as means ± standard deviation (SD).

2.5. Toxicity Evaluation through Zebrafish Model

2.5.1. Zebrafish Husbandry and Maintenance

Adult wild-type zebrafish was maintained at $26.0 \pm 1.0\,°C$ in a recirculating aquaria system under a photoperiod of 14:10 h (light:dark) at the zebrafish facility of the International Iberian Nanotechnology Laboratory (INL, Braga, Portugal), to be used as breeding stock. Spawners were screened for diseases and had never been previously exposed to a chemical insult. The animals were fed ad libitum twice a day with commercial flakes and supplemented with live brine shrimp eggs. For egg production, zebrafish males and females at a sex ratio of 2:1 were placed into 30 L aquarium coupled with a bottom-open net cage. Spawning was triggered once the light onset in the following morning. Newly fertilized eggs were collected and washed, and viable zygotes were selected for experiments

2.5.2. Zebrafish Embryotoxicity Test

Nanotoxicity and bioactivity of CL (free and loaded in PCL NPs) were assessed using the zebrafish embryotoxicity test (OECD, FET test guideline 236 [56]) as described in [59,60]. During the experimental period, the resulting embryos were kept at $28.0 \pm 1.0\,°C$ with 14:10 light:dark cycle. At selected time-points (Table 1), lethal and sub-lethal effects were assessed via morphological and physiological developmental features analysis that are characteristic of zebrafish embryogenesis [61]. FET experiments were considered valid for a mortality rate inferior to 25% in the control group. All experiments were carried out in strict accordance with the Council of Europe, Directive 2010/63/EU (revised Directive 86/609/EEC), on the protection of experimental animals, including the fact that these were carried out up to 80 h post-fertilization (h_{pf}).

Table 1. Zebrafish developmental parameters assessed during embryogenesis.

	Time-Point Evaluation (h_{pf}—i.e., Hours Post-Fertilization)		
	8 h_{pf}	32 h_{pf}	56 h_{pf}
Developmental parameters	Epiboly Yolk volume Egg volume	Head-trunk index Cardiac frequency Yolk volume Egg volume Pupil surface Eye surface Spontaneous movements	Cardiac frequency Yolk volume Yolk extension Hatching

Toxicity Assessment

To study the toxicity of PQ, GA, and CL, two types of FET assays were carried out to assess long and short insults. In the long experiments (Figure 1A), zebrafish embryos were exposed to the test conditions from 2 until 80 h_{pf}. For this period, four time-points during embryonic development were considered: 8, 32, 56, and 80 h_{pf}. Each experiment was performed with 40 embryos per condition. In the short exposure (Figure 1B), zebrafish embryos with 2, 4, 8, and 24 h_{pf} were exposed for 24 h after which were analyzed in the microscope. Collected data were of 20 embryos per condition. During the experiments, all embryos were constantly checked for mortality. The incubation medium was renewed daily to ensure oxygenation. Mortality was defined as an embryo that lacked cardiac function, blood circulation motility, and/or was in a certain state of degradation. The parameters studied and respective timepoints are listed in Table 1.

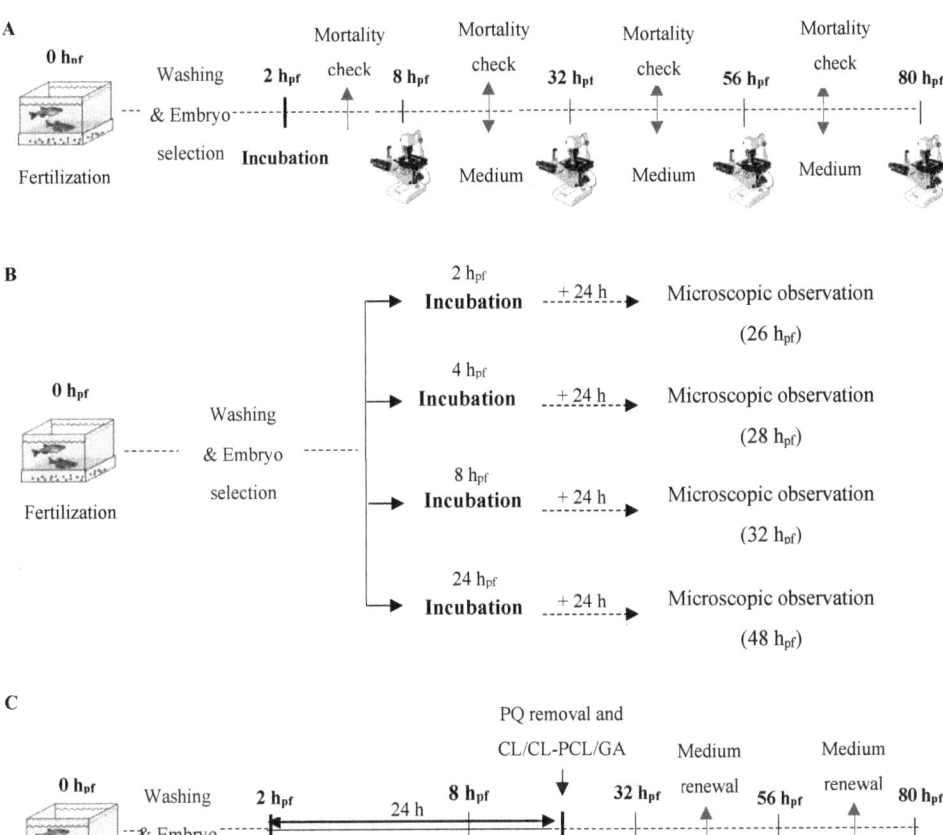

Figure 1. Experimental design of zebrafish embryotoxicity test (FET): (**A**) long exposure; (**B**) short exposure; and (**C**) post-incubation assay. h_{pf}—hours post-fertilization; PQ—paraquat; GA—gallic acid (GA), CL—*Curcuma longa* extract (CL) and CL-PCL—CL-poly-ε-caprolactone nanoparticles.

Bioactivity Assessment

With the goal of evaluating the potential protective effect of CL extract and CL-PCL nanoparticles, post-incubation tests were carried out (Figure 1C). In these experiments, 2 h_{pf} zebrafish embryos were exposed to PQ for 24 h. Zebrafish embryos were kept until 80 h_{pf} with daily medium renovation and successive mortality checks. Each experiment was performed with 10 embryos, in quadruplicates.

2.6. Statistical Analysis

Microscopy images were measured and analyzed by Fiji software. Statistical analysis was performed using Statistica software (StatSoft v.8, Tulsa, OK, USA). Prior to data analysis, all assumptions were met, testing for normality (Shapiro–Wilk test) and homogeneity of variances (Levene's test). Results were presented as mean ± standard deviation (SD) and p value < 0.05 was considered as statistically significant. Details on specific statistical analysis are referred in the legend of the figures. All data were plotted by GraphPad Prism software (GraphPad Inc., v.6, San Diego, CA, USA).

3. Results

3.1. CL Extract and Nanoparticles Characterization

The chemical profile of methanolic CL extract was investigated in order to measure the concentration of total CC and other curcuminoids. The chromatograms of commercial standard curcumin and our CL extract have similar composition (Figure 2). As previously reported [62], based on the spectral data and retention times obtained, the major compounds found were curcumin and, at lower amounts, the curcuminoids bisdemethoxycurcumin and demothoxycurcumin. From the DAD data collected, the CL extract did not show other significant compounds besides these. Our CL extract contains a total of 459 µg/mL of CC and its curcuminoids. For clarity, the HPLC results relating to the period between 38 and 44 min are presented in Supplementary Information (Figure S3).

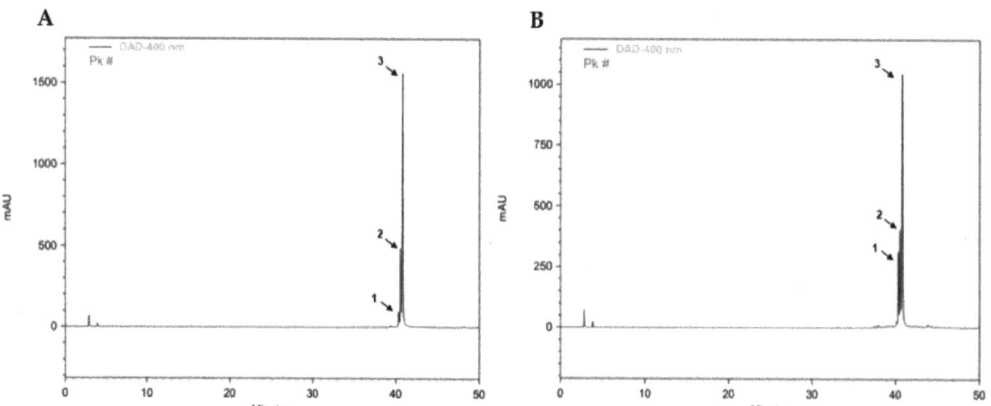

Figure 2. High-performance liquid chromatography (HPLC) analysis of commercial standard curcumin (**A**) and *Curcuma longa* (CL) methanolic extract (**B**). The major compounds present were curcumin (3), demethoxycurcumin (2), and bisdemothoxycurcumin (1).

The prepared nanoparticles were characterized in terms of size distribution, PdI, and zeta potential. Analyzing Figure 3, notorious differences were observed between empty and CL-PCL nanoparticles. The former presented higher sizes and PdI values (Figure 3A), with significant variations in long-term (one-way ANOVA, $F(3,14) = 106.240$, $p < 0.001$). Furthermore, surface charge of the developed empty nanoparticles showed a significant decrease along time: at day 0, measured zeta potential was -8 mV and, 4 weeks later, -18 mV (Figure 3B) (one-way ANOVA, $F(3,14) = 409.630$, $p < 0.001$). Regarding the CL-PCL nanoparticles, it was observed that 4 weeks after production, the nanoformulation presented values of size and PdI of 279 nm and 0.26, respectively (Figure 3C). Additionally, it presented a highly negative surface charge at weeks 2, 3, and 4 (>20 mV) (Figure 3D). Significant differences were found in the size and PdI value at weeks 2 and 4 (one-way ANOVA, size: $F(3,12) = 10.890$, $p < 0.01$; PdI value: $F(3,12) = 11.400$, $p < 0.01$), and in the zeta potential at week 0 (one-way ANOVA, $F(3,12) = 57.770$, $p < 0.001$).

3.2. Bioactivity Validation in Zebrafish
3.2.1. Toxicity Profiling

In the present study, the short and long terms toxicity of PQ, GA, CL, and CL nanoparticles exposure to zebrafish embryos were evaluated. Regarding PQ long exposure results, notorious effects were observed among the independent variables evaluated (Table 2). The longer exposure of zebrafish embryos to 0.64 µg/mL of PQ caused a significant decrease on pupil surface (ANCOVA, $F(4,41) = 7.882$, $p < 0.001$) and a significant decrease on yolk extension (ANCOVA, $F(4,43) = 0.998$, $p < 0.05$). Concerning the neuro-motor coordination indicator parameters, this herbicide negatively affected zebrafish embryos in a concentration dependent manner. A significant decrease in the cardiac frequency was registered (one-away ANOVA, $F(4,40) = 7.746$, $p < 0.001$), and an increase in the percentage of embryos with spontaneous movements (SM) and of larvae exhibiting free-swimming (FS) (SM: chi-square test, $\chi^2 = 11.53$, $p < 0.001$; FS: chi-square test, $\chi^2 = 16.34$, $p < 0.01$) were recorded. In comparison with the control group (that is, 0.00 µg/mL of PQ), 0.64 µg/mL of PQ triggered an increase of 20 and 25% in the number of SM and FS, respectively. Moreover, it was observed that increasing PQ (nominal) test concentration caused toxicity in a time-dependent manner. The exposure to the highest (nominal) test concentration of PQ (i.e., 1.28 µg/mL) resulted in a reduction of 55% in embryo survival (Figure S1A).

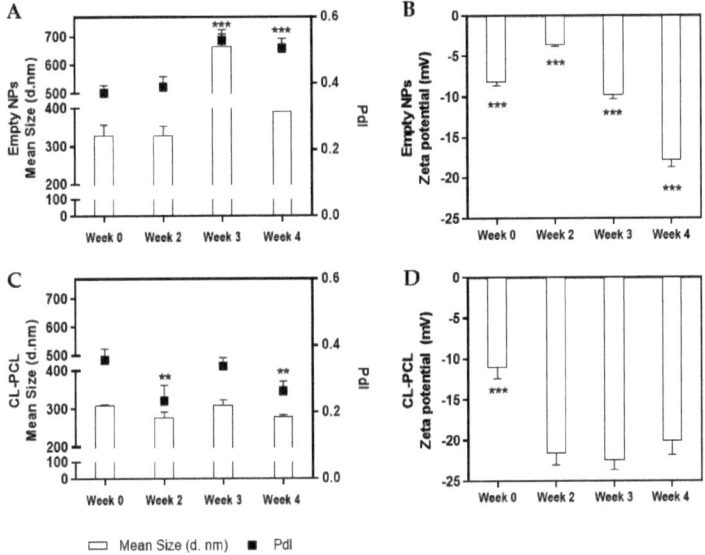

Figure 3. Polymeric nanoparticles characterization: (**A**) mean size (d. nm) and polydispersity index (PdI) of empty nanoparticles (NPs); (**B**) zeta potential of empty NPs; (**C**) mean size (d. nm) and PdI of *Curcuma longa* extract (CL)-poly-ε-caprolactone (CL-PCL) NPs; and (**D**) zeta potential of CL-PCL NPs at different time points. Loaded nanoparticles present values of mean size and PdI lower than the empty ones and are slightly more negatively charged. Both formulations are stable for at least 4 weeks after preparation. Results are expressed as mean ± SD. ** $p < 0.01$; *** $p < 0.001$, one-way ANOVA followed by Student–Newman–Keuls post hoc test.

A well-known antioxidant compound—GA [63] was considered to validate its potential to mitigate the oxidative stress. Accordingly, for GA (Table 2), it was observed that the range of the tested (nominal) concentrations did not interfere with the epibolic arc development and embryos straightening. At 32 h_{pf}, however, the embryos exposed to 25 µg/mL GA presented a significant reduction of their yolk volume (ANCOVA, $F(3,25) = 2.113$, $p < 0.05$). Furthermore, it was observed that, although the highest (nominal) test concentration of GA (50 µg/mL) did not interfere with the percentage of zebrafish embryos with SM, effects on the cardiac frequency (one-way ANOVA, $F(3,16) = 9.143$, $p < 0.001$) and on the number of larvae having FS (one-way ANOVA, $F(3,8) = 2.976$, $p < 0.05$) were observed. A volume of 50 µg/mL of GA, in comparison to the untreated embryos, triggered a decrease of approximately 17% on the cardiac frequency and, on the other hand, an increase of 18% to 59% on the FS behavior. In addition, it was recorded that an increased concentration of GA resulted in reduced zebrafish embryonic mortality (Figure S1B).

Table 2. Zebrafish embryotoxicity test—long exposure toxicity assessment. (+) stands for statistically significant effect and (−) for non-statistically significant effect. h_{pf}—hours post-fertilization. CL—*Curcuma longa* extract.

	Independent Variables	h_{pf}	Paraquat 0.64 µg/mL	Gallic Acid 25 µg/mL	Gallic Acid 50 µg/mL	CL 5 µg/mL
Morphometric parameters	Epibolic arc	8	−	−	−	−
	Yolk volume	8	−	−	−	−
	Head-trunk index	32	−	−	−	−
	Yolk volume	32	−	+	−	−
	Pupil surface	32	+	−	−	−
	Yolk volume	56	−	−	−	−
	Eye surface	56	−	−	−	−
	Yolk extension	56	+	−	−	−
Neuro-motor parameters	Spontaneous movements	32	+	+	−	−
	Cardiac frequency	56	+	−	+	−
	Free-swimming	80	+	+	+	+

Regarding CL long exposure of zebrafish embryos (Table 2), it is possible to infer that 5 µg/mL of CL did not induce a delay on the embryonic development, did not affect the structures rich in lipids as the yolk, eye, and pupil and did not interfere with the SM and cardiac frequency. However, this extract rich in polyphenolic compounds significantly reduced FS ability (one-way ANOVA, $F(3,8) = 14.21$, $p < 0.01$). Moreover, it was observed that increasing CL concentration (tested at 1.5, 2.5, 5, and 10 µg/mL) led to a decrease in mortality. At 80 h_{pf}, zebrafish embryos exposed to 5 µg/mL showed 9% cumulative survival (Figure S1C).

Additionally, the results obtained for the short exposure of zebrafish embryos to 5 µg/mL of CL and CL-PCL nanoparticles are summarized in Table 3. In general, 5 µg/mL of either CL Free, CL-PCL, or empty nanoparticles did not interfere with the embryos straightening. However, it was observed that the unformulated drug and the polymer used to encapsulate the plant extract caused a significant increase in the yolk volume (CL Free: ANCOVA, 2 h_{pf}: $F(2,59) = 14.38$, $p < 0.001$; 8 h_{pf}: $F(2,59) = 4.18$, $p < 0.05$; Empty NPs: ANCOVA, 4 h_{pf}: $F(3,75) = 8.59$, $p < 0.001$; 8 h_{pf}: $F(3,82) = 5.30$, $p < 0.01$). Concerning the results for zebrafish embryonic cardiac frequency, it was observed that 5 µg/mL of CL Free significantly affected cardiac frequency. The non-treated embryos incubated at 8 h_{pf} showed 2.3 beats/s but, when exposed to 5 µg/mL of CL Free, the cardiac frequency decreased to 1.9 beats/s (one-way ANOVA, 2 h_{pf}: $F(2,37) = 10.26$, $p < 0.001$; 8 h_{pf}: $F(2,36) = 15.40$, $p < 0.001$; 24 h_{pf}: $F(2,36) = 6.238$, $p < 0.01$). At 8 and 24 h_{pf}, zebrafish embryos treated with CL-PCL nanoparticles did not show signs of cardiotoxicity. Nevertheless, the empty nanoparticles caused a reduction in zebrafish cardiac frequency, being statistically significant for those incubated at 4, 8, and 24 h_{pf} (Table 3). Furthermore, the effects of CL short exposure on the SM of zebrafish embryos were also recorded. CL Free promoted, in a concentration dependent manner, an increase in the percentage of embryos with SM, which was verified to be statistically significant for embryos incubated at 24 h_{pf} with 5 µg/mL (one-way ANOVA, 24 hpf: $F(2,7) = 5.331$, $p < 0.05$). In contrast, zebrafish embryos exposed to an increasingly higher (nominal) test concentration of CL-PCL nanoparticles showed a decrease in SM. At 4 h_{pf}, the number of muscular contractions was significantly reduced from 53 to 33% (one-way ANOVA: 4 h_{pf}: $F(3,9) = 8.987$, $p < 0.01$). PCL did not affect the muscular contractions of zebrafish embryos. Furthermore, the short exposure of 5 µg/mL of CL extract (free and encapsulated) seems to have a protective effect on zebrafish embryos survival, as a decrease in mortality rate was observed (Figure S2).

Table 3. Zebrafish embryotoxicity test—short exposure toxicity assessment. (+) stands for statistically significant effect and (−) for non-statistically significant effect. h_{pf}, hours post-fertilization; CL Free, *Curcuma longa* extract; CL-PCL NPs, *Curcuma longa* extract—poly-ε-caprolactone nanoparticles and empty NPs, empty nanoparticles.

	Independent Variables	h_{pf}	CL Free (5 µg/mL)	CL-PCL NPs (5 µg/mL)	Empty NPs (5 µg/mL)
Morphometric parameters	Yolk volume	2	+	−	−
		4	−	−	+
		8	+	−	+
		24	−	−	−
	Head-trunk index	2	−	−	−
		4	−	−	−
		8	−	−	−
		24	−	−	−
	Pupil surface	2	−	−	−
		4	−	−	−
		8	−	+	−
		24	−	+	−
Neuro-motor parameters	Cardiac frequency	2	+	+	−
		4	−	+	+
		8	+	−	+
		24	+	−	+
	Spontaneous movements	2	−	−	−
		4	−	+	−
		8	−	−	−
		24	+	−	−

3.2.2. Bioactivity Profiling

The putative protective activity of CL, Free and loaded onto PCL nanoparticles, was evaluated by a FET assay using embryos exposed to PQ. According to the results obtained, 5 µg/mL of CL (free and encapsulated), 25 and 50 µg/mL of GA, and 0.64 µg/mL of PQ were the selected (nominal) test concentrations for a post-incubation test (exposure to PQ then incubation with GA). Results obtained are illustrated in Table S2. The post-incubation of GA potentiated the survival of zebrafish embryos and was effective in minimizing the negative effects of PQ in the zebrafish muscular contraction at 32 h_{pf}. However, it was not able to mitigate cardiotoxicity caused by PQ exposure (Table S2).

Table 4 and Figure 4 illustrate the results related to the medicinal potential of CL extract (free and encapsulated) against PQ toxicity. According to Figure 4A, it was observed that 13% of the embryos exposed to PQ (and non-treated with any other compound) died, while in conditions where embryos were treated with 5 µg/mL of CL Free and CL-PCL, the mortality rate was significantly lower (i.e., 3% and 0%, respectively). Analyzing the zebrafish embryos cardiac frequency (Figure 4B), it was observed that the toxic effect instigated by PQ exposure was quite remarkable. At 32 h_{pf}, the heartbeat of the embryos incubated with PQ decreased significantly to 1.18 beat/s, while in the control group, the embryos presented an average of 1.66 beats/s (Figure 4B). CL Free and the formulated drug counteracted the cardiotoxic effect of PQ by increasing the heart beating of the zebrafish embryos for 1.48 and 1.57 beats/s, respectively (Figure 4B). At 56 h_{pf}, the cardiac frequency of zebrafish embryos exposed to PQ was still lower (2.48 beats/s), compared to the control group (2.71 beats/s). However, those treated with CL-PCL nanoparticles exhibited an increased cardiac frequency (2.64 beats/s) (Figure 4B). Differences reported were statistically significant at both time points (32 h_{pf}: one-way ANOVA, $F(7,74) = 24.33$, $p < 0.001$; 56 h_{pf}: one-way ANOVA, $F(7,69) = 10.11$, $p < 0.001$).

Table 4. Zebrafish embryotoxicity test—post-incubation experiment. (+) stands for statistically significant effect and (−) for non-statistically significant effect. h_{pf}, hours post-fertilization; CL Free—*Curcuma longa* extract (5 µg/mL); CL-PCL NPs—*Curcuma longa* -poly-ε-caprolactone nanoparticles (5 µg/mL) and PQ—paraquat (0.64 µg/mL).

	Independent Variables	h_{pf}	CL Free	CL-PCL NPs	Empty NPs	PQ ↓ H₂O	PQ ↓ CL Free	PQ ↓ CL-PCL	PQ ↓ Empty NPs
						Post-Incubation Conditions			
Morphometric parameters	Epibolic arc	8	−	−	−	−	−	−	−
	Yolk volume	8	−	−	−	+	+	−	−
	Head-trunk index	32	−	−	−	−	−	−	+
	Yolk volume	32	−	−	−	−	−	−	−
	Pupil surface	32	−	−	−	−	−	−	−
	Yolk volume	56	−	−	−	−	−	−	−
	Eye surface	56	−	−	−	−	−	−	−

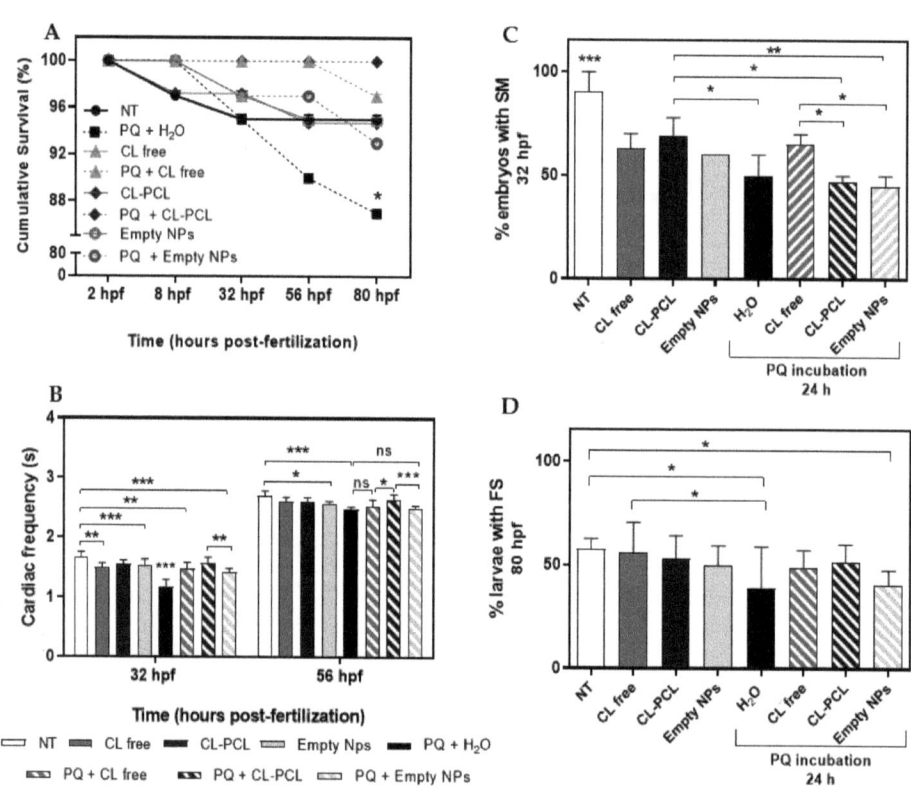

Figure 4. The effects of 24 h insult of paraquat (PQ) followed by a long exposure of *Curcuma longa* extract (CL) Free and encapsulated (CL-PCL) on (**A**) cumulative survival at 8, 32, 56 to and 80 h post-fertilization (h_{pf}); (**B**) cardiac frequency (s) at 32 and 56 h_{pf}; (**C**) spontaneous movements (SM) at 32 h_{pf}, and (**D**) free swimming (FS) at 80 h_{pf} of zebrafish embryos. At 2 h_{pf}, the embryos were first exposed for 24 h to 0.64 µg/mL of PQ, followed by an insult of 5 µg/mL of CL free, CL-PCL and empty NPs. CL (free and formulated) reduced the mortality rate. Furthermore, the treatment with CL Free and CL-PCL increased not only the heart frequency of zebrafish embryos exposed to PQ but also increased the number of embryos with SM and the larvae with FS behavior. Results are expressed as mean ± SD. ns—not statistically different, * $p < 0.05$; ** $p < 0.01$; *** $p < 0.001$; one-way ANOVA, followed by Newman–Keuls and Fisher LSD post hoc tests.

Furthermore, at 32 h_{pf}, the percentage of zebrafish embryos with SM decreased from 90% at the control group, to 50% for those exposed to PQ. Nevertheless, the treatment with free CL attenuated the toxic effect on muscular contractions by increasing in 15% the number of embryos with SM (Figure 4C). Differences among conditions were demonstrated to be statistically significant (one-way ANOVA, $F(7,16) = 13.14$, $p < 0.001$). At 80 h_{pf}, the effect of 24 h exposure of 0.64 µg/mL of PQ was still notorious (Figure 4D). CL Free and CL-PCL showed to be efficient in reverting the effects caused by PQ. The treatment with these substances increased the larvae FS behavior from 39.2% to 48.6% and 51.4%, respectively (Figure 4D). Differences between conditions were statistically significant (one-way ANOVA, $F(7,24) = 1.350$, $p < 0.001$).

Regarding the morphometric parameters (Table 4), none of the tested conditions seemed to interfere with the zebrafish embryonic development at 8 h_{pf}, as well as with the pupil and eye development (32 and 56 h_{pf}, respectively). In the early stages of development, zebrafish embryos exposed to PQ showed a significant decrease on yolk volume, compared to the control group (ANCOVA, $F(7,150) = 3.025$, $p < 0.01$). However, the formulated drug counteracts the damage caused by increasing the volume of this structure, and with the progression of the embryonic development, differences on yolk volume tended to dissipate. Aside from this, it was also observed that the post-incubation of CL Free and CL-PCL attenuated the production of reactive oxygen species (ROS) (data not shown).

4. Discussion

Oxidative stress plays an important role in mitochondrial activity and consequently on neurodegeneration [3,9,10]. To date, the strategies developed for the treatment of disorders related to neuronal degeneration and cognitive deterioration have been ineffective [9,10,64]. Therefore, the search for new candidates with high antioxidant and anti-inflammatory potential that interact with a wide diversity of molecular targets and pathways has been constant [11,13]. The powdered rhizomes of CL have been used for centuries as a medicinal herb with a wide range of biological applications [16,18,22]. This plant and, in particular, its yellow-colored constituent (CC), has been shown to be very effective in the attenuation of the symptoms associated with neurodegenerative disorders [12,23–25].

In this work, in order to potentiate CL activity, polymeric nanoparticles were developed by solvent displacement method, which is a simple, economic, and widely employed technique [25,54,65,66]. Biodegradable polymers have provided numerous avenues to improve drug development and, consequently, the therapeutic efficacy of medicinal entities [67]. PCL is an example of a nontoxic, biodegradable, and biocompatible polymer widely used and approved by the U.S. Food and Drug Administration (FDA) [25,54,55,67]. Considering the obtained results, CL-PCL nanoparticles remain stable for at least four weeks after their production. Comparing CL loaded and empty formulations, large differences were observed in terms of size, homogeneity, and charge potential. CL-loaded formulations are smaller, with broad size distribution and highly negative surface charge. The inclusion of the plant extract seems to potentiate the stabilization of nanoparticles, perhaps by the addition of a negative charge or promoting exposure of negative groups. Physicochemical features of nanoparticles also determine how they interact with cells and tissues. The more negative surface charge of loaded nanoparticles confers more stability also over time, as clearly observed, and it does not cause relevant cytotoxicity issues, which would be the case if the nanoparticles presented cationic charge. More homogeneity in size distribution and slightly smaller particles, when loaded with CL, are also advantageous in terms of predictability of cell response and also facilitate the selection of administration route. In a previous study [25], we optimized the PCL nanoformulations to obtain >70% curcumin encapsulation and also included a release profile of PCL loaded nanoparticles. It reveals that there is sustained release and that, after 168 h, approximately 75% of curcumin is released from the nanoparticles. Regarding the toxicological profile, the free and formulated compounds were demonstrated to be generally inert. As reported by [68], the exposure of zebrafish embryos to these conditions did not induce a delay on development and in toxicity at the level of cardiac function, SM and FS behavior.

The brain and heart are the organs with most demanding energetic necessities and are totally dependent upon oxidative phosphorylation to supply the huge amount of ATP required [69]. In contrast, PQ interferes with electrons transference by activity inhibition of the complex I of mitochondrial respiratory chain, resulting in uncontrolled ROS production [6,70]. Furthermore, the ability of ROS to affect both mitochondrial function and nucleic acids results in critical cellular alterations. This chain of events promotes the phosphorylation of proteins intrinsically associated with neurodegeneration [71], and the reduction of cardiac activity [69]. The obtained results demonstrate that the exposure to PQ caused negative effects in the structures and organs related to zebrafish lipid metabolism. Reference [72] reported that PQ exposure on Kunming mice resulted in lipid peroxidation; increase in ROS levels; and damage in the biliary, gastrointestinal, and nervous systems, in addition to lungs, kidneys, and the liver. Our own results (Figure S4) show that a 24 h incubation with PQ leads to an increase in ROS levels on zebrafish embryos that is maintained for another 24 h, while there is a dose-dependent increase upon exposure to H_2O_2. Furthermore, our data suggest that PQ affects neuro-motor coordination, such as trunk contractions and pectoral fin movements, and lead to signs of cardiotoxicity, as a reduced cardiac frequency was observed. These results are concordant with other reports [53,73,74]. PQ exposure on zebrafish results in decreased sensorimotor reflexes, spontaneous movements, and distance swum. Furthermore, the PQ-treated group showed anxiety-like behavior by swimming near the walls [73,74]. In *Drosophila melanogaster*, this herbicide caused alternation of motor-related behavioral patterns, including resting tremors, rotation of the body, and postural instability [68]. According to our results, CL improved survival rate of zebrafish embryos and counteracted the reduction in cardiac frequency as well as the negative effects on the neuro-motor parameters. In Figure S5, it is noticeable that free and formulated curcumin can reduce high ROS levels induced by 24 h contact with PQ. This protective effect of CL seems to be associated with the capacity to restore the endogenous antioxidant system by increasing levels of catalase, superoxide dismutase and glutathione, decreased oxidative markers levels, reduced lipid peroxidation, and counteracted cardiac cell damage [18,75]. Comparing the CL-PCL nanoparticles with the CL Free, the nanosystems have shown to be, in most cases, more competent in mitigating the oxidative damage. This effect may be due to the fact that the encapsulation of CL enhances its bioavailability and pharmacokinetics [25,65,76]. Furthermore, CL was more efficient in attenuating the toxic effect of PQ than GA.

5. Conclusions

Our data support the notion that CL possesses antioxidant activity, as well as cardio and neuroprotective properties, validating its potential to minimize damages caused by the herbicide PQ. Delivery of this plant extract incorporated in polymeric nanoparticles also conferred protection against PQ-induced toxicity, in FET assays. The embryos treated with free or formulated CL showed a decrease in mortality rate and cardiotoxicity caused by exposure to the tested herbicide, and the heartbeat of zebrafish embryos increased to values close to those of the control group. Additionally, CL free and CL-PCL minimize the toxic effect of PQ on neuro-motor parameters, as the number of embryos with SM and the number of larvae with FS increased with both treatments. Furthermore, the present work corroborates the potential of zebrafish embryos as a model to assess drug discovery and nanotoxicity.

Supplementary Materials: The following supporting information can be downloaded at: https://www.mdpi.com/article/10.3390/polym14183773/s1, Figure S1: Effects of (A) paraquat; (B) gallic acid and (C) *Curcuma longa* extract on zebrafish embryos at different hours post-fertilization; Figure S2: Short exposure effect of *Curcuma longa* extract (CL) free (A) and loaded to poly-ε-caprolactone nanoparticles (CL-PCL NPs) (B), on survival of zebrafish embryos; Figure S3: Detailed analysis of high-performance liquid chromatography (HPLC) analysis of commercial standard curcumin

(A) and *Curcuma longa* (CL) methanolic extract (B) relating to the period between 38 and 44 min, clearly showing the major compounds present were curcumin (3), demethoxycurcumin (2) and bisdemothoxycurcumin (1); Figure S4. Titration of reactive oxygen species (ROS) induced by paraquat (PQ) and hydrogen peroxide (H_2O_2) exposure on zebrafish embryos. At 2 hpf, the embryos were exposed to 0.64 µg/mL of PQ or 0.17–3.4 µg/mL of H_2O_2 for 24 h and, after this incubation time, the media were replaced by freshwater. ROS levels were quantified with DCFH-DA probe (excitation wavelength 485 nm and emission wavelength 538 nm), measured 24 h after incubation (time-point 24 h) and one day after its removal (time-point 48 h); Figure S5. The effects of *Curcuma longa* extract (CL) exposure on paraquat (PQ)-induced ROS generation in zebrafish embryos. Zebrafish embryos were exposed to PQ for 24 h and treated with CL free, CL-poly-ε-caprolactone (CL-PCL) nanoparticles and gallic acid (GA). The concentrations tested were: 0.64 µg/mL PQ; 5 µg/mL CL free, CL-PCL and empty NPs and 25 µg/mL of GA. ROS measurement was performed one day after PQ removal (time-point 48 h); Table S1: Zebrafish embryotoxicity test—short exposure toxicity assessment; Table S2: Zebrafish embryotoxicity test—post-incubation experiment.

Author Contributions: Conceptualization, A.C.P.D. and A.C.G.; methodology, A.C.P.D., A.C.G. and M.P.S.; validation, A.C.P.D., A.C.G. and M.P.S.; formal analysis, A.T. and I.P.; investigation, A.T., I.P. and M.P.S.; resources, A.C.P.D., A.C.G. and B.E.; data curation, A.T., I.P. and M.P.S.; writing—original draft preparation, A.T.; writing—review and editing, A.C.P.D., A.C.G. and M.P.S.; supervision, A.C.P.D., A.C.G. and M.P.S.; project administration, A.C.P.D. and A.C.G.; funding acquisition, A.C.P.D. and A.C.G. All authors have read and agreed to the published version of the manuscript.

Funding: Marisa P. Sarria was supported by Marie Skłodowska—Curie Actions from European Union's 7th Framework Programme for Research, Technological Development and Demonstration under Grant Agreement 600375. This work was supported by the strategic programme UID/BIA/04050/2019 funded by national funds through the Fundação para a Ciência e a Tecnologia I.P. Marisa P. Sárria was supported by Marie Curie COFUND funding from the European Union's 7th Framework Programme for research, technological development, and demonstration under grant agreement 600375. Begoña Espiña is supported by the co-funding from Northern Regional Operational Program through the Project NORTE-45-2015-02 Nanotechnology based functional solutions. B.E. acknowledges the financial support of the project SbDtoolBox—Nanotechnology-based tools and tests for Safer-by-Design nanomaterials, with the reference n° NORTE-01-0145-FEDER-000047, funded by Norte2020—North-Regional Operational Programme under the PORTUGAL 2020 Partnership Agreement, through the European Regional Development Fund (ERDF).

Institutional Review Board Statement: Ethical review and approval were waived for this study as it is not required according to Portuguese or European laws. As mentioned in the Section 2.5.1. of Material and Methods, tests on fish embryonic stages (OECD, FET test guideline 236) were conducted in strict conformity to current guidelines defined at Council of Europe's Directive 2010/63/EU (revised Directive 86/609/EEC) on protection of experimental animals, setting the regulatory limit of exposure at the free-living stage (that is, end of embryogenesis). As the last endpoint tested (80 h post-fertilization (hpf)) precedes this developmental age, a statutory ethical consent is not required.

Informed Consent Statement: Not applicable.

Data Availability Statement: Data will be made available upon reasonable request.

Conflicts of Interest: The authors declare no conflict of interest.

References

1. Costa, L.G.; Cole, T.B.; Dao, K.; Chang, Y.C.; Coburn, J.; Garrick, J.M. Effects of air pollution on the nervous system and its possible role in neurodevelopmental and neurodegenerative disorders. *Pharmacol. Ther.* **2020**, *210*, 107523. [CrossRef] [PubMed]
2. Poprac, P.; Jomova, K.; Simunkova, M.; Kollar, V.; Rhodes, C.J.; Valko, M. Targeting Free Radicals in Oxidative Stress-Related Human Diseases. *Trends Pharmacol. Sci.* **2017**, *38*, 592–607. [CrossRef] [PubMed]
3. Sharifi-Rad, M.; Anil Kumar, N.V.; Zucca, P.; Varoni, E.M.; Dini, L.; Panzarini, E.; Rajkovic, J.; Tsouh Fokou, P.V.; Azzini, E.; Peluso, I.; et al. Lifestyle, Oxidative Stress, and Antioxidants: Back and Forth in the Pathophysiology of Chronic Diseases. *Front. Physiol.* **2020**, *11*, 694. [CrossRef] [PubMed]
4. Zheng, F.; Goncalves, F.M.; Abiko, Y.; Li, H.; Kumagai, Y.; Aschner, M. Redox toxicology of environmental chemicals causing oxidative stress. *Redox Biol.* **2020**, *34*, 101475. [CrossRef] [PubMed]

5. Baltazar, M.T.; Dinis-Oliveira, R.J.; de Lourdes Bastos, M.; Tsatsakis, A.M.; Duarte, J.A.; Carvalho, F. Pesticides exposure as etiological factors of Parkinson's disease and other neurodegenerative diseases—A mechanistic approach. *Toxicol. Lett.* **2014**, *230*, 85–103. [CrossRef]
6. Bastias-Candia, S.; Zolezzi, J.M.; Inestrosa, N.C. Revisiting the Paraquat-Induced Sporadic Parkinson's Disease-Like Model. *Mol. Neurobiol.* **2019**, *56*, 1044–1055. [CrossRef]
7. Xiao, J.; Dong, X.; Zhang, X.; Ye, F.; Cheng, J.; Dan, G.; Zhao, Y.; Zou, Z.; Cao, J.; Sai, Y. Pesticides Exposure and Dopaminergic Neurodegeneration. *Expo. Health* **2021**, *13*, 295–306. [CrossRef]
8. Wang, Q.; Ren, N.; Cai, Z.; Lin, Q.; Wang, Z.; Zhang, Q.; Wu, S.; Li, H. Paraquat and MPTP induce neurodegeneration and alteration in the expression profile of microRNAs: The role of transcription factor Nrf2. *NPJ Parkinsons Dis.* **2017**, *3*, 31. [CrossRef]
9. Angelova, P.R.; Abramov, A.Y. Role of mitochondrial ROS in the brain: From physiology to neurodegeneration. *FEBS Lett.* **2018**, *592*, 692–702. [CrossRef]
10. Singh, A.; Kukreti, R.; Saso, L.; Kukreti, S. Oxidative Stress: A Key Modulator in Neurodegenerative Diseases. *Molecules* **2019**, *24*, 1583. [CrossRef]
11. Amato, A.; Terzo, S.; Mule, F. Natural Compounds as Beneficial Antioxidant Agents in Neurodegenerative Disorders: A Focus on Alzheimer's Disease. *Antioxidants* **2019**, *8*, 608. [CrossRef]
12. Benameur, T.; Soleti, R.; Panaro, M.A.; La Torre, M.E.; Monda, V.; Messina, G.; Porro, C. Curcumin as Prospective Anti-Aging Natural Compound: Focus on Brain. *Molecules* **2021**, *26*, 4794. [CrossRef] [PubMed]
13. Uttara, B.; Singh, A.V.; Zamboni, P.; Mahajan, R.T. Oxidative stress and neurodegenerative diseases: A review of upstream and downstream antioxidant therapeutic options. *Curr. Neuropharmacol.* **2009**, *7*, 65–74. [CrossRef] [PubMed]
14. Cragg, G.M.; Newman, D.J. Natural products: A continuing source of novel drug leads. *Biochim. Biophys. Acta* **2013**, *1830*, 3670–3695. [CrossRef] [PubMed]
15. Kirichenko, T.V.; Sukhorukov, V.N.; Markin, A.M.; Nikiforov, N.G.; Liu, P.Y.; Sobenin, I.A.; Tarasov, V.V.; Orekhov, A.N.; Aliev, G. Medicinal Plants as a Potential and Successful Treatment Option in the Context of Atherosclerosis. *Front. Pharmacol.* **2020**, *11*, 403. [CrossRef]
16. Kocaadam, B.; Sanlier, N. Curcumin, an active component of turmeric (*Curcuma longa*), and its effects on health. *Crit. Rev. Food Sci. Nutr.* **2017**, *57*, 2889–2895. [CrossRef]
17. Prasad, S.; Gupta, S.C.; Tyagi, A.K.; Aggarwal, B.B. Curcumin, a component of golden spice: From bedside to bench and back. *Biotechnol. Adv.* **2014**, *32*, 1053–1064. [CrossRef]
18. Abrahams, S.; Haylett, W.L.; Johnson, G.; Carr, J.A.; Bardien, S. Antioxidant effects of curcumin in models of neurodegeneration, aging, oxidative and nitrosative stress: A review. *Neuroscience* **2019**, *406*, 1–21. [CrossRef]
19. Jaroonwitchawan, T.; Chaicharoenaudomrung, N.; Namkaew, J.; Noisa, P. Curcumin attenuates paraquat-induced cell death in human neuroblastoma cells through modulating oxidative stress and autophagy. *Neurosci. Lett.* **2017**, *636*, 40–47. [CrossRef]
20. Farhood, B.; Mortezaee, K.; Goradel, N.H.; Khanlarkhani, N.; Salehi, E.; Nashtaei, M.S.; Najafi, M.; Sahebkar, A. Curcumin as an anti-inflammatory agent: Implications to radiotherapy and chemotherapy. *J. Cell. Physiol.* **2019**, *234*, 5728–5740. [CrossRef]
21. Tomeh, M.A.; Hadianamrei, R.; Zhao, X. A Review of Curcumin and Its Derivatives as Anticancer Agents. *Int. J. Mol. Sci.* **2019**, *20*, 1033. [CrossRef] [PubMed]
22. Santos-Parker, J.R.; Strahler, T.R.; Bassett, C.J.; Bispham, N.Z.; Chonchol, M.B.; Seals, D.R. Curcumin supplementation improves vascular endothelial function in healthy middle-aged and older adults by increasing nitric oxide bioavailability and reducing oxidative stress. *Aging* **2017**, *9*, 187–208. [CrossRef] [PubMed]
23. Askarizadeh, A.; Barreto, G.E.; Henney, N.C.; Majeed, M.; Sahebkar, A. Neuroprotection by curcumin: A review on brain delivery strategies. *Int. J. Pharm.* **2020**, *585*, 119476. [CrossRef]
24. Fernandes, M.; Lopes, I.; Magalhaes, L.; Sarria, M.P.; Machado, R.; Sousa, J.C.; Botelho, C.; Teixeira, J.; Gomes, A.C. Novel concept of exosome-like liposomes for the treatment of Alzheimer's disease. *J. Control. Release* **2021**, *336*, 130–143. [CrossRef]
25. Marslin, G.; Sarmento, B.F.; Franklin, G.; Martins, J.A.; Silva, C.J.; Gomes, A.F.; Sarria, M.P.; Coutinho, O.M.; Dias, A.C. Curcumin Encapsulated into Methoxy Poly(Ethylene Glycol) Poly(epsilon-Caprolactone) Nanoparticles Increases Cellular Uptake and Neuroprotective Effect in Glioma Cells. *Planta Med.* **2017**, *83*, 434–444. [PubMed]
26. Guo, C.; Ma, J.; Zhong, Q.; Zhao, M.; Hu, T.; Chen, T.; Qiu, L.; Wen, L. Curcumin improves alcoholic fatty liver by inhibiting fatty acid biosynthesis. *Toxicol. Appl. Pharmacol.* **2017**, *328*, 1–9. [CrossRef]
27. Kheiripour, N.; Plarak, A.; Heshmati, A.; Asl, S.S.; Mehri, F.; Ebadollahi-Natanzi, A.; Ranjbar, A.; Hosseini, A. Evaluation of the hepatoprotective effects of curcumin and nanocurcumin against paraquat-induced liver injury in rats: Modulation of oxidative stress and Nrf2 pathway. *J. Biochem. Mol. Toxicol.* **2021**, *35*, e22739. [CrossRef]
28. Soleimani, V.; Sahebkar, A.; Hosseinzadeh, H. Turmeric (*Curcuma longa*) and its major constituent (curcumin) as nontoxic and safe substances: Review. *Phytother. Res.* **2018**, *32*, 985–995. [CrossRef]
29. Dei Cas, M.; Ghidoni, R. Dietary Curcumin: Correlation between Bioavailability and Health Potential. *Nutrients* **2019**, *11*, 2147. [CrossRef]
30. Ma, Z.; Wang, N.; He, H.; Tang, X. Pharmaceutical strategies of improving oral systemic bioavailability of curcumin for clinical application. *J. Control. Release* **2019**, *316*, 359–380. [CrossRef]

31. Kharat, M.; Du, Z.; Zhang, G.; McClements, D.J. Physical and Chemical Stability of Curcumin in Aqueous Solutions and Emulsions: Impact of pH, Temperature, and Molecular Environment. *J. Agric. Food Chem.* **2017**, *65*, 1525–1532. [CrossRef] [PubMed]
32. Lao, C.D.; Ruffin, M.T.T.; Normolle, D.; Heath, D.D.; Murray, S.I.; Bailey, J.M.; Boggs, M.E.; Crowell, J.; Rock, C.L.; Brenner, D.E. Dose escalation of a curcuminoid formulation. *BMC Complement. Altern. Med.* **2006**, *6*, 10. [CrossRef] [PubMed]
33. Anand, P.; Kunnumakkara, A.B.; Newman, R.A.; Aggarwal, B.B. Bioavailability of Curcumin: Problems and Promises. *Mol. Pharm.* **2007**, *4*, 807–818. [CrossRef] [PubMed]
34. Ding, S.; Zhang, N.; Lyu, Z.; Zhu, W.; Chang, Y.-C.; Hu, X.; Du, D.; Lin, Y. Protein-based nanomaterials and nanosystems for biomedical applications: A review. *Mater. Today* **2021**, *43*, 166–184. [CrossRef]
35. La Barbera, L.; Mauri, E.; D'Amelio, M.; Gori, M. Functionalization strategies of polymeric nanoparticles for drug delivery in Alzheimer's disease: Current trends and future perspectives. *Front. Neurosci.* **2022**, *16*, 939855. [CrossRef]
36. Ko, C.N.; Zang, S.; Zhou, Y.; Zhong, Z.; Yang, C. Nanocarriers for effective delivery: Modulation of innate immunity for the management of infections and the associated complications. *J. Nanobiotechnol.* **2022**, *20*, 380. [CrossRef] [PubMed]
37. Patra, J.K.; Das, G.; Fraceto, L.F.; Campos, E.V.R.; Rodriguez-Torres, M.D.P.; Acosta-Torres, L.S.; Diaz-Torres, L.A.; Grillo, R.; Swamy, M.K.; Sharma, S.; et al. Nano based drug delivery systems: Recent developments and future prospects. *J. Nanobiotechnol.* **2018**, *16*, 71. [CrossRef]
38. Azandaryani, A.H.; Kashanian, S.; Jamshidnejad-Tosaramandani, T. Recent Insights into Effective Nanomaterials and Biomacromolecules Conjugation in Advanced Drug Targeting. *Curr. Pharm. Biotechnol.* **2019**, *20*, 526–541. [CrossRef]
39. Lam, P.L.; Wong, W.Y.; Bian, Z.; Chui, C.H.; Gambari, R. Recent advances in green nanoparticulate systems for drug delivery: Efficient delivery and safety concern. *Nanomedicine* **2017**, *12*, 357–385. [CrossRef]
40. Mitchell, M.J.; Billingsley, M.M.; Haley, R.M.; Wechsler, M.E.; Peppas, N.A.; Langer, R. Engineering precision nanoparticles for drug delivery. *Nat. Rev. Drug Discov.* **2021**, *20*, 101–124. [CrossRef] [PubMed]
41. Zhang, Y.; Chan, H.F.; Leong, K.W. Advanced materials and processing for drug delivery: The past and the future. *Adv. Drug Deliv. Rev.* **2013**, *65*, 104–120. [CrossRef] [PubMed]
42. Ban, C.; Jo, M.; Park, Y.H.; Kim, J.H.; Han, J.Y.; Lee, K.W.; Kweon, D.H.; Choi, Y.J. Enhancing the oral bioavailability of curcumin using solid lipid nanoparticles. *Food Chem.* **2020**, *302*, 125328. [CrossRef] [PubMed]
43. Rizzo, L.Y.; Golombek, S.K.; Mertens, M.E.; Pan, Y.; Laaf, D.; Broda, J.; Jayapaul, J.; Mockel, D.; Subr, V.; Hennink, W.E.; et al. In Vivo Nanotoxicity Testing using the Zebrafish Embryo Assay. *J. Mater. Chem. B* **2013**, *1*, 3918–3925. [CrossRef]
44. Lieschke, G.J.; Currie, P.D. Animal models of human disease: Zebrafish swim into view. *Nat. Rev. Genet.* **2007**, *8*, 353–367. [CrossRef] [PubMed]
45. Lu, S.; Hu, M.; Wang, Z.; Liu, H.; Kou, Y.; Lyu, Z.; Tian, J. Generation and Application of the Zebrafish heg1 Mutant as a Cardiovascular Disease Model. *Biomolecules* **2020**, *10*, 1542. [CrossRef]
46. Zhang, A.; Wu, M.; Tan, J.; Yu, N.; Xu, M.; Liu, W.; Zhang, Y. Establishment of a zebrafish hematological disease model induced by 1,4-benzoquinone. *Dis. Models Mech.* **2019**, *12*, dmm037903. [CrossRef]
47. Bowley, G.; Kugler, E.; Wilkinson, R.; Lawrie, A.; van Eeden, F.; Chico, T.J.A.; Evans, P.C.; Noel, E.S.; Serbanovic-Canic, J. Zebrafish as a tractable model of human cardiovascular disease. *Br. J. Pharmacol.* **2022**, *179*, 900–917. [CrossRef]
48. MacRae, C.A.; Peterson, R.T. Zebrafish as tools for drug discovery. *Nat. Rev. Drug Discov.* **2015**, *14*, 721–731. [CrossRef]
49. McCutcheon, V.; Park, E.; Liu, E.; Sobhebidari, P.; Tavakkoli, J.; Wen, X.Y.; Baker, A.J. A Novel Model of Traumatic Brain Injury in Adult Zebrafish Demonstrates Response to Injury and Treatment Comparable with Mammalian Models. *J. Neurotrauma* **2017**, *34*, 1382–1393. [CrossRef] [PubMed]
50. Oliveira, A.C.N.; Sarria, M.P.; Moreira, P.; Fernandes, J.; Castro, L.; Lopes, I.; Corte-Real, M.; Cavaco-Paulo, A.; Real Oliveira, M.; Gomes, A.C. Counter ions and constituents combination affect DODAX: MO nanocarriers toxicity in vitro and in vivo. *Toxicol. Res.* **2016**, *5*, 1244–1255. [CrossRef]
51. Annunziato, M.; Eeza, M.N.H.; Bashirova, N.; Lawson, J.; Matysik, J.; Benetti, D.; Grosell, M.; Stieglitz, J.D.; Alia, A.; Berry, J.P. An integrated systems-level model of the toxicity of brevetoxin based on high-resolution magic-angle spinning nuclear magnetic resonance (HRMAS NMR) metabolic profiling of zebrafish embryos. *Sci. Total Environ.* **2022**, *803*, 149858. [CrossRef] [PubMed]
52. Hermsen, S.A.; van den Brandhof, E.J.; van der Ven, L.T.; Piersma, A.H. Relative embryotoxicity of two classes of chemicals in a modified zebrafish embryotoxicity test and comparison with their in vivo potencies. *Toxicol. Vitro* **2011**, *25*, 745–753. [CrossRef] [PubMed]
53. Nellore, J.; Nandita, P. Paraquat exposure induces behavioral deficits in larval zebrafish during the window of dopamine neurogenesis. *Toxicol. Rep.* **2015**, *2*, 950–956. [CrossRef]
54. Oliveira, A.I.; Pinho, C.; Fonte, P.; Sarmento, B.; Dias, A.C.P. Development, characterization, antioxidant and hepatoprotective properties of poly(E-caprolactone) nanoparticles loaded with a neuroprotective fraction of Hypericum perforatum. *Int. J. Biol. Macromol.* **2018**, *110*, 185–196. [CrossRef] [PubMed]
55. Woodruff, M.A.; Hutmacher, D.W. The return of a forgotten polymer—Polycaprolactone in the 21st century. *Prog. Polym. Sci.* **2010**, *35*, 1217–1256. [CrossRef]
56. OECD. *Test No. 236: Fish Embryo Acute Toxicity (FET) Test*; OECD Guidelines for the Testing of Chemicals, Section 2; OECD Publishing: Paris, France, 2013. [CrossRef]

57. Beekhuijzen, M.; de Koning, C.; Flores-Guillen, M.E.; de Vries-Buitenweg, S.; Tobor-Kaplon, M.; van de Waart, B.; Emmen, H. From cutting edge to guideline: A first step in harmonization of the zebrafish embryotoxicity test (ZET) by describing the most optimal test conditions and morphology scoring system. *Reprod. Toxicol.* **2015**, *56*, 64–76. [CrossRef]
58. Selderslaghs, I.W.; Van Rompay, A.R.; De Coen, W.; Witters, H.E. Development of a screening assay to identify teratogenic and embryotoxic chemicals using the zebrafish embryo. *Reprod. Toxicol.* **2009**, *28*, 308–320. [CrossRef]
59. Machado, S.; Gonzalez-Ballesteros, N.; Goncalves, A.; Magalhaes, L.; Sarria Pereira de Passos, M.; Rodriguez-Arguelles, M.C.; Castro Gomes, A. Toxicity in vitro and in Zebrafish Embryonic Development of Gold Nanoparticles Biosynthesized Using Cystoseira Macroalgae Extracts. *Int. J. Nanomed.* **2021**, *16*, 5017–5036. [CrossRef]
60. Sárria, M.P.; Vieira, A.; Lima, Â.; Fernandes, S.P.; Lopes, I.; Gonçalves, A.; Gomes, A.C.; Salonen, L.M.; Espiña, B. Acute ecotoxicity assessment of a covalent organic framework. *Environ. Sci. Nano* **2021**, *8*, 1680–1689. [CrossRef]
61. Kimmel, C.B.; Ballard, W.W.; Kimmel, S.R.; Ullmann, B.; Schilling, T.F. Stages of embryonic development of the zebrafish. *Dev. Dyn.* **1995**, *203*, 253–310. [CrossRef]
62. Chang, H.B.; Chen, B.H. Inhibition of lung cancer cells A549 and H460 by curcuminoid extracts and nanoemulsions prepared from *Curcuma longa* Linnaeus. *Int. J. Nanomed.* **2015**, *10*, 5059–5080.
63. Bai, J.; Zhang, Y.; Tang, C.; Hou, Y.; Ai, X.; Chen, X.; Zhang, Y.; Wang, X.; Meng, X. Gallic acid: Pharmacological activities and molecular mechanisms involved in inflammation-related diseases. *Biomed. Pharmacother.* **2021**, *133*, 110985. [CrossRef] [PubMed]
64. Correia, S.C.; Santos, R.X.; Perry, G.; Zhu, X.; Moreira, P.I.; Smith, M.A. Mitochondrial importance in Alzheimer's, Huntington's and Parkinson's diseases. *Adv. Exp. Med. Biol.* **2012**, *724*, 205–221.
65. Gou, M.; Men, K.; Shi, H.; Xiang, M.; Zhang, J.; Song, J.; Long, J.; Wan, Y.; Luo, F.; Zhao, X.; et al. Curcumin-loaded biodegradable polymeric micelles for colon cancer therapy in vitro and in vivo. *Nanoscale* **2011**, *3*, 1558–1567. [CrossRef] [PubMed]
66. López-Córdoba, A.; Medina-Jaramillo, C.; Piñeros-Hernandez, D.; Goyanes, S. Cassava starch films containing rosemary nanoparticles produced by solvent displacement method. *Food Hydrocoll.* **2017**, *71*, 26–34. [CrossRef]
67. Lukasiewicz, S.; Mikolajczyk, A.; Blasiak, E.; Fic, E.; Dziedzicka-Wasylewska, M. Polycaprolactone Nanoparticles as Promising Candidates for Nanocarriers in Novel Nanomedicines. *Pharmaceutics* **2021**, *13*, 191. [CrossRef]
68. Orellana-Paucar, A.M.; Serruys, A.S.; Afrikanova, T.; Maes, J.; De Borggraeve, W.; Alen, J.; Leon-Tamariz, F.; Wilches-Arizabala, I.M.; Crawford, A.D.; de Witte, P.A.; et al. Anticonvulsant activity of bisabolene sesquiterpenoids of *Curcuma longa* in zebrafish and mouse seizure models. *Epilepsy Behav.* **2012**, *24*, 14–22. [CrossRef]
69. Genge, C.E.; Lin, E.; Lee, L.; Sheng, X.; Rayani, K.; Gunawan, M.; Stevens, C.M.; Li, A.Y.; Talab, S.S.; Claydon, T.W.; et al. The Zebrafish Heart as a Model of Mammalian Cardiac Function. *Rev. Physiol. Biochem. Pharmacol.* **2016**, *171*, 99–136.
70. Tanner, C.M.; Kamel, F.; Ross, G.W.; Hoppin, J.A.; Goldman, S.M.; Korell, M.; Marras, C.; Bhudhikanok, G.S.; Kasten, M.; Chade, A.R.; et al. Rotenone, paraquat, and Parkinson's disease. *Environ. Health Perspect.* **2011**, *119*, 866–872. [CrossRef]
71. Federico, A.; Cardaioli, E.; Da Pozzo, P.; Formichi, P.; Gallus, G.N.; Radi, E. Mitochondria, oxidative stress and neurodegeneration. *J. Neurol. Sci.* **2012**, *322*, 254–262. [CrossRef]
72. Gao, L.; Yuan, H.; Xu, E.; Liu, J. Toxicology of paraquat and pharmacology of the protective effect of 5-hydroxy-1-methylhydantoin on lung injury caused by paraquat based on metabolomics. *Sci. Rep.* **2020**, *10*, 1790. [CrossRef] [PubMed]
73. Joseph, T.P.; Jagadeesan, N.; Sai, L.Y.; Lin, S.L.; Sahu, S.; Schachner, M. Adhesion Molecule L1 Agonist Mimetics Protect Against the Pesticide Paraquat-Induced Locomotor Deficits and Biochemical Alterations in Zebrafish. *Front. Neurosci.* **2020**, *14*, 458. [CrossRef] [PubMed]
74. Pinho, B.R.; Reis, S.D.; Hartley, R.C.; Murphy, M.P.; Oliveira, J.M.A. Mitochondrial superoxide generation induces a parkinsonian phenotype in zebrafish and huntingtin aggregation in human cells. *Free Radic. Biol. Med.* **2019**, *130*, 318–327. [CrossRef] [PubMed]
75. Correa, F.; Buelna-Chontal, M.; Hernandez-Resendiz, S.; Garcia-Nino, W.R.; Roldan, F.J.; Soto, V.; Silva-Palacios, A.; Amador, A.; Pedraza-Chaverri, J.; Tapia, E.; et al. Curcumin maintains cardiac and mitochondrial function in chronic kidney disease. *Free Radic. Biol. Med.* **2013**, *61*, 119–129. [CrossRef] [PubMed]
76. Alkhader, E.; Roberts, C.J.; Rosli, R.; Yuen, K.H.; Seow, E.K.; Lee, Y.Z.; Billa, N. Pharmacokinetic and anti-colon cancer properties of curcumin-containing chitosan-pectinate composite nanoparticles. *J. Biomater. Sci. Polym. Ed.* **2018**, *29*, 2281–2298. [CrossRef]

Article

Spray-Drying Microencapsulation of Pink Guava (*Psidium guajava*) Carotenoids Using Mucilage from *Opuntia ficus-indica* Cladodes and Aloe Vera Leaves as Encapsulating Materials

María Carolina Otálora [1,*], Andrea Wilches-Torres [1] and Jovanny A. Gómez Castaño [2,*]

1. Grupo de Investigación en Ciencias Básicas (NÚCLEO), Facultad de Ciencias e Ingeniería, Universidad de Boyacá, Tunja 050030, Boyacá, Colombia; andreawilches@uniboyaca.edu.co
2. Grupo Química-Física Molecular y Modelamiento Computacional (QUIMOL®), Facultad de Ciencias, Universidad Pedagógica y Tecnológica de Colombia (UPTC), Avenida Central del Norte, Tunja 050030, Boyacá, Colombia
* Correspondence: marotalora@uniboyaca.edu.co (M.C.O.); jovanny.gomez@uptc.edu.co (J.A.G.C.)

Citation: Otálora, M.C.; Wilches-Torres, A.; Gómez Castaño, J.A. Spray-Drying Microencapsulation of Pink Guava (*Psidium guajava*) Carotenoids Using Mucilage from *Opuntia ficus-indica* Cladodes and Aloe Vera Leaves as Encapsulating Materials. *Polymers* **2022**, *14*, 310. https://doi.org/10.3390/polym14020310

Academic Editors: Lorenzo Antonio Picos Corrales, Angel Licea-Claverie and Grégorio Crini

Received: 14 December 2021
Accepted: 11 January 2022
Published: 13 January 2022

Publisher's Note: MDPI stays neutral with regard to jurisdictional claims in published maps and institutional affiliations.

Copyright: © 2022 by the authors. Licensee MDPI, Basel, Switzerland. This article is an open access article distributed under the terms and conditions of the Creative Commons Attribution (CC BY) license (https://creativecommons.org/licenses/by/4.0/).

Abstract: In this work, the capacity of the mucilage extracted from the cladodes of *Opuntia ficus-indica* (OFI) and aloe vera (AV) leaves as wall material in the microencapsulation of pink guava carotenoids using spray-drying was studied. The stability of the encapsulated carotenoids was quantified using UV–vis and HPLC/MS techniques. Likewise, the antioxidant activity (TEAC), color (CIELab), structural (FTIR) and microstructural (SEM and particle size) properties, as well as the total dietary content, of both types of mucilage microcapsules were determined. Our results show that the use of AV mucilage, compared to OFI mucilage, increased both the retention of β-carotene and the antioxidant capacity of the carotenoid microcapsules by around 14%, as well as the total carotenoid content (TCC) by around 26%, and also favors the formation of spherical-type particles (Ø ≅ 26 µm) without the apparent damage of a more uniform size and with an attractive red-yellow hue. This type of microcapsules is proposed as a convenient alternative means to incorporate guava carotenoids, a natural colorant with a high antioxidant capacity, and dietary fiber content in the manufacture of functional products, which is a topic of interest for the food, pharmaceutical, and cosmetic industries.

Keywords: guava; carotenoids; microencapsulation; mucilage; spray-drying; aloe vera; *Opuntia ficus-indica*

1. Introduction

Guava (*Psidium guajava* L.) is a tropical fruit native to Central America, northern South America, and the Caribbean that has spread to many other tropical and subtropical regions, including southern North America, southwestern Europe, tropical Africa, Oceania, and south and southeast Asia. Its annual production is estimated at more than 55 million tons worldwide, with India being the largest producer at around 45% of the total production [1].

Guava is a rich source of vitamin C and dietary fiber (hemicellulose, pectin, cellulose, and lignin), and has a lower content of other micronutrients such as vitamins A, B1 (thiamine), B2 (riboflavin), B3 (niacin), and B5 (pantothenic acid). In addition, it is a good source of phosphorus, calcium, iron, potassium, and sodium [2,3]. A single guava fruit can provide as much as 250% of the required daily value of vitamin C [3]. Furthermore, this fruit has been recognized as a potential source of functional compounds with high antioxidant capacities, such as tannins, phenols, triterpenes, flavonoids, and carotenoids [4,5].

Pink guava, a pink-fleshed fruit known as the "apple of the tropics" due to its attractive color, aroma, and flavor characteristics, has been considered an ideal source of β-carotene, γ-carotene, β-cryptoxanthin, rubixanthin, lutein, cryptoflavin, phytofluene neochrome, and lycopene [6,7]. Among these, lycopene ($C_{40}H_{56}$) is the aliphatic hydrocarbon-type carotenoid responsible for the reddish hue of the pink guava pulp. This compound is also attributed to the beneficial effects associated with human health related to the prevention of cardiovascular and degenerative diseases, cancer, diabetes, and inflammation, among others [8–10]. However, lycopene is prone to undergoing isomerization and oxidation reactions when handled in isolation in the presence of oxidants (air and water), light, and temperature, which leads to a significant loss of its bioactive and physicochemical properties [11]. These factors, in addition to its poor solubility in aqueous media and ensuing low bioavailability, restrict its application as a functional bio-ingredient [12].

Microencapsulation using biopolymers as structural materials has been used in the food, pharmaceutical, and cosmetic industries. This process is both an economic and natural strategy used to safely contain and deliver sensitive active or volatile compounds while isolating them from external factors such as temperature, oxygen, light, humidity, pH variations, macromolecules, and metabolites. This ensures their bioavailability and antioxidant capacity; however, to ensure the effectiveness and stability of the encapsulation, a critical aspect that must be properly designed is the structuring of the biopolymeric wall [13,14].

Natural hydrocolloids such as mucilage, gums, and gelatins have been shown to be functional biopolymers that possess an intrinsic antioxidant capacity and high dietary fiber content. These compounds also contain suitable physicochemical properties for use as encapsulating wall materials [15–19]. The mucilage extracted from the cladodes of *Opuntia ficus-indica* (OFI) is a heteropolysaccharide with thickening, binding, emulsifying, stabilizing, and gelling properties. It is rich in dietary fiber (73.9 g/100 g of powdered mucilage), which has been used as a wall material in different encapsulation formulations [20,21]. OFI mucilage has a molar mass of 3×10^6 gmol^{-1} and is composed of L-arabinose (24.6–42.0%), D-galactose (21.0–40.1%), D-xylose (22.0–22.2%), L-rhamnose (7.0–13.1%), and α-D-(1 → 4) galacturonic acid (8.0–12.7%) [16]. Likewise, the mucilage extracted from the hydroparenchyma of aloe vera (AV) leaves is a polysaccharide that is considered an excellent wall material to be applied in the encapsulation of active ingredients due to its thickening and emulsifying properties [15,16]. This biopolymer is rich in dietary fiber (37.0 g/100 g of powdered mucilage) and is mainly composed of acemannan, a polysaccharide with a molecular weight of around 40–50 kDa. It contains large amounts of partially acetylated mannose units (>60%), glucose (~20%), and galactose (<10%) [16].

To date, there are few studies focused on the effect of encapsulation on the bioavailability and conservation of the antioxidant capacity of the carotenoids extracted from the guava fruit, using different wall biomaterials. Leite et al. recently demonstrated that lipid core nanocapsules, using polysorbate 80 coated poly-ε-caprolactone as the wall material, could be efficiently applied to stabilize the lycopene-rich extract of red guava. The lipid core nanocapsules optimized the stability of lycopene at 5 °C for 7 months, improved its toxicity against MCF-7 cancer cells, inhibited the production of intracellular ROS and NF-κB in human microglial cells, and did not impact the integrity of the erythrocyte membrane [13]. Chaves et al. more recently prepared microcapsules loaded with guava pulp by means of the spray-drying (SD) technique, using a mixture of inulin and maltodextrin as functional encapsulating material [22]. They found that microcapsule formulation with a higher percentage of inulin resulted in a better retention and stability of the antioxidant activity of the pulp over time. It also resulted in a higher retention of the carotenoid content and a more stable microstructure. Previously, Osorio et al. encapsulated the aqueous extract of pink-fleshed guava fruit by SD using maltodextrin, arabic gum, and their mixtures as wall materials [23]. SEM observations verified the production of spherical microencapsulates, while thermal analyses revealed a higher thermal stability of the maltodextrin particles.

In this study, we investigated the influence on the physicochemical and antioxidant properties of pink guava carotenoids given by the mucilage of OFI cladodes and AV leaves when used as the wall material in microencapsulation processes using the SD technique. The structural (FTIR), microstructural (SEM and particle size), and thermal (DSC/TGA) properties, as well as the total dietary fiber content, were determined for each type of microcapsule formulation, i.e., SD-OFI and SD-AV. For this study, Colombian pink guavas were used, whose current production is estimated at around 550,000 tons per year. This makes Colombia rank 17th among the world producers of this fruit [1].

2. Materials and Methods

2.1. Reagents

Acetonitrile and methanol solvents (HPLC grade), as well as acetone and ethanol solvents (analytical grade), were purchased from Merck (Darmstadt, Germany). 2,2'azinobis (3-ethylbenzthiazoline-6-sulfonic acid) (ABTS) was purchased from Sigma Aldrich (St. Louis, MO, USA).

2.2. Vegetal Materials

Guava fresh fruits (*Psidium guajava* L.) of the Palmira ICA-1 regional variety were purchased at a local supermarket in the city of Tunja in the department of Boyacá, Colombia. The fruits were selected based on their uniform size, firmness, color, and absence of defects (spots, depressions, or cracks). The fruits were washed with distilled water and manually peeled to obtain the pulp (edible part) and seeds. The edible portion was immediately chopped and homogenized in a blender at minimum power for 1 min. Then, the pulp was sieved to remove the seeds, maintained in an amber bottle to protect against photodegradation, and stored at 4 °C until its analysis. The guava pulp presented a pH of 3.85 ± 0.02 and soluble solids content of 8.00 ± 0.03 °Brix, determined with a digital pH meter (ORION™ Versa Star™, Thermo Scientific Inc., Waltham, MA, USA) and a digital hand refractometer (Boeco model 32395, Hamburg, Germany), respectively.

2.3. Characterization of Vegetal Material

The guava pulp was frozen at −80 °C in an ultra-low temperature freezer (Buzzer, model MDF–86V188E, Shanghai, China) for 48 h. Then, the samples were freeze-dried in a Freezone 4.5 L freeze dryer (Labconco, Kansas City, MO, USA) at −84 °C in a vacuum of 0.13 mbar for 48 h. After freeze-drying, the samples were triturated using a food processor and stored in amber bottles until further analysis. The non-microencapsulated lyophilized pulp of guava (LGP) was designated as a control sample.

2.4. Extraction of Wall Materials

For the extraction of mucilage from cladodes of *Opuntia ficus-indica* and aloe vera leaves, the methodology reported by Quinzio et al. [24] and Otálora et al. [16] was used.

The cladodes and aloe vera leaves were washed with distilled water at room temperature, after which the epidermis of each plant material was carefully separated from the inner pulp (i.e., the medulla) using a Teflon knife.

The medulla obtained from the cladodes was cut into small pieces and placed in a 1000 mL beaker. Distilled water at room temperature was added to the beaker in a 1:2 v/v (medulla:water) ratio. The mixture was left for 24 h and manually squeezed to extract the gel. Similarly, the medulla obtained from aloe vera leaves was also cut into small pieces and manually squeezed to extract the gel.

Each of the gels obtained from the cladodes and aloe vera leaves were filtered through a nylon cloth, separately. Then, 95% ethanol was added to each of the filtered gels in a ratio of 3:1 (ethanol:centrifuged gel) at room temperature. The mixture was then stirred manually with a glass rod until the appearance of a white-milky gel (precipitated mucilage) was present.

The mucilage gels obtained from the cladodes and aloe vera leaves were placed in Petri dishes and dried in an oven (UM 400, Memmert, Schwabach, Germany) for 24 h at 60 and 105 °C, respectively. The dried materials were manually macerated in a porcelain mortar until a fine powder was obtained. The OFI (obtained from cladodes) and AV (obtained from aloe vera leaves) powdered mucilages were placed in separate high-density polyethylene bags and stored in a desiccator at room temperature with a relative humidity of 30% until use.

2.5. Preparation of Microcapsules

A mass of 1.2 g of mucilage obtained from Opuntia ficus-indica cladodes and 0.4 g of mucilage obtained from aloe vera leaves were each dissolved separately in 100 mL of distilled water at 18 °C. To ensure complete solubilization, both solutions were constantly stirred at 300 rpm for 2 h using a magnetic stirrer (C-MAG HS 7 S000, IKA, Staufen im Breisgau, Germany). The lyophilized guava pulp was mixed separately with each mucilage solution in a ratio of 1:10 w/v (10 g:100 mL), respectively. In each case, the feed mixture was kept under constant magnetic stirring at room temperature until homogeneity was achieved. The total solids content of the feed mixes was 5.55% for SD-AV and 4.34% for SD-OFI. Each mixture was then fed into a mini spray dryer (B-290, Büchi Labortechnik, Switzerland) with aspiration maintained at 100% (35 m^3/h) to maximize the separation rate of the cyclone [25] and a compressed air pressure of 40 bar. The spray drier used a nozzle with an internal diameter of 0.7 mm, a feed flow of 350 mL/h, and an inlet air temperature of 120 °C. The two microencapsulates that were obtained from this procedure, SD-OFI (pulp/mucilage obtained from cladodes *Opuntia ficus-indica*) and SD-AV (pulp/mucilage obtained from aloe vera leaves), were stored in the dark at −20 °C for subsequent analysis and use.

2.6. Characterization of the Guava Pulp Microcapsules

2.6.1. Carotenoids Quantification

The lyophilized guava pulp (LGP) and microcapsules (SD-OFI and SD-AV) were accurately weighed and dispersed in acetone, stirred for 1 min at room temperature, and filtered with a Millipore membrane (0.45 μm). UV-Vis analysis was run at 450 nm in order to quantify the β-carotene equivalent present in each of the samples (LGP, SD-OFI, and SD-AV) using a UV–vis spectrophotometer (V530, Jasco, Hachioji, Tokyo, Japan). The total carotenoid content was determined from the standard curve of β-carotene. The retention of total carotenoid in the microencapsulated was expressed as μg β-carotene/g of sample.

2.6.2. Lycopene and β-Carotene Analysis by HPLC–MS

The HPLC–MS analyses of the lyophilized pulp of guava (LGP), SD-OFI, and SD-AV microencapsulates were performed using an Acquity ultraperformance liquid chromatography (UPLC) system equipped with a Xevo TQD Mass Spectrometer (Waters Corp., Milford, MA, USA) equipped with an electrospray ionization (ESI) probe that was operated in positive ion mode. An Acquity UPLC® T3 HSS C18 analytical column (2.1 mm × 100 mm, 1.7 μm particle size) was used for the analysis of analytes present in each sample with the column temperature being maintained at 40 °C and using a flow rate of 0.4 mL min^{-1}. The solvent mixture was composed of acetonitrile (solvent A) and methanol (solvent B). The gradient elution program was set as follows: 0–5 min, 100% A; 5–8.5 min, 80% B and 20% A; and 8.5–12 min, 100% A. The injection volume was set at 2 μL. The MS parameters were as follows: capillary voltage was set at 2.5 kV while block and desolvation temperatures were set at 150 °C and 400 °C, respectively. Desolvation gas flow rate was set to 800 L h^{-1} and cone gas was set at 50 L h^{-1}. Cone voltages were set to 68 and 64 V and collision energies were set to 60 and 52 eV for lycopene and β-carotene, respectively.

2.6.3. Trolox Equivalent Antioxidant Capacity (TEAC)

Trolox equivalent antioxidant capacity (TEAC) was measured using the method reported by Re et al. [26]. The lyophilized guava pulp (LGP) and microcapsules (SD-OFI and SD-AV) were accurately weighed and dispersed in methanol, stirred at room temperature, and filtered with a Millipore membrane. The liquid samples were mixed with an ABTS$^{\bullet+}$ solution and its absorbance was read at 734 nm using a UV–vis spectrophotometer (V530, Jasco, Hachioji, Tokyo, Japan). Results were expressed in mmol Trolox equivalents/kg of sample (TEAC).

2.6.4. Color Parameters

The CIELab parameters (L^*, a^*, b^*) of lyophilized guava pulp (LGP) and microencapsulates SD-OFI and SD-AV samples were measured using a colorimeter (CM-5, Konica Minolta Sensing, Inc., Osaka, Japan). The other color parameters, chroma (C^*ab) and hue (hab), were calculated using Equations (1) and (2).

$$C^*_{ab} = [(a^*)^2 + (b^*)^2]^{1/2} \tag{1}$$

$$h_{ab} = \arctan[b^*/a^*] \tag{2}$$

2.7. Fourier-Transform Infrared (FTIR) Spectroscopy

The infrared spectra of the powdered mucilages of *Opuntia ficus-indica* (OFI) and aloe vera (AV) as well as microencapsulates SD-OFI and SD-AV were recorded on a Shimadzu Prestigie 21 spectrophotometer (Duisburg, Germany) equipped with a Michelson-type interferometer, a KBr/Ge beam-splitter, a ceramic lamp, and a DLATGS detector. The FTIR spectra were measured in the range of 4500–500 cm^{-1} with a resolution of 3.0 cm^{-1} and 30 cumulative scans using the attenuated total reflectance/reflection (ATR) technique.

2.8. Microstructural Characterization

2.8.1. Scanning Electron Microscopy (SEM)

The microscopic morphology of the powdered mucilages (OFI and AV) as well as the microencapsulates (SD-OFI and SD-AV) were evaluated by scanning electron microscopy (SEM) using EVO MA 10-Carl Zeiss equipment (Oberkochen, Germany) operating at 20 kV. All samples were coated by gold–palladium sputtering before their examination.

2.8.2. Particle Size

The size distributions and average diameters of the SD-OFI and SD-AV microencapsulates were determined using a laser diffraction particle size analyzer Mastersizer 3000 system equipped with a Hydro MV dispersion unit (Malvern Panalytical Ltd., Malvern, UK). A particle refractive index of 1.52, dispersant refractive index of 1.33, and an obscuration range of 5.18–5.32% were applied. Volume weighted mean diameter D[4,3] and area-volume mean diameter D[3,2] were obtained. The span of the volume-based distribution using droplet size was calculated considering an average diameter equivalent to 90%, 50%, and 10% of the cumulative volume.

2.9. Thermal Characterization

Thermogravimetric analysis (TGA)/differential scanning calorimetry (DSC) of the *Opuntia ficus-indica* and aloe vera powdered mucilages, as well as microencapsulates, SD-OFI, and SD-AV were performed on a TA Instrument (SDT Q600 V20.9 Build 20, New Castle, DE, USA). Argon was used as a purge gas (100 mL/min). The dried samples of OFI and AV mucilages were placed in aluminum pans and heated from 20 to 600 °C at a heating rate of 10 °C/min.

2.10. Dietary Fiber Content

Total dietary fiber (TDF) contents in SD-FI and SD-AV microencapsulated samples were determined using a total dietary fiber test kit (TDF-100A), provided by Sigma Aldrich (St. Louis, MO, USA), which is based on the enzymatic–gravimetric method AOAC 985.29 [27].

2.11. Statistical Analysis

Physicochemical data, as presented in Table 1, were reported as the mean ± standard deviation ($n = 3$). The Kruskal–Wallis test was performed to identify differences among the means using InfoStat/P version 1.1 statistical software. Differences at probability level $p < 0.05$ were considered significant.

Table 1. Physicochemical characterization of lyophilized guava pulp (LGP) and guava pulp microcapsules (SD-AV and SD-OFI). Different letters in the same row and column for each parameter indicate a statistical difference ($p < 0.05$) between samples.

Parameter	LGP	SD-AV	SD-OFI
TCC [1]	190.9 ± 0.2 [a]	42.6 ± 0.2 [b]	31.4 ± 0.3 [c]
TEAC [2]	32.2 ± 0.3 [a]	26.8 ± 0.2 [b]	23.2 ± 0.2 [c]
Total dietary fiber [3]	-	22.8 ± 0.1 [b]	32.1 ± 0.1 [a]
Color parameters			
L^* (luminosity)	70.83 ± 0.02 [b]	68.94 ± 0.01 [b]	77.80 ± 0.02 [a]
a^*	13.07 ± 0.04 [a]	7.48 ± 0.02 [b]	5.89 ± 0.02 [c]
b^*	24.27 ± 0.01 [b]	31.61 ± 0.03 [a]	22.39 ± 0.01 [b]
C_{ab}^* (chroma)	27.29 ± 0.02 [b]	31.78 ± 0.03 [a]	22.56 ± 0.02 [b]
h_{ab}^* (hue)	61.55 ± 0.03 [b]	77.27 ± 0.01 [a]	77.19 ± 0.01 [a]

[1] TCC is represented as µg β-carotene/g of sample in dry base. [2] TEAC is represented as mmol Trolox equivalents/kg of sample in dry base. [3] Expressed as g/100 g.

3. Results and Discussion

3.1. Total Carotenoid Content, Antioxidant Capacity, Dietary Fiber Content, and Color Parameters

Table 1 shows the physicochemical parameters of total carotenoid content (TCC), antioxidant capacity (TEAC), dietary fiber content, and color (CIE*Lab*) of the pink guava pulp microcapsules (SD-OFI and SD-AV), together with the respective values for the lyophilized unencapsulated sample (LGP).

As shown in Table 1, spray-drying microencapsulation of the guava pulp using AV and OFI mucilage as the wall material led to a decrease in the total carotenoid contents of around 78% and 84%, respectively, when compared to the unencapsulated (LGP) sample. This poor carotenoid retention after atomization can be attributed to the high temperature to which these thermolabile compounds were exposed during the spray-drying microencapsulation. These results agree with recent studies reported in the review by Santos [28]. Likewise, the stability of microencapsulated carotenoids may have been influenced by the temperature at which they were stored until their later physicochemical characterization. The final load of total carotenoids in the aloe mucilage microcapsule (SD-AV) resulted in a slightly higher amount (3.6%) compared to the cactus mucilage microcapsule (SD-OFI). This difference was correlated with the amount of total dissolved solids in the feed mixes (5.55% for SD-AV and 4.34% for SD-OFI) introduced into the spray-drying equipment. The total solids content impacts the viscosity of the feed mix, which influences the size of the atomized droplets and particles, and thus modifies the retention rate of carotenoids [28]. The slight increase in the total solid concentration in the SD-AV feed mixture led to an increase in its viscosity, which reduced the circulation of the pulp core within the droplets. It also reduced the formation time of the semi-permeable membrane around the core, resulting in a decrease in the loss of carotenoids by a migration towards the surface of the microparticle. Another phenomenon that could have contributed to the higher presence of carotenoids in the SD-AV microcapsule is the greater degree of interaction of the pulp biomolecules with the macromolecules of the AV mucilage [29].

Despite the significant total loss of carotenoids that occurred during the spray-drying of the guava pulp microencapsulation, the antioxidant capacity (TEAC) measured in SD-AV and OFI-AV microcapsules presented losses of only around 17% and 28%, respectively, compared to the LGP sample (Table 1). This result indicates that the main active principles responsible for the antioxidant capacities (of both the mucilage biopolymer and the guava pulp) were largely preserved during the spray-drying microencapsulation process. Since the amount of guava pulp in both feed mixes was the same, the significantly ($p < 0.05$) higher TEAC value in the SD-AV microcapsules compared to the SD-OFI microcapsules (26.8 vs. 23.2 mmol equivalents of Trolox/kg of sample on a dry basis, respectively) can be attributed to the cooperative effect resulting from the lycopene content of pink guava and phenolic content of AV mucilage [30] (Section 3.3). Similar core-wall contributing effects on antioxidant capacities have been observed in microcapsules of betaxanthin from *Opuntia megacantha* fruits using a mixture of maltodextrin and cactus cladode mucilage as the encapsulating agent [31].

The color parameters listed in Table 1 show how the luminosity value (L^*) (i.e., the light/bright appearance) for the SD-OFI microcapsules was higher than the values obtained in both the lyophilized guava pulp and the aloe vera microcapsule. This is possibly due to the whiter color coming from the higher ratio in weight of OFI mucilage in relation to the guava pulp (pulp:mucilage ratio of 1:0.12 for SD-OFI vs. 1:0.04 for SD-AV) used in the feed mixture for this microencapsulation. The range of values of the a^* and b^* parameters indicate that the color of both the microcapsules and the lyophilized guava pulp was framed in red-yellow hues. The chroma parameter (Cab^*) was significantly lower ($p < 0.05$) for the SD-OFI microcapsules (i.e., lower a^* and b^* values) when compared to the SD-AV microcapsules. This behavior could be correlated with a higher degradation rate of lycopene (pink pigment), which would coincide with the lower antioxidant capacity of the SD-OFI microcapsule. As shown below, this greater loss of lycopene in the SD-OFI microcapsules was confirmed by HPLC-MS quantification (Section 3.2) and could be correlated with their more cracked microscopic morphology (Section 3.4). Finally, the parameter (h_{ab}^*) agreed with a yellowish hue for both types of microcapsules (77.19 for SD-OFI and 77.27 for SD-AV). This was associated with the average size determined for these particles (25.9 μm for SD-OFI and 26.4 μm for SD-AV, see Section 3.4). This result coincides with the reported effect of increasing the value of h_{ab}^* as a function of a decrease in the particle size [31].

As reported by other studies [22,23], the nature and concentrations of the wall materials are variables that play an important role in the color parameters of microcapsules produced by spray-drying. Shishir et al. observed a decrease in the a^*/b^* value for microcapsules produced with DE 10 maltodextrins as the encapsulating material [22]. Meanwhile, Osorio et al. reported an increase in color parameters ($+a^*$, $+b^*$) in the production of microcapsules using DE 19–20 maltodextrin as the encapsulating agent [23]. In the present work, the concentration of OFI and AV mucilages did influence the chromatic parameters (Cab^*) of the microcapsules. However, the difference in the color parameters between the microecapsulation studies using maltodextrin and those carried out in the present work may have been related to the fact that maltodextrin is a dextrinization of starch, in that it influences the dextrose equivalent (DE), while that mucilage is a polysaccharide.

The total dietary fiber content of 9.3 g/100 g of sample was greater in the SD-OFI microcapsules when compared to the SD-AV microcapsules. This behavior could be associated with the concentration of the mucilage, and therefore, the total dietary fiber content in the feed-mixture, since the amount of guava pulp was the same. Camacho, et al. [32] reported a total dietary fiber content in pink guava pulp, Palmira ICA-1 regional variety, of 5.42 ± 0.07%. Otálora et al. [16] determined a total dietary fiber content *Opuntia ficus indica* and aloe vera powder mucilage of 73.9 g/100 g of sample and 37.0 g/100 g of sample, respectively. Therefore, both types of microcapsules could be incorporated into food products with health benefits for consumers.

3.2. HPLC-MS Identification Analysis

The determination of the contents of lycopene and β-carotene in the SD-OFI and SD-AV guava pulp microcapsules and the lyophilized guava pulp (LGP) sample was carried out using HPLC-MS, the chromatograms of which are shown in Figure 1.

The quantification of β-carotene by HPLC-MS shows a minimal variation in its content between the three types of samples analyzed (179 for LGP, 181 for SD-OFI, and 211 mg/kg dry sample for SD-AV). This suggests a small effect on this component during the spray-drying process. The slightly lower β-carotene content in the LGP sample compared to the SD-AV and SD-OFI microcapsules may have be related to the protective effect against oxidative processes provided by the mucilage biomaterial. The microscopic structure of LGP is characterized by an irregular and angular morphology (Section 3.4) that could contribute to the diffusion of oxygen (β-carotene degradation factor) in contrast to the structure of the mucilage microparticles that provide a barrier effect against oxidation of the carotenoids loaded in their interior [29,33]. The SD-OFI microcapsules showed a lower level of β-carotene retention compared to the SD-AV microcapsules which was attributed to the greater internal porosity and lower wall thickness of the OFI mucilage wall (*vide infra*). This structural difference was what allowed for a greater degree of oxygen diffusion. The more irregular microstructure of the SD-OFI encapsulating matrix may be have been related to the higher viscosity of the emulsion entering the spray-drying equipment (see Section 3.1). This can cause a longer time for the formation of the OFI mucilage film around the droplets as well as a greater exposure to heat during the drying process [34,35]. Similar behavior was observed in lemon essential oil microcapsules using a mixture of mesquite gum and chia mucilage as a wall material [36].

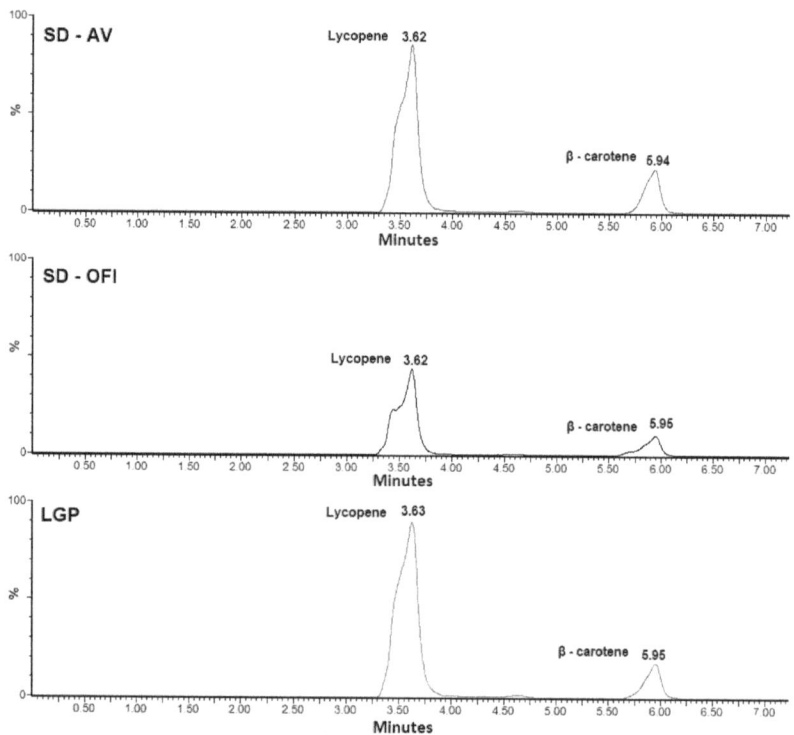

Figure 1. HPLC-MS chromatograms of SD-OFI and SD-AV guava pulp microcapsules and lyophilized guava pulp (LGP).

In contrast, significantly lower amounts of lycopene were found in the SD-OFI (4155 mg/kg dry sample) and SD-AV (4104 mg/kg dry sample) microcapsules compared to the amount determined in the lyophilized pulp sample (9281 mg/kg). This showed a loss of approximately 55% of the lycopene content by thermal effects (Section 3.1) during the guava pulp microencapsulation process by spray-drying in both cases.

3.3. Fourier Transform Infrared Spectroscopy (FTIR)

The FTIR spectra of lyophilized guava pulp, SD-OFI and SD-AV microcapsules, as well as the respective spectra of the free OFI and AV mucilage, are presented in Figure 2a,b.

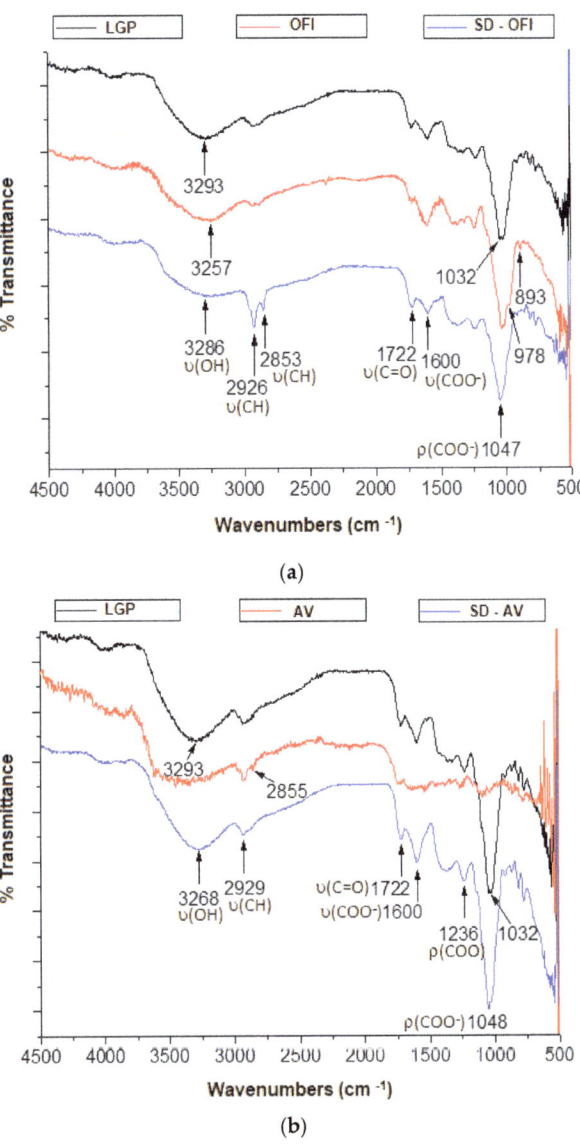

Figure 2. FTIR spectra of lyophilized guava pulp (LGP), mucilages ((**a**) OFI and (**b**) AV) and microcapsules (SD-OFI and SD-AV).

The main infrared absorbances observed in the microcapsules are attributable to the most representative functional groups present both in the OFI or AV mucilage as well as in the guava pulp. The band at 3268 and 3286 cm^{-1} observed in the FTIR spectra of the SD-AV and SD-OFI microcapsules, respectively, was attributed to a combined contribution of the hydroxyl group from both R-OH and the C(O)-OH moieties involved in intramolecular hydrogen bonds. The absorptions observed at 2929 (SD-AV) and at 2926 and 2853 cm^{-1} (SD-OFI) were assigned to the aliphatic C–H stretch. Meanwhile, the signals at 1722 and 1600 cm^{-1}, observed in both microcapsule FTIR spectra, were attributed to the carbonyl (C=O) and COO- stretching modes, respectively, of the D-galactopyranosyluronic acid residues. The position of the carbonyl signal at 1722 cm^{-1} in both microcapsule spectra was associated with the interaction of the carbonyl of the acetyl residue in mucilage and the C=O stretching groups in the hemicellulose, pectin, and lignin structures in the pulp of guava. A similar behavior was observed in gallic acid microcap19sules using aloe vera as a wall material [15]. On the other hand, the intense signals at 1047 (SD-OFI) and 1048 cm^{-1} (SD-AV) were related to the compound's movement between the polysaccharide skeleton and the stretching vibration of the C–O flexion, indicating the presence of alcohols, ethers, esters, and carboxylic acids which are mainly linked to phenolic acids and flavonoids [16,37]. These spectroscopic results suggest that guava molecular structures and wall materials are preserved during the microencapsulation process and that no new chemical bonds were formed. In other words, the carotenoids present in the guava pulp were physically trapped in the AV and OFI mucilage matrix, and only hydrogen bonds and van der Waals interactions were formed [38].

3.4. Microscopy Morphology and Particle Size

SEM micrographs of the OFI and AV mucilages are presented in Figure 3, while the SEM micrographs for the SD-OFI and SD-AV microcapsules are shown in Figure 4.

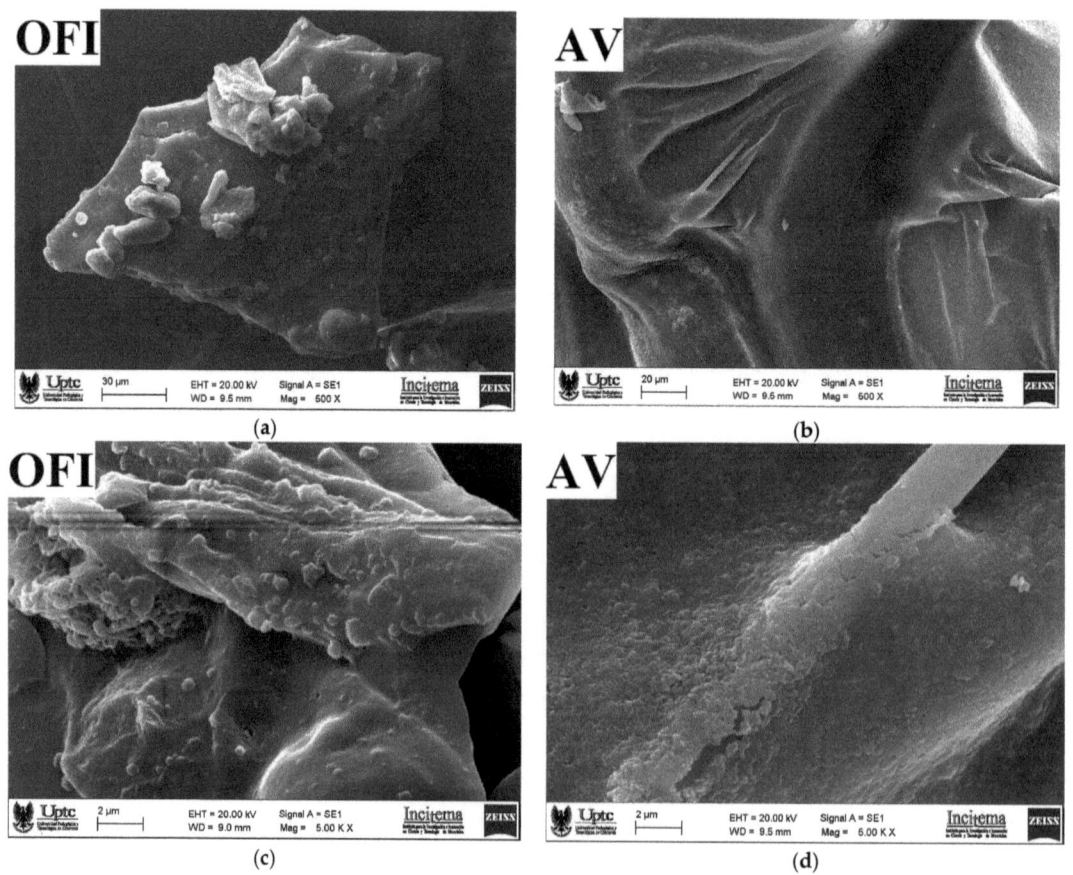

Figure 3. SEM micrograph images of the surface at 500× (**a**,**b**) and 5000× (**c**,**d**) of mucilages of OFI and AV in powder, respectively.

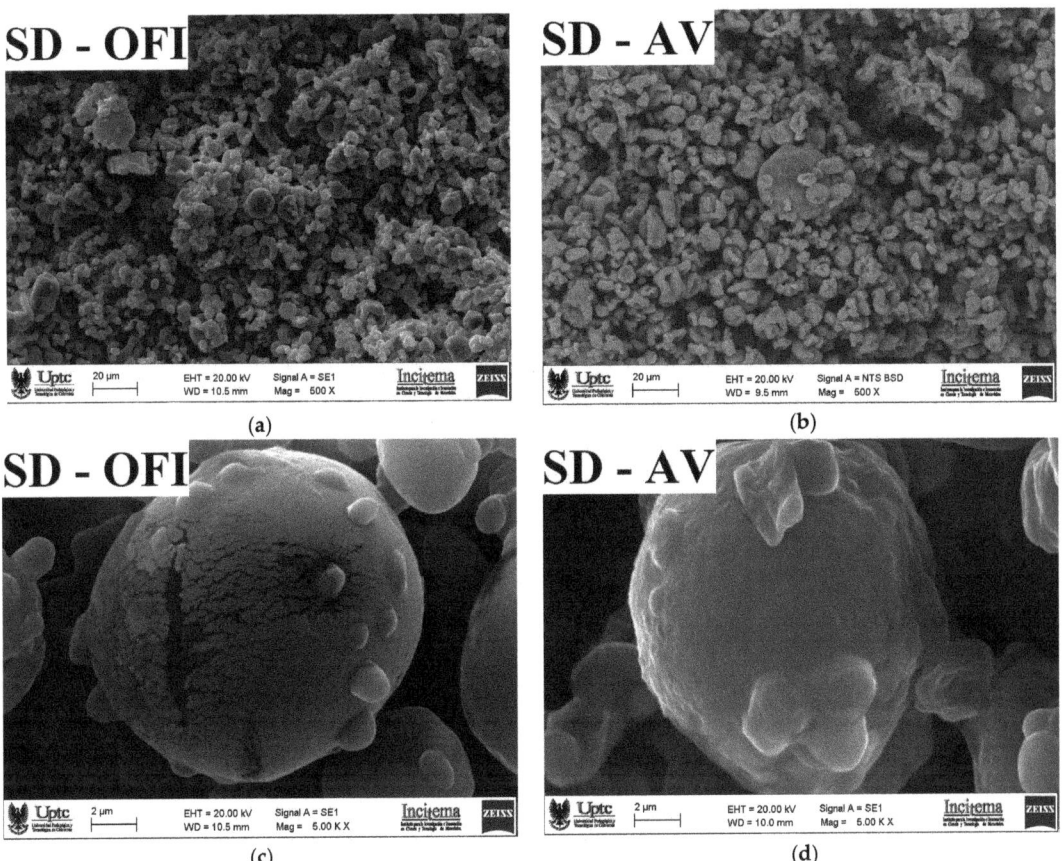

Figure 4. SEM micrograph images of the surface at 500× (**a,b**) and at 5000× (**c,d**) of SD-SD–OFI and SD-AV microcapsules, respectively.

As shown in Figure 3a, the surface structure of the OFI mucilage, captured with a magnification of 500×, was characterized by an irregular, compact, dense, and cracked morphology. In comparison, the surface structure of the AV mucilage (Figure 3b) had a rough, scaly, and porous morphology. The images of the mucilage which were taken with a magnification of 5000×, revealed the presence of small particles, possibly corresponding to protein aggregates adhered to the carbohydrate blocks in the OFI sample (Figure 3c). The images of the AV sample (Figure 3d) show a heterogeneous and slightly rough morphology with some cavities. These types of morphologies have been associated with the conditions used during the extraction and flocculation processes of the mucilage during ethanol precipitation as well as the drying conditions of the sample [39,40]

The SEM micrograph of the surface structure of the SD-OFI microcapsule, observed with a magnification of 500× (Figure 4a), shows an agglomeration effect (adherence) between the particles. This is consistent with a phenomenon of attraction by electrostatic and van der Waals forces that is characteristic of samples with a high amount of carbohydrates [15]. A similar morphology was observed in microcapsules of betaxanthins from orange *Opuntia megacantha* fruits using a mixture of maltodextrin-cactus cladode mucilage as encapsulating agents [41]. On the other hand, the surface of the SD-AV microcapsule (Figure 4b) showed less agglomerate formation, which can be attributed to a greater resistance to the hot air flow during the drying process [41].

The SEM micrographs observed with a magnification of 5000× show particles that were spherical with a cracked and dented morphology in the SD-OFI sample. The cracking and bulging irregularities in these microcapsules can be attributed to the desorption of the air by the droplets and the shrinkage of the particles during the drying process [20]. Similar morphology has been reported for Lacto-bacillus acidophilus La-05 microcapsules protected with flaxseed mucilage and a soluble protein [42]. On the other hand, the SD-AV microcapsules showed an irregularly particle shape (spherical-type), and most of them without apparent damage. This morphology is responsible for discouraging the entry of oxygen into the capsule, thus increasing the stability of the bioactive (β-carotene) compound [43]. However, some of the particles presented a dented morphology with roughness surfaces, which was attributed to the shrinkage of the particles during the drying process typical of microcapsules produced by the spray-drying process [19].

The particle size distribution and the average diameter of the SD-OFI and SD-AV microcapsules are shown in Figure 5. Both types of microcapsules showed a bimodal behavior, i.e., two maxima, with an average maximum particle size of 25.9 (SD-OFI) and 26.4 μm (SD-AV). A similar particle size, around 25.3 μm, was observed in Lactobacillus casei microcapsules using Alyssum homolocarpum mucilage and inulin as wall materials [44]. This mean particle diameter for the SD-OFI and SD-AV microcapsules was related to the size of the droplets formed from the input emulsions during the spray-drying and the viscoelastic characteristics of the mucilage wall materials [45]. The span values of the SD-OFI and SD-AV microcapsules were 1.31 and 1.56, respectively, indicating a uniform particle size distribution and greater homogeneity in both cases. These results are in agreement with those reported by Campo et al. [46], who observed an increase in span values with the addition of high amounts of chia oil in chia seed mucilages nanoparticles.

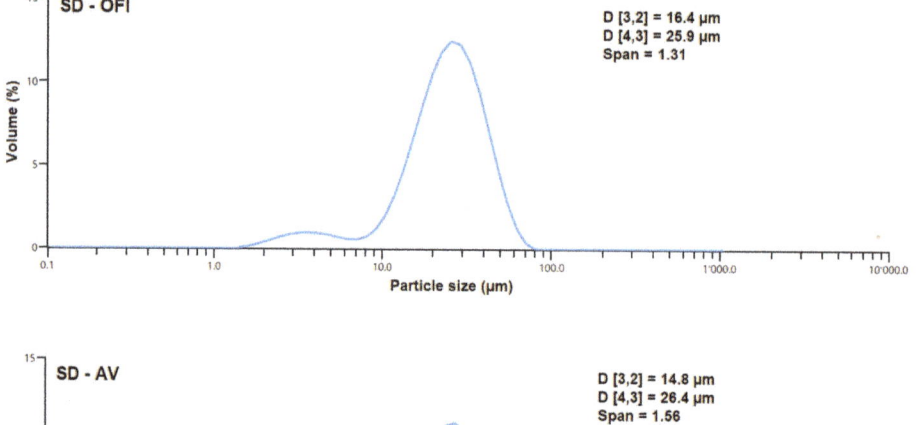

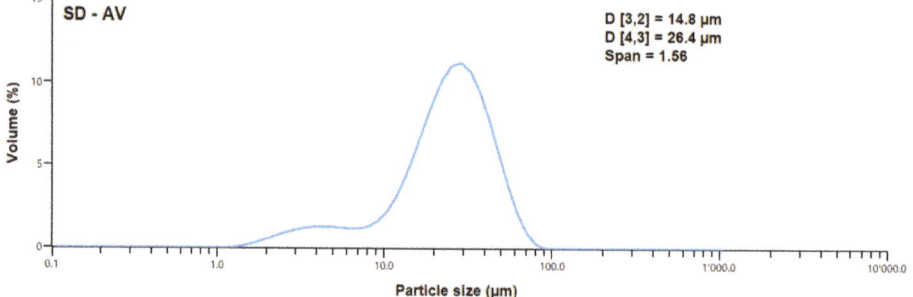

Figure 5. Particle size distribution plots of SD-OFI and SD-AV microcapsules. D[3,2] is area-volume mean diameter and D[4,3] is volume weighted mean diameter calculated by the software of the equipment used.

3.5. Thermal Properties

The thermal behavior of the OFI and AV mucilages, as well as the SD-OFI and SD-AV microcapsules, are shown in Figure 6.

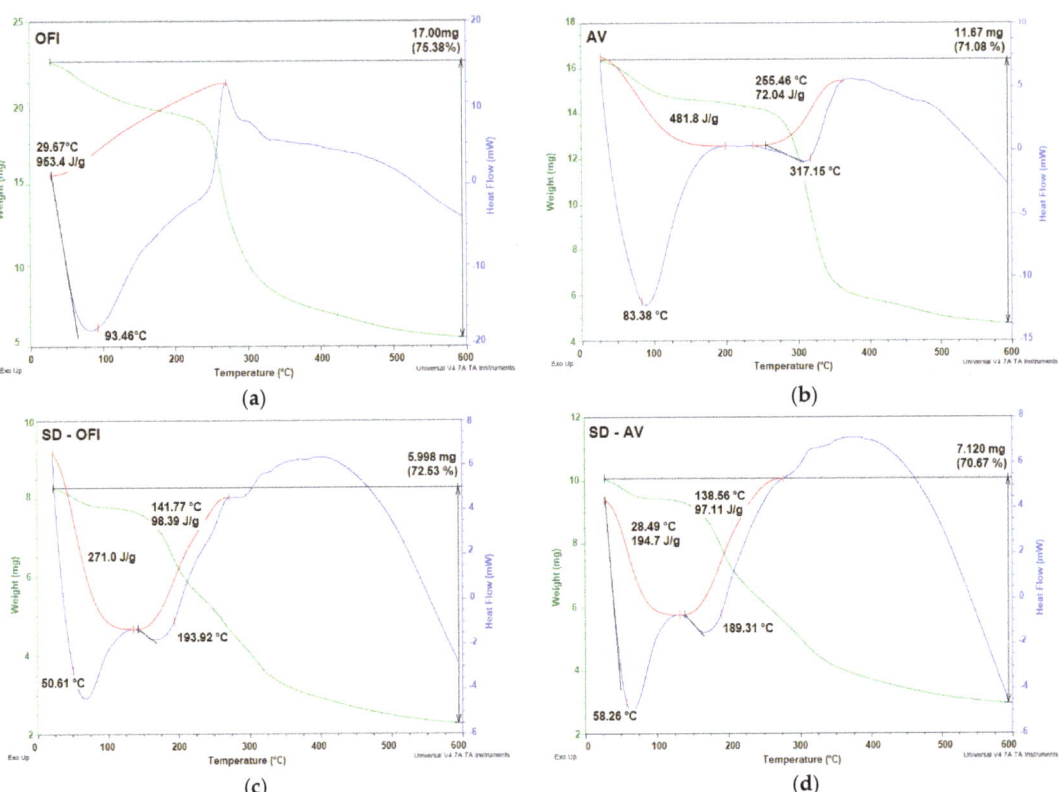

Figure 6. TGA/DSC thermograms of OFI (a) and AV (b) mucilages, and SD-OFI (c) and SD-AV (d) microcapsules.

The thermograms of the OFI and AV mucilages showed the typical thermal characteristics reported for these biomaterials [16]. Both SD-OFI and SD-AV microcapsule samples showed similar thermal behavior. The thermogram of the SD-OFI and SD-AV microcapsule materials was characterized by two endothermic events. The first event occurred between 30 and 150 °C (Tg of 50.61 °C for SD-OFI and 58.26 °C for SD-AV) with a weight loss of less than 5%. This was attributed to the evaporation of the water adsorbed and structurally incorporated into theses samples. The presence of water in these samples was attributed to the hydrophilic nature of the functional groups of the polysaccharides present in the OFI and AV mucilages. The Tg values represent the interaction and the crosslinking density between the components of the guava pulp and the wall materials, as well as the stiffness, structure of the polymer chain, and molecular weight of the species contained in the microcapsules [47]. The second thermal event occurred between 150 and 200 °C (with a peak of 193.92 °C for SD-OFI and 189.31 °C for SD-AV), with a mass loss of 67.53% (SD-OFI) and 65.67% (SD-AV). This was attributed to the decomposition/volatilization of the microcapsule material. Similar thermal behavior was observed in annatto extract microcapsules using Psyllium mucilage as a wall material [48]. In general terms, our thermal analysis revealed that SD-OFI microcapsules were slightly more thermally stable than SD-AV microcapsules due to their higher melting and degradation temperature.

4. Conclusions

The present study revealed the possibility of microencapsulating pink guava carotenoids by spray-drying (SD) using natural hydrocolloids (mucilages) extracted from cladodes of *Opuntia ficus-indica* (OFI) and aloe vera (AV) leaves as wall materials. This microcapsule formulation was possible given the good emulsifying capacity of the OFI and AV mucilages, allowing at the same time to protect against oxidation and by-pass the hydrophobic nature of guava carotenoids. The stability of the feed spray-drying emulsion was determined by the properties of the microcapsules. Meanwhile the concentration of the encapsulating agent impacted the characteristics of the microcapsules. The SD-AV microcapsules presented a higher content of 11.2 µm/g-dry-sample of carotenoids and 14% more antioxidant capacity with an irregular (almost spherical) particle morphology without apparent damage. In contrast, the SD-OFI microcapsules showed a greater thermal stability and dietary fiber content (29% higher). Red guava pulp microcapsules using OFI and AV mucilages as encapsulating material can be a valued source of bio-functional natural ingredients (colorant with anti-radical power) to be incorporated into food products with benefits for the health of the consumer. These microcapsules may have important commercial applications in the future in the health-promoting natural additive market. For instance, they can be applied as food fortification ingredients, due to the wall material rich in dietary fiber that can provide therapeutic effects different from those provided by carotenoids (i.e., an antioxidant capacity). An important direction of this research was the acquisition of polymers from the recovery of food by-products, a current trend in microencapsulation by spray-drying.

Author Contributions: Data curation, M.C.O. and J.A.G.C.; formal analysis, J.A.G.C. and M.C.O.; investigation, A.W.-T. and M.C.O.; project administration, M.C.O.; writing (original draft preparation), M.C.O.; writing (review and editing), J.A.G.C. and M.C.O. All authors have read and agreed to the published version of the manuscript.

Funding: This work was funded by the Universidad de Boyacá and the Universidad Pedagógica y Tecnológica de Colombia (UPTC) through the interinstitutional project SGI 2384 of the Vicerrectoría de Investigaciones of the Universidad Pedagógica y Tecnológica de Colombia.

Institutional Review Board Statement: Not applicable.

Informed Consent Statement: Not applicable.

Data Availability Statement: The data presented in this study are available on request from the corresponding author.

Acknowledgments: The authors greatly acknowledge the financial support provided by the Universidad de Boyacá and the Universidad Pedagógica y Tecnológica de Colombia (UPTC).

Conflicts of Interest: The authors declare that there is no conflict of interest.

References

1. Tridge.com. Available online: https://www.tridge.com/intelligences/guava/production (accessed on 11 November 2021).
2. Chang, Y.P.; Woo, K.K.; Gnanaraj, C. Pink guava. In *Valorization of Fruit Processing By-Products*; Galanakis, C.M., Ed.; Elsevier Inc.: Amsterdam, The Netherlands, 2020; pp. 227–252. [CrossRef]
3. Nutritiondata.self.com. Available online: https://nutritiondata.self.com/facts/fruits-and-fruit-juices/1927/2 (accessed on 11 November 2021).
4. Chang, S.K.; Alasalvar, C.; Shanhidi, F. Superfruits: Phytochemicals, antioxidant efficacies and health effects—A comprehensive review. *Crit. Rev. Food Sci. Nutr.* **2018**, *59*, 1580–1604. [CrossRef]
5. Kong, K.W.; Ismail, A. Lycopene content and lipophilic antioxidant capacity of by-products from *Psidium guajava* fruits produced during puree production industry. *Food Bioprod. Process.* **2011**, *89*, 53–61. [CrossRef]
6. Thaipong, K.; Boonprako, U.; Crosby, K.; Cisneros-Zevallos, L.; Hawkins, D. Comparison of ABTS, DPPH, FRAP, and ORAC assays for estimating antioxidant activity from guava fruit extracts. *J. Food Compost. Anal.* **2006**, *19*, 669–675. [CrossRef]
7. Koguishi de Brito, C.A.; Becker, P.; de Souza, J.C.; André, H.M. In vitro antioxidant capacity, phenolic, ascorbic acid and lycopene content of guava (*Psidium guajava* L.) juices and nectars. *Bol. Cent. Pesqui. Process. Aliment.* **2009**, *27*, 175–182. [CrossRef]

8. Nagarajan, J.; Ramanan, R.N.; Raghunandan, M.E.; Galanakis, C.M.; Krishnamurthy, N.P. Carotenoids. In *Nutraceutical and Functional Food Components. Effects of Innovative Processing Techniques*; Galanakis, C.M., Ed.; Elsevier Inc.: Amsterdam, The Netherlands, 2017; pp. 259–296. [CrossRef]
9. Santos, W.N.L.; Sauthier, M.C.S.; Santos, A.M.P.; Santana, D.A.; Azevedo, R.S.A.; Caldas, J.C. Simultaneous determination of 13 phenolic bioactive compounds in guava (*Psidium guajava* L.) by HPLC-PAD with evaluation using PCA and Neural Network Analysis (NNA). *Microchem. J.* **2017**, *133*, 583–592. [CrossRef]
10. Vasconcelos, A.G.; Amorim, A.G.N.; dos Santos, R.C.; de Souza, J.M.T.; Souza, L.K.M.; Araújo, T.S.L.; Nicolau, L.A.D.; Carvalho, L.L.; de Aquino, P.E.A.; da Silva Martins, C.; et al. Lycopene rich extract from red guava (*Psidium guajava* L.) displays anti-inflammatory and antioxidant profile by reducing suggestive hallmarks of acute inflammatory response in mice. *Int. Food Res. J.* **2017**, *99*, 959–968. [CrossRef] [PubMed]
11. Souza, A.L.R.; Hidalgo-Chávez, D.W.; Pontes, S.M.; Gomes, F.S.; Cabral, L.M.C.; Tonon, R.V. Microencapsulation by spray drying of a lycopene-rich tomato concentrate: Characterization and stability. *LWT—Food Sci. Technol.* **2018**, *91*, 286–292. [CrossRef]
12. Rehman, A.; Tong, Q.; Jafari, S.M.; Assadpour, E.; Shehzad, Q.; Aadil, R.M.; Iqbal, M.W.; Rashed, M.M.A.; Mushtaq, B.S.; Ashraf, W. Carotenoid-loaded nanocarriers: A comprehensive review. *Adv. Colloid Interface Sci.* **2020**, *275*, 102048. [CrossRef] [PubMed]
13. Vasconcelos, A.G.; Valim, M.O.; Amorim, A.G.N.; do Amaral, C.P.; de Almeida, M.P.; Borges, T.K.S.; Socodato, R.; Portugal, C.C.; Brand, G.D.; Mattos, S.J.C.; et al. Cytotoxic activity of poly-ε-caprolactone lipid-core nanocapsules loaded with lycopene-rich extract from red guava (*Psidium guajava* L.) on breast cancer cells. *Int. Food Res. J.* **2020**, *136*, 109548. [CrossRef] [PubMed]
14. Gheonea, I.; Aprodu, I.; Cîrciumaru, A.; Rapeanu, G.; Bahrim, G.E.; Stanciuc, N. Microencapsulation of lycopene from tomatoes peels by complex coacervation and freeze-drying: Evidences on phytochemical profile, stability and food applications. *J. Food Eng.* **2021**, *288*, 110166. [CrossRef]
15. Medina-Torres, L.; Núñez-Ramírez, D.M.; Calderas, F.; González-Laredo, R.F.; Minjares-Fuentes, R.; Valadez-García, M.A.; Bernad-Bernad, M.J.; Manero, O. Microencapsulation of gallic acid by spray drying with aloe vera mucilage (aloe barbadensis miller) as wall material. *Ind. Crops Prod.* **2019**, *138*, 111461. [CrossRef]
16. Otálora, M.C.; Wilches-Torres, A.; Gómez Castaño, J.A. Extraction and Physicochemical Characterization of Dried Powder Mucilage from *Opuntia ficus-indica* Cladodes and Aloe Vera Leaves: A Comparative Study. *Polymers* **2021**, *13*, 1689. [CrossRef]
17. Soukoulis, C.; Gaiani, C.; Hoffmann, L. Plant seed mucilage as emerging biopolymer in food industry applications. *Curr. Opin. Food Sci.* **2018**, *22*, 28–42. [CrossRef]
18. De Campo, C.; Dick, M.; dos Santos, P.P.; Costa, T.M.H.; Paese, K.; Guterres, S.S.; Rios, A.O.; Flôres, S.H. Zeaxanthin nanoencapsulation with Opuntia monacantha mucilage as structuring material: Characterization and stability evaluation under different temperatures. *Colloids Surf. A Physicochem. Eng. Asp.* **2018**, *558*, 410–421. [CrossRef]
19. Soto-Castro, D.; Gutiérrez Miguel Chávez, M.; León-Martínez, F.; Santiago-García, P.A.; Aragón-Lucero, I.; Antonio-Antonio, F. Spray drying microencapsulation of betalain rich extracts from Escontria chiotilla and Stenocereus queretaroensis fruits using cactus mucilage. *Food Chem.* **2019**, *272*, 715–722. [CrossRef]
20. Otálora, M.C.; Gómez Castaño, J.A.; Wilches-Torres, A. Preparation, study and characterization of complex coacervates formed between gelatin and cactus mucilage extracted from cladodes of *Opuntia ficus-indica*. *LWT—Food Sci. Technol.* **2019**, *112*, 108234. [CrossRef]
21. Medina-Torres, L.; García-Cruz, E.E.; Calderas, F.; González Laredo, R.F.; Sánchez-Olivares, G.; Gallegos-Infante, J.A.; Rocha-Guzmán, N.E.; Rodríguez-Ramírez, J. Microencapsulation by spray drying of gallic acid with nopal mucilage (*Opuntia ficus indica*). *LWT—Food Sci. Technol.* **2013**, *50*, 642–650. [CrossRef]
22. Shishir, M.R.I.; Taip, F.S.; Aziz, N.A.; Talib, R.A. Physical Properties of Spray-dried Pink Guava (*Psidium Guajava*) Powder. *Agric. Agric. Sci. Procedia* **2014**, *2*, 74–81. [CrossRef]
23. Osorio, C.; Forero, D.P.; Carriazo, J.G. Characterization and performance assessment of guava (*Psidium guajava* L.) microencapsulates obtained by spray-drying. *Int. Food Res. J.* **2011**, *44*, 1174–1181. [CrossRef]
24. Quinzio, C.; Corvalán, M.; López, B.; Iturriaga, L. Studying stability against coalescence in tuna mucilage emulsions. *Acta Hortic.* **2009**, *811*, 427–431. [CrossRef]
25. B-290 Mini Spray Dryer Operation Manual 093001N. Available online: https://static1.buchi.com/sites/default/files/downloads/B290_OM_en_I_0.pdf?cf595fc09d939d0eb8f2bee907c35bca8feeee47 (accessed on 7 June 2021).
26. Re, R.; Pellegrini, N.; Proteggente, A.; Pannala, A.; Yang, M.; Rice-Evans, C. Antioxidant activity applying an improved ABTS radical cation decolorization assay. *Free Radic. Biol. Med.* **1999**, *26*, 1231–1237. [CrossRef]
27. Cunniff, P. Enzymatic-gravimetric method. In *Official Methods of Analysis of AOAC International*, 16th ed.; AOAC: Gaithersburg, MD, USA, 1997.
28. Santos, P.D.F.; Rubio, F.T.V.; da Silva, M.P.; Pinho, L.S.; Carmen Sílvia Favaro-Trindade, C.S. Microencapsulation of carotenoid-rich materials: A review. *Food Res. Int.* **2021**, *147*, 110571. [CrossRef] [PubMed]
29. Quinzio, C.; Ayunta, C.; Alancay, M.; de Mishima, B.L.; Iturriaga, L. Physicochemical and rheological properties of mucilage extracted from *Opuntia ficus indica* (L. Miller). Comparative study with guar gum and xanthan gum. *J. Food Meas. Charact.* **2018**, *12*, 459–470. [CrossRef]

30. Medina-Torres, L.; Calderas, F.; Minjares, R.; Femenia, A.; Sanchez-Olivares, G.; Gonzalez-Laredo, F.R.; Santiago-Adame, R.; Ramirez-Nuñez, D.M.; Rodríguez-Ramírez, J.; Manero, O. Structure preservation of Aloe vera (barbadensis Miller) mucilage in a spray drying process. *LWT—Food Sci. Technol.* **2016**, *66*, 93–100. [CrossRef]
31. Otálora, M.C.; Carriazo, J.G.; Osorio, C.; Nazareno, M.A. Encapsulation of cactus (*Opuntia megacantha*) betaxanthins by ionic gelation and spray drying: A comparative study. *Int. Food Res. J.* **2018**, *111*, 423–430. [CrossRef] [PubMed]
32. Camacho, M.I.; Espinal, M.; García, J.; Jiménez, L.; Silva, K.; Restrepo, P. Desarrollo de productos enriquecidos con fibra de guayaba. In *Desarrollo de Productos Funcionales Promisorios a Partir de la Guayaba (Psidium guajava L.) para el Fortalecimiento de la Cadena Productiva*, 1st ed.; Morales, A.L., Melgarejo, L.M., Eds.; Panamericana Formas e Impresos S.A.: Bogotá, Colombia, 2010; pp. 155–174.
33. Haas, K.; Obernberger, J.; Zehetner, E.; Kiesslich, A.; Volkert, M.; Jaeger, H. Impact of powder particle structure on the oxidation stability and color of encapsulated crystalline and emulsified carotenoids in carrot concentrate powders. *J. Food Eng.* **2019**, *263*, 398–408. [CrossRef]
34. Soottitantawat, A.; Bigeard, F.; Yoshii, H.; Furuta, T.; Ohkawara, M.; Linko, P. Influence of emulsion and powder size on the stability of encapsulated D-limonene by spray drying. *Innov. Food Sci. Emerg. Technol.* **2005**, *6*, 107–114. [CrossRef]
35. Cortés-Camargo, S.; Cruz-Olivares, J.E.; Barragán-Huerta, B.; Dublán-García, O.; Román-Guerrero, A.; Pérez-Alonso, C. Microencapsulation by spray drying of lemon essential oil: Evaluation of mixtures of mesquite gum–nopal mucilage as new wall materials. *J. Microencapsul.* **2017**, *34*, 395–407. [CrossRef]
36. Cortés-Camargo, S.; Acuña-Avila, P.E.; Rodríguez-Huezo, M.E.; Román-Guerrero, A.; Varela-Guerrero, V.; Pérez-Alonso, C. Effect of chia mucilage addition on oxidation and release kinetics of lemon essential oil microencapsulated using mesquite gum—Chia mucilage mixtures. *Int. Food Res. J.* **2019**, *116*, 1010–1019. [CrossRef]
37. Carvalho Gualberto, N.; Santos de Oliveira, C.; Pedreira Nogueira, J.; Silva de Jesus, M.; Santos Araujo, H.C.; Rajan, M.; Santos Leite Neta, M.T.; Narain, N. Bioactive compounds and antioxidant activities in the agro-industrial residues of acerola (*Malpighia emarginata* L.), guava (*Psidium guajava* L.), genipap (*Genipa americana* L.) and umbu (*Spondias tuberosa* L.) fruits assisted by ultrasonic or shaker extraction. *Int. Food Res. J.* **2021**, *147*, 110538. [CrossRef]
38. Sun, X.; Xu, Y.; Zhao, L.; Yan, H.; Wang, S.; Wang, D. The stability and bioaccessibility of fucoxanthin in spray-dried microcapsules based on various biopolymers. *RSC Adv.* **2018**, *8*, 35139–35149. [CrossRef]
39. Jiang, Y.; Du, J.; Zhang, L.; Li, W. Properties of pectin extracted from fermented and steeped hawthorn wine pomace: A comparison. *Carbohydr. Polym.* **2018**, *197*, 174–182. [CrossRef] [PubMed]
40. Andrade, L.A.; Aparecida de Oliveira Silva, D.; Nunes, C.A.; Pereira, J. Experimental techniques for the extraction of taro mucilage with enhanced emulsifier properties using chemical characterization. *Food Chem.* **2020**, *327*, 127095. [CrossRef] [PubMed]
41. Cervantes-Martínez, C.V.; Medina-Torres, L.; González-Laredo, R.F.; Calderas, F.; Sánchez-Olivares, G.; Herrera-Valencia, E.E.; Gallegos Infante, J.A.; Rocha-Guzman, N.E.; Rodríguez-Ramírez, J. Study of spray drying of the Aloe vera mucilage (Aloe vera barbadensis Miller) as a function of its rheological properties. *LWT—Food Sci. Technol.* **2014**, *55*, 426–435. [CrossRef]
42. Bustamante, M.; Villarroel, M.; Rubilar, M.; Shene, C. Lactobacillus acidophilusLa-05 encapsulated by spray drying: Effect of mucilage and protein from flaxseed (*Linum usitatissimum* L.). *LWT—Food Sci. Technol.* **2015**, *62*, 1162–1168. [CrossRef]
43. Ortiz-Basurto, R.I.; Rubio-Ibarra, M.E.; Ragazzo-Sanchez, J.A.; Beristain, C.I.; Jimenez Fernandez, M. Microencapsulation of Eugenia uniflora L. juice by spray drying using fructans with different degrees of polymerisation. *Carbohydr. Polym.* **2017**, *175*, 603–609. [CrossRef] [PubMed]
44. Homayouni-Rad, A.; Mortazavian, A.M.; Mashkani, M.G.; Hajipour, N.; Pourjafar, H. Effect of Alyssum homolocarpum mucilage and inulin microencapsulation on the survivability of Lactobacillus casei in simulated gastrointestinal and high temperature conditions. *Biocatal. Agric. Biotechnol.* **2021**, *35*, 102075. [CrossRef]
45. Carneiro, H.C.F.; Tonon, R.V.; Grosso, C.R.F.; Hubinger, M.D. Encapsulation efficiency and oxidative stability of flaxseed oil microencapsulated by spray drying using different combinations of wall materials. *J. Food Eng.* **2013**, *115*, 443–451. [CrossRef]
46. Campo, C.; Santos, P.P.; Costa, T.M.H.; Paese, K.; Guterres, S.S.; Rios, A.O.; Flôres, S.H. Nanoencapsulation of chia seed oil with chia mucilage (*Salvia hispanica* L.) as wall material: Characterization and stability evaluation. *Food Chem.* **2017**, *234*, 1–9. [CrossRef]
47. Otálora, M.C.; Carriazo, J.G.; Iturriaga, L.; Nazareno, M.A.; Osorio, C. Microencapsulation of betalains obtained from cactus fruit (*Opuntia ficus-indica*) by spray drying using cactus cladode mucilage and maltodextrin as encapsulating agents. *Food Chem.* **2015**, *187*, 174–181. [CrossRef]
48. Monge Neto, A.A.; Fonseca Tomazini, L.; Gouveia Mizuta, A.; Gomes Correa, R.C.; Scaramal Madrona, G.; Faria de Moraes, F.; Peralta, R.M. Direct microencapsulation of an annatto extract by precipitation of psyllium husk mucilage polysaccharides. *Food Hydrocoll.* **2021**, *112*, 106333. [CrossRef]

Article

Encapsulation of Blackberry Phenolics and Volatiles Using Apple Fibers and Disaccharides

Mirela Kopjar [1,*], Ivana Buljeta [1], Mario Nosić [1], Ivana Ivić [1], Josip Šimunović [2] and Anita Pichler [1]

[1] Faculty of Food Technology Osijek, Josip Juraj Strossmayer University of Osijek, F. Kuhača 18, 31000 Osijek, Croatia; ivana.buljeta@ptfos.hr (I.B.); m.nosic@yahoo.com (M.N.); ivana.ivic@ptfos.hr (I.I.); anita.pichler@ptfos.hr (A.P.)

[2] Department of Food, Bioprocessing and Nutrition Sciences, North Carolina State University, Raleigh, NC 27695, USA; simun@ncsu.edu

* Correspondence: mirela.kopjar@ptfos.hr; Tel.: +385-3122-4309

Citation: Kopjar, M.; Buljeta, I.; Nosić, M.; Ivić, I.; Šimunović, J.; Pichler, A. Encapsulation of Blackberry Phenolics and Volatiles Using Apple Fibers and Disaccharides. *Polymers* **2022**, *14*, 2179. https://doi.org/10.3390/polym14112179

Academic Editors: Angel Licea-Claverie, Grégorio Crini and Lorenzo Antonio Picos Corrales

Received: 19 April 2022
Accepted: 25 May 2022
Published: 27 May 2022

Publisher's Note: MDPI stays neutral with regard to jurisdictional claims in published maps and institutional affiliations.

Copyright: © 2022 by the authors. Licensee MDPI, Basel, Switzerland. This article is an open access article distributed under the terms and conditions of the Creative Commons Attribution (CC BY) license (https://creativecommons.org/licenses/by/4.0/).

Abstract: The objective of this study was to determine the effect of disaccharides on the encapsulation of the phenolics and volatiles of blackberry juice with the use of apple fiber. For this purpose, apple fiber/blackberry microparticles were prepared as the control, as well as microparticles additionally containing disaccharides, i.e., sucrose or trehalose. Fiber:disaccharide ratios were 1:0.5, 1:1, and 1:2. Formulated microparticles were characterized for total phenolics, proanthocyanidins, individual phenolics, antioxidant activity, flavor profiles, and color parameters. Both applied disaccharides affected the encapsulation of phenolics and volatiles by the apple fibers. Control microparticles had a higher content of phenolics than microparticles with disaccharides. Comparing disaccharides, the microparticles with trehalose had a higher content of phenolics than the ones containing sucrose. The amount of proanthocyanidins in the control microparticles was 47.81 mg PB2/100 g; in trehalose, the microparticles ranged from 39.88 to 42.99 mg PB2/100 g, and in sucrose, the microparticles ranged from 12.98 to 26.42 mg PB2/100 g, depending on the fiber:disaccharide ratio. Cyanidin-3-glucoside was the dominant anthocyanin. Its amount in the control microparticles was 151.97 mg/100 g, while in the trehalose microparticles, this ranged from 111.97 to 142.56 mg /100 g and in sucrose microparticles, from 100.28 to 138.74 mg /100 g. On the other hand, microparticles with disaccharides had a higher content of volatiles than the control microparticles. Trehalose microparticles had a higher content of volatiles than sucrose ones. These results show that the formulation of microparticles, i.e., the selection of carriers, had an important role in the final quality of the encapsulates.

Keywords: apple fiber; trehalose; sucrose; blackberry phenolics; blackberry volatiles

1. Introduction

Color and flavor are very important quality parameters of foods, and they are very often the driving forces for consumers' acceptance of particular food products. Nowadays, the utilization of different food additives is on the rise in order to enhance products with the use of color and/or flavor compounds. Furthermore, these compounds can have important health benefits, i.e., by making food additives functional. Fruits are rich in different bioactive and flavor compounds that make them very popular among consumers, but on the other hand, they are quite fragile. The high post-harvest respiration rate of fruits results in their nutritional and microbiological deterioration and limited shelf life. All of this leads to a loss of quality and health benefits. Traditionally, to avoid these negative aspects of post-harvest respiration, fruits are converted into different products (such as frozen, dried, canned products, jams, jellies, and juices) which can be available to consumers all year long. Nowadays, one of the emerging trends in the food industry is the formulation of delivery systems that can effectively preserve both phenolic and volatile compounds and efficiently incorporate them into food products. For that purpose, the selection of carriers/polymers for these compounds is of major importance. Nowadays, sustainability is quite an important subject in the food industry. Understanding the impact of synthetic materials on the environment directed the food industry toward natural polymers that can be beneficial as efficient eco-friendly materials [1]. The production of different products from apples leads to the formation of a considerable amount of its by-product, i.e., apple pomace, which can be used as a sustainable source for the production of valuable polymers such as apple fibers [2]. Dietary fibers have many health benefits, which makes them an excellent choice as delivery systems of selected active compounds. Some health benefits of fibers that have been highlighted in studies over the years are the lowering of blood cholesterol and sugar levels as well as a decrease in the risk of developing coronary heart diseases, hypertension, diabetes, obesity, and gastrointestinal disorders [3,4]. Besides health benefits, fibers possess some technological and functional attributes that allow them to be used as thickeners, oil texturizers, and moisture retention agents during the preparation of different types of products, including yogurts, ice creams, sauces, dressings, beverages, meat, and bakery products [5–8]. The apple fibers that were chosen for this experiment in addition to cellulose, hemicellulose, and pectin are also a rich source of bioactive compounds [9]. Some progress has already been made in investigations concerning the interactions of dietary fibers and phenolics and volatiles. Hydroxytyrosol and 3,4-dihydroxyphenylglycol, which are phenolics with important biological activity, were complexed with strawberry dietary fibers in order to formulate effective functional food ingredients that could potentially promote intestinal health [10]. Next to strawberry dietary fibers, these phenolics were also complexed with pectin, with the aim of formulating complexes that could potentially be used for colon targeting [11,12]. Pectin and cellulose as cell wall models can be used as carriers of anthocyanins and phenolic acids since they can bind those phenolics [13,14]. Not all phenolics equally bind to dietary fibers. For example, procyanidins bound to components of the apple cell wall, while hydroxycinnamic acids and epicatechin exhibited the opposite trend [15]. The polysaccharides of the apple cell wall affected the antioxidant activity of quercetin [16], while fiber from onions did not [17]. Cellulose can not only be used as an efficient carrier of raspberry phenolics, but also of raspberry volatiles; thus, the obtained complexes can be used as colorants and flavorings in the food industry [18,19]. In our study, we selected blackberries as a source of phenolic and volatile compounds. Blackberries are known as good sources of phenolics with high antioxidant potential, which are responsible for diverse health benefits [20,21]. In addition to their appealing color, blackberries are also known for their pleasant flavor [22,23].

Even with the existence of many different encapsulation techniques [24,25], and although higher unit costs for freeze-drying are a disadvantage, this technique is still widely used to obtain high-value food products, especially when the goal is to have higher polyphenol retention [26,27]. The conditions of the selected encapsulation technique, the chemical and physical properties of the carriers, and active ingredients play a major role in encapsulation efficiency [24]. In order to improve the encapsulation of the phenolics and volatiles of blackberry juice with apple fibers by freeze-drying, two disaccharides were selected, namely sucrose and trehalose. As far as we know, this combination of carriers has not been studied. Sucrose is a commonly used saccharide in fruit product formulations, while the use of trehalose in formulations of fruit products is increasing in frequency. The positive influence of trehalose on volatiles and phenolics in different freeze-dried fruit products has been proven in several studies [28–39], therefore we wanted to determine whether a similar effect could be achieved in combination with apple fibers. An additional benefit of trehalose is its slow digestion, which results in a lower glycemic index, with a lower insulin release compared to sucrose [40]. Likewise, trehalose has substantially lower cariogenic potential compared to sucrose, and it does not have laxative effects similar to the other low-cariogenic bulk sweeteners [41].

The results of our previous research [42] showed that apple fibers could be good carriers of blackberry polyphenols and volatiles. The aim of this study was to evaluate the effect of the addition of disaccharides on the encapsulation of the phenolic and volatile compounds of blackberry juice with apple fiber. In order to investigate whether disaccharides can improve the encapsulation of the mentioned compounds with apple fiber, sucrose and trehalose were used in fiber:disaccharides ratios of 1:0.5, 1:1, and 1:2. All formulated microparticles were evaluated for total phenolics, proanthocyanidins, individual phenolic content, antioxidant activity, color parameters, and flavor profile.

2. Materials and Methods

2.1. Chemicals

Apple fibers in powder form was obtained from Biesterfeld AG (Zagreb, Croatia). 2,2-diphenyl-1-picrylhydrazyl (DPPH), 2,2'-azino-bis(3-ethylbenzothiazoline-6-sulfonic acid) diammonium salt (ABTS), trolox, 4-(dimethylamino)-cinnamaldehyde (DMAC), chlorogenic, caffeic, p-coumaric, gallic and ellagic acids, rutin, quercetin, phloretin, phlorizin, and myrtenol were purchased from Sigma-Aldrich (St. Louis, MO, USA). Potassium persulfate, Folin-Chiocalteu reagent, and sodium carbonate were obtained from Kemika (Zagreb, Croatia). Neocuproine, 2,4,6-tri(2-pyridyl)-s-triazine (TPTZ), and copper (II) chloride were purchased from Acros Organics (Geel, Belgium). Methanol (HPLC grade) was acquired from J.T. Baker (Deventer, The Netherlands), and orthophosphoric acid (HPLC grade > 85%) was obtained from Fisher Scientific (Loughborough, UK). Iron (III) chloride hexahydrate, sodium acetate, ethanol, ammonium acetate, and starch were bought from Gram-mol (Zagreb, Croatia). Cyanidin 3-glucoside was obtained from Extrasynthese (Genay, France), and 3,5-dinitrosalicylic acid was from Alfa Aesar (Kandel Germany). Sucrose was obtained from Fluka (Buchs, Switzerland). Trehalose was a donation from Hayashibara Company (Nagase group, Okayama, Japan).

2.2. Formulation of Microparticles

Blackberry juice was prepared by pressing the berries through the press and additionally filtrating them through cheesecloth to remove solids in order to obtain clear juice. Subsequently, the obtained juice was heated at 90 °C for 3 min to inactivate enzymes. Complexation was performed by mixing selected ingredients: apple fibers and blackberry juice for the control sample, and apple fibers, disaccharides, and blackberry juice for the microparticles, with the addition of disaccharides. For the formulation of the control sample, apple fibers (3 g) were added to blackberry juice (50 mL). The obtained mixture was homogenized by stirring on a magnetic stirrer for 20 min. Microparticles with the added disaccharides (sucrose or trehalose) were prepared by the addition of apple fibers and disaccharides into blackberry juice, followed by the previously described homogenization process. For these microparticles, the fiber:disaccharides ratios were 1:0.5, 1:1, and 1:2. Prepared homogenized mixtures of the control sample and the microparticles with the added disaccharides were transferred to cuvettes and centrifuged for 15 min (4000 rpm). Phenolics and volatiles from blackberry juice adsorbed onto the apple fibers or apple fibers/disaccharides represented the wet precipitate, which was further used to obtain dried microparticles. The fluid part that contained unadsorbed blackberry phenolics and volatiles was discarded. Freeze-drying was performed in a laboratory freeze-dryer (Christ Freeze Dryer, Alpha 1–4, Osterode am Harz, Germany). Firstly, before freeze-drying, wet precipitates were frozen (−18 °C). Conditions for freeze-drying were set as follows: freezing temperature was −55 °C, the temperature of sublimation was from −35 °C to 0 °C, and the vacuum level was 0.220 mbar. Temperatures of the isothermal desorption varied from 0 °C to 21 °C under the vacuum of 0.060 mbar. The complete process of freeze-drying lasted for 12 h.

2.3. Extraction of Microparticles

Extraction of microparticles was necessary in order to evaluate the total phenolics, proanthocyanidins, individual phenolics, and antioxidant activity. A total of 0.2 g of microparticles was weighted, and 10 mL of acidified methanol (HCl:methanol ratio was 1:99) was added. Obtained mixtures were mixed and homogenized on a multi-speed vortex.

These homogenized mixtures were left in the dark for 24 h and then filtered. The resulting extracts were used for the mentioned analyses.

2.4. Evaluation of the Total Phenolics

The total phenolic content in the samples was determined by the Folin–Ciocalteu method [43]. In brief, in a glass tube, 0.2 mL of the extract and 1.8 mL of deionized water were added and mixed. To this mixture, 10 mL (1:10) of the Folin–Ciocalteu reagent and 8 mL of sodium carbonate (7.5%) were added and mixed. The complete mixture was incubated at room temperature in the dark for 2 h, and then the absorbance was measured at 765 nm on a UV/Vis spectrophotometer (Cary 60 UV-Vis, Agilent Technologies, Santa Clara, CA, USA). The measurements were performed in triplicate, and the results were expressed as g of gallic acid equivalents per 100 g of microparticles (g GAE/100 g).

2.5. Evaluation of Total Proanthocyanidins

Total proanthocyanidin content was determined by the application of the colorimetric DMAC method [44]. Briefly, in a glass tube, an aliquot of the extract and a 4-(dimethylamino)-cinnamaldehyde solution were added and mixed. The complete mixture was incubated for 30 min, and then the absorbance was measured at 640 nm. The measurements were performed in triplicate, and the results were expressed as mg of procyanidin B2 equivalents per 100 g of microparticles (mg PB2E/100 g).

2.6. Evaluation of Antioxidant Activity

The ABTS assay followed the method of Arnao et al. [45], with some modifications. Briefly, 0.1 mL of extract was mixed with 3 mL of ABTS reagent and then left in the dark. The antioxidant activity of the samples was also determined by the radical scavenging activity method using 2,2-diphenyl-1-picrylhydrazyl (DPPH) [46]. Cupric reducing antioxidant capacity assay was carried out according to the method of Apak et al. [47]. Moreover, the antioxidant capacity of the samples was determined according to the Benzie and Strain method [48], with some modifications. For each method for blank, the extract was replaced with water, and all measurements were performed in triplicate. Antioxidant activities evaluated by ABTS, DPPH, FRAP, and CUPRAC methods were calculated from the calibration curve, with Trolox as the standard and expressed as µmol TE/100 g.

2.7. Identification and Quantification of Phenolic Compounds

Extracts were filtered through 0.2 µm PTFE membranes and evaluated using an HPLC (high-performance liquid chromatography) system 1260 Infinity II (Agilent technology, Santa Clara, CA, USA). The apparatus was equipped with a quaternary pump, a vial sampler, DAD detector, and a Poroshell 120 EC C-18 column (4.6 × 100 mm, 2.7 µm). The method used to evaluate the phenolic compounds was previously described in a study by Ivić et al. [49]. Calibration curves were obtained for cyanidin-3-glucoside, quercetin, chlorogenic acid, ellagic acid, rutin, caffeic acid, p-coumaric acid, gallic acid, phloretin, and phlorizin hydrate. Cyanidin-3-dioxalylglucoside and two derivates of hydroxycinnamic acids were tentatively identified by comparing their peak spectrum with standards and literature data, and were then quantified using cyanidin-3-glucoside and chlorogenic acid, respectively.

2.8. Color Measurement and Color Change

Chromometer Minolta CR-400 (Minolta; Osaka, Japan) was used for the evaluation of the color parameters of the formulated microparticles by Lab system. The measured color parameters were as follows: L^* (lightness; 0 is black and 100 is white), a^* (redness (+) and greenness (−)), b^* (yellowness (+) and blueness (−)), C^* (the color saturation value-chroma), and $°h$ (the hue angle). The L^*, a^*, and b^* values were used for the calculation of total color change, ΔE.

2.9. Evaluation of Volatiles

Volatiles from the formulated microparticles were extracted by solid-phase microextraction (SPME). A total of 0.3 g of the sample, 4.7 g of water, and 1 g of NaCl were weighed into a 10 mL glass vial. For the extraction of volatiles, SPME fiber coated with a divinylbenzene/carboxen/polydimethylsiloxane (DVB/CAR/PDMS) sorbent (50/30 μm, StableFlex™, Supelco, Bellefonte, PA, USA) was used. The method was described by Vukoja et al. [18]. Compounds were confirmed by matching their mass spectra with the National Institute of Standards and Technology mass spectral database (NIST, East Amwell Township, NJ, USA) and through their retention time (RT) and retention index (RI). Two repetitions were made for each sample. Quantification was conducted with myrtenol as an internal standard, and results were expressed as μg/kg.

2.10. Fourier Transform Infrared with Attenuated Total Reflection (FTIR-ATR) Spectroscopy Analysis

The FTIR-ATR (Cary 630, Agilent, Santa Clara, CA, USA) technique was used to obtain the infrared (IR) spectra of the microparticles. The observed IR spectra are the absorbance of different microparticles versus the wavenumber range, from 4000 cm^{-1} to 600 cm^{-1}.

2.11. Statistical Analysis

A comparison of the formulated microparticles was conducted by analysis of variance (ANOVA) and Fisher's least significant difference (LSD), with the significance defined at $p < 0.05$. Additionally, principal component analysis (PCA) on the volatile compounds was conducted. The software program STATISTICA 13.1 (StatSoft Inc., Tulsa, OK, USA) was used for the statistical analyses. Results were presented as the mean values ± standard deviation.

3. Results

3.1. Phenolic Compounds, Antioxidant Activities, and Color of Formulated Microparticles

Apple fibers, apple fiber/blackberry juice microparticles as control sample, and microparticles with the added disaccharides are presented in Figure 1.

Results of the total phenolics (TP) and proanthocyanidin (PA) content of the formulated microparticles are presented in Table 1. The highest TP content had microparticles that were formulated only with apple fiber and blackberry juice (1.35 GAE g/100 g). The addition of disaccharides during the formulation of the microparticles caused a decrease of TP, i.e., adsorption of phenolics onto the apple fiber was lower in comparison to the apple fiber without the added disaccharides. An increase of added disaccharides had a negative effect on the adsorption of phenolics, and there was no change among microparticles formulated with sucrose and trehalose. Considering only proanthocyanidins, a different trend was observed as compared to total phenolics. The microparticles prepared only with apple fiber had the highest PA content (47.81 mg/100 g). However, unlike for TP, the difference between disaccharide types was observed. The addition of trehalose during formulation had a more positive impact on the adsorption of PA onto the apple fiber than the addition of sucrose. When trehalose was applied during the formulation of the microparticles, higher PA content was adsorbed onto the apple fiber, and with the increase in the amount of trehalose, a decrease of PA content occurred (42.99 mg/100 g, 40.67 mg/100 g, and 39.88 mg/100 g). The same tendency, i.e., a decrease in the adsorption of proanthocyanidins with the increase of sucrose occurred, but PA content was significantly lower (26.42 mg/100 g, 21.41 mg/100 g, and 12.98 mg/100 g).

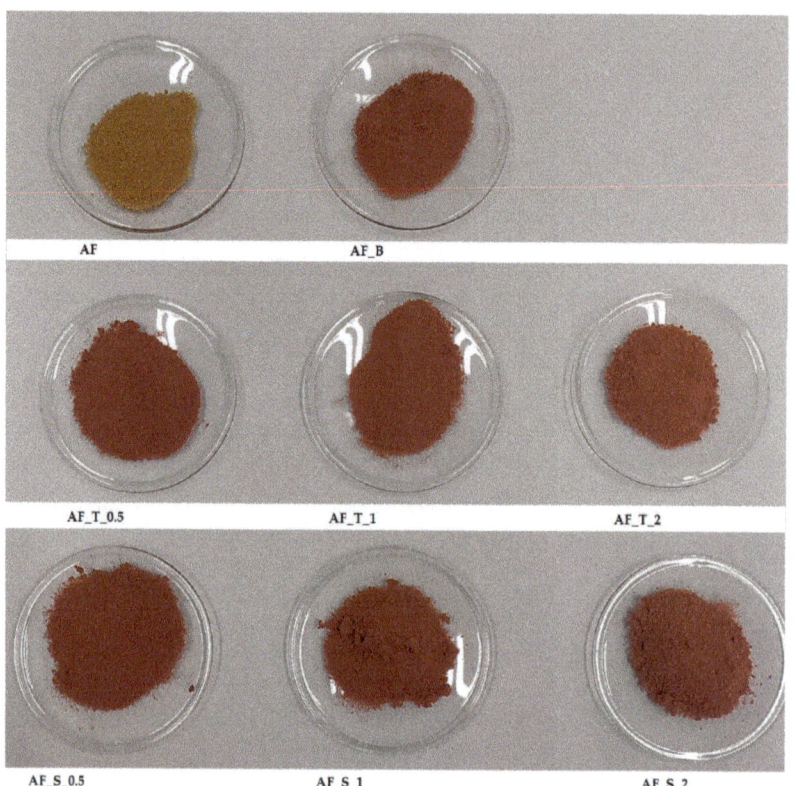

Figure 1. Presentation of formulated microparticles. AF_B—apple fiber/blackberry microparticles (control sample); AF—apple fiber; S—sucrose; T—trehalose; 0.5—ratio of apple fibers:disaccharide (1:0.5); 1—ratio of apple fibers:disaccharide (1:1); 2—ratio of apple fibers:disaccharide (1:2).

Table 1. Total phenolics (TP), proanthocyanidins (PA), and antioxidant activity of formulated microparticles.

Samples	TP	PA	DPPH	ABTS	FRAP	CUPRAC
AF_B	1.35 ± 0.01 [a]	47.81 ± 0.12 [a]	53.18 ± 0.02 [a]	65.65 ± 0.32 [a]	9.13 ± 0.66 [a]	555.39 ± 4.24 [a]
AF_S_0.5	1.29 ± 0.02 [b]	26.42 ± 0.36 [e]	48.05 ± 1.01 [c]	55.98 ± 0.57 [b]	7.55 ± 0.78 [b]	494.75 ± 7.29 [c]
AF_S_1	1.15 ± 0.02 [d]	21.41 ± 0.18 [f]	47.80 ± 0.35 [c]	46.95 ± 0.98 [d]	7.04 ± 0.53 [c]	451.29 ± 7.21 [d]
AF_S_2	0.95 ± 0.02 [f]	12.98 ± 0.59 [g]	42.16 ± 0.51 [e]	41.74 ± 0.74 [e]	5.98 ± 0.41 [d]	378.74 ± 6.85 [e]
AF_T_0.5	1.25 ± 0.01 [b]	42.99 ± 0.60 [b]	51.23 ± 0.01 [b]	54.22 ± 0.92 [b]	8.06 ± 0.58 [a]	551.79 ± 7.77 [a]
AF_T_1	1.15 ± 0.01 [c]	40.67 ± 0.19 [c]	47.93 ± 1.42 [c,d]	51.51 ± 0.35 [c]	7.65 ± 0.69 [b]	506.46 ± 0.38 [b]
AF_T_2	1.02 ± 0.02 [d]	39.88 ± 0.07 [d]	45.55 ± 0.72 [d]	45.98 ± 0.25 [d]	6.41 ± 0.43 [c]	449.94 ± 1.42 [d]

AF_B—apple fiber/blackberry microparticles (control sample); AF—apple fiber; S—sucrose; T—trehalose; 0.5—ratio of apple fibers:disaccharide (1:0.5); 1—ratio of apple fibers:disaccharide (1:1); 2—ratio of apple fibers:disaccharide (1:2). TP—expressed as g GAE/100 g; PA—expressed as mg PB2/100 g; antioxidant activity—expressed as μmol TE/100 g. Values in the same column marked with different superscripts (a–g) are statistically different.

Individual phenolic compounds of blackberry juice and apple fiber are presented in Table 2. The type of disaccharides and their ratio to apple fiber affected the adsorption of individual phenolic compounds (Table 3). Blackberry juice contained quercetin, rutin, cyanidin-3-glucoside, cyanidin-3-dioxalylglucoside, ellagic acid, caffeic acid, chlorogenic acid, p-coumaric acid, and gallic acid, while in microparticles, anthocyanins, quercetin, ellagic acid, and chlorogenic acid were determined with the two derivatives of hydrocinnamic acids—phloretin and phlorizin hydrate; these two derivatives originated from apple fibers. Cyanidin-3-glucoside was determined in the highest amount in the apple fiber/blackberry microparticles (151.97 mg/100 g). With the increased addition of disaccharides, lower amounts of this anthocyanin was detected, which means that the addition of disaccharides during the formulation of microparticles negatively affected the adsorption of this compound onto the apple fiber. Regardless of the disaccharide type, with the increased amounts of added disaccharides, a decrease in cyanidin-3-glucoside content was observed. When the fiber:disaccharides ratios were 1:0.5 and 1:2, higher anthocyanin content was observed in the trehalose microparticles, while the opposite results were observed for the third type of microparticles. The highest amount of cyanidin-3-dioxalylglucoside was also observed in apple fiber/blackberry microparticles (36.71 mg/100 g). Similarly, for cyanidin-3-glucoside, the increased addition of disaccharides lowered the amount of this anthocyanin. Similar results were also obtained for quercetin. The highest amount of this phenolic (92.96 mg/100 g) was determined in the apple fiber/blackberry microparticles. Other microparticles had lower quercetin content. In this case, when the fiber:disaccharide ratios were 1:0.5 and 1:2, higher quercetin content was observed in the trehalose microparticles, while there was no difference for the third type of microparticles. Ellagic acid was also detected in the highest amount in the apple fiber/blackberry microparticles (28.95 mg/100 g). Regarding microparticles with disaccharides, the same trend as for quercetin was observed, thus the trehalose microparticles had a slightly higher content of this phenolic acid. Chlorogenic acid as well as two derivatives of hydroxycinnamic acid were also determined in the microparticles. All phenolic acids were also determined in the highest amount in the apple fiber/blackberry microparticles, but the differences between those microparticles and the microparticles with disaccharides were not so high.

Table 2. Individual phenolic compounds (mg/100 g) of blackberry juice and apple fiber.

Blackberry Juice		Apple Fiber	
Cyanidin 3-glucoside	339.8 ± 0.40	Phloretin	17.64 ± 0.48
Cyanidin 3-dioxalylglucoside	118.8 ± 0.03	Phlorizin hydrate	78.21 ± 0.60
Ellagic acid	27.35 ± 0.00	Chlorogenic acid	51.64 ± 2.70
Caffeic acid	3.8 ± 0.00	HC-1	17.36 ± 0.13
Chlorogenic acid	31.55 ± 0.03	HC-2	15.72 ± 0.02
p-Coumaric acid	41.1 ± 0.00	Quercetin	130.60 ± 5.38
Gallic acid	36.3 ± 0.01	Rutin	9.54 ± 1.14
Quercetin	22.7 ± 0.05		
Rutin	3.7 ± 0.00		

HC-1 and HC-2—two derivatives of hydroxycinnamic acid.

Table 3. Individual phenolic content (mg/100 g) in formulated microparticles.

	AF_B	AF_S_0.5	AF_S_1	AF_S_2	AF_T_0.5	AF_T_1	AF_T_2
C-3-G	151.97 ± 0.11 [a]	138.74 ± 0.42 [c]	127.01 ± 2.88 [d]	100.28 ± 0.28 [g]	142.56 ± 1.06 [b]	120.17 ± 1.28 [e]	111.97 ± 4.39 [f]
C-3-DG	36.71 ± 0.46 [a]	32.58 ± 0.31 [b]	30.61 ± 0.32 [c]	24.90 ± 0.38 [f]	32.20 ± 0.07 [b]	27.98 ± 0.36 [d]	26.20 ± 0.54 [e]
Q	92.96 ± 1.14 [a]	81.72 ± 0.36 [c]	75.99 ± 1.02 [d]	62.93 ± 0.12 [f]	83.69 ± 0.00 [b]	74.64 ± 1.19 [d]	68.27 ± 1.44 [e]
EA	28.95 ± 0.01 [a]	24.35 ± 0.09 [c]	23.43 ± 0.47 [d]	20.20 ± 0.09 [e]	26.32 ± 0.26 [b]	22.89 ± 0.71 [d]	22.77 ± 0.65 [d]
ChA	23.63 ± 0.28 [a]	21.58 ± 0.10 [c]	21.67 ± 0.20 [c]	19.25 ± 0.01 [d]	22.32 ± 0.15 [b]	20.89 ± 0.25 [c]	19.66 ± 0.32 [d]
HC-1	21.24 ± 0.04 [a]	19.59 ± 0.00 [c]	19.91 ± 0.12 [c]	18.08 ± 0.02 [e]	20.65 ± 0.07 [b]	19.78 ± 0.07 [c]	19.16 ± 0.22 [d]
HC-2	17.82 ± 0.04 [a]	16.61 ± 0.00 [d]	17.16 ± 0.07 [b,c]	15.83 ± 0.02 [e]	17.40 ± 0.00 [b]	16.93 ± 0.09 [c]	16.48 ± 0.09 [d]
P	16.98 ± 0.03 [e]	14.45 ± 0.01 [c]	10.89 ± 0.27 [b]	9.09 ± 0.01 [a]	16.47 ± 0.09 [e]	15.39 ± 0.38 [d]	14.84 ± 0.61 [c]
Ph	22.47 ± 0.07 [e]	20.33 ± 0.49 [d]	15.45 ± 0.07 [c]	13.39 ± 0.37 [b]	15.83 ± 0.08 [c]	13.45 ± 0.45 [b]	11.21 ± 0.28 [a]

AF_B—apple fiber/blackberry microparticles (control sample); AF—apple fiber; S—sucrose; T—trehalose; 0.5—ratio of apple fibers:disaccharide (1:0.5); 1—ratio of apple fibers:disaccharide (1:1); 2—ratio of apple fibers:disaccharide (1:2). C-3-G—cyanidin-3-glucoside; C-3-DG—cyanidin-3-dioxalylglucoside; Q—quercetin; EA—ellagic acid; ChA—chlorogenic acid; HC-1 and HC-2—two derivatives of hydroxycinnamic acid; P—phloretin; Ph—phlorizin hydrate. Values in the same row marked (a–g) with different superscripts are statistically different.

Antioxidant activities of the formulated microparticles were evaluated by the application of the DPPH, ABTS, FRAP, and CUPRAC methods (Table 1). For all four methods, it was observed that apple fiber microparticles had the highest antioxidant activities, while all the other microparticles with added disaccharides had lower values. In fact, by comparing the results of antioxidant activities with the results of phenolics, it can be concluded that antioxidant activity followed the same trend. Higher values for microparticles with trehalose addition were obtained with the FRAP and especially CUPRAC methods, which can be correlated with the higher amount of proanthocyanidins adsorbed onto those microparticles.

Furthermore, the L*, a*, and b* color parameters were recorded in order to establish whether disaccharides had influenced the color of formulated microparticles (Table 4). The L* value of disaccharide microparticles was higher than that of microparticles formulated only with apple fiber. In addition, it was observed that with the increase in the amount of disaccharides, the L* values increased. When the fiber:disaccharides ratios were 1:0.5 and 1:1, the L* values were equal regardless of the disaccharide type. The microparticles with the highest amount of trehalose had a higher L* value than the corresponding sucrose microparticles. The a* and b* values of the disaccharide microparticles were higher than those for microparticles formulated only with apple fiber. Slightly higher values were observed for microparticles formulated with sucrose. Based on L*, a*, and b*, the color change (ΔE) was calculated for the microparticles with disaccharides relative to the apple fiber microparticles. The color change increased with the increase in disaccharides, and the highest difference between microparticles was observed when they were added in the highest amount. The °h and C* values were also determined. Slight changes in these parameters were observed in microparticles with added disaccharides as compared to the apple fiber microparticles.

Table 4. Color parameters of formulated microparticles.

Samples	L*	a*	b*	ΔE	°h	C*
AF_B	48.13 ± 0.01 [e]	19.06 ± 0.07 [c]	11.23 ± 0.02 [c,d]		29.81 ± 0.07 [c]	22.59 ± 0.07 [c,d]
AF_S_0.5	48.73 ± 0.06 [d]	20.08 ± 0.06 [a]	11.53 ± 0.02 [c]	0.83	29.88 ± 0.10 [c]	23.15 ± 0.04 [b]
AF_S_1	50.86 ± 0.03 [c]	20.11 ± 0.05 [a]	11.80 ± 0.03 [b]	2.84	30.40 ± 0.10 [b]	23.32 ± 0.03 [b]
AF_S_2	52.75 ± 0.04 [b]	20.37 ± 0.02 [a]	12.54 ± 0.04 [a]	4.87	31.62 ± 0.11 [a]	23.93 ± 0.01 [a]
AF_T_0.5	48.69 ± 0.03 [d]	19.46 ± 0.02 [b]	11.12 ± 0.03 [d]	0.59	29.73 ± 0.04 [c]	22.43 ± 0.03 [d]
AF_T_1	50.30 ± 0.07 [c]	19.63 ± 0.02 [b]	11.49 ± 0.01 [c]	2.19	30.36 ± 0.02 [b]	22.74 ± 0.01 [c]
AF_T_2	55.14 ± 0.04 [a]	17.98 ± 0.02 [d]	10.25 ± 0.05 [e]	7.26	29.69 ± 0.11 [c]	20.70 ± 0.03 [e]

AF_B—apple fiber/blackberry microparticles (control sample); AF—apple fiber; S—sucrose; T—trehalose; 0.5—ratio of apple fibers:disaccharide (1:0.5); 1—ratio of apple fibers:disaccharide (1:1); 2—ratio of apple fibers:disaccharide (1:2). Values in the same column marked (a–e) with different superscripts are statistically different.

3.2. Flavor Profile of Formulated Microparticles

Seventeen volatiles were determined in the microparticles, while 16 were identified in blackberry juice (minus hexanal in compared microparticles) and 9 in apple fiber (minus linalool oxide, guaiacol, phenethyl alcohol, menthol, nerol, and eugenol as compared to microparticles). Apple fiber/blackberry microparticles were used as a control for the comparison of volatiles in microparticles with disaccharides, i.e., the effect of disaccharides on the adsorption of volatiles (Tables 5 and 6). Disaccharide type and their ratio to the fiber during formulation had an effect on the adsorption of volatile compounds. Two aldehydes were determined—hexanal and heptanal. Both of them were determined in higher amounts in the microparticles with added disaccharides than on apple fiber. With the increase in disaccharides, an increase in their amount in microparticles occurred, with trehalose having a more positive effect. Phenethyl and perillyl alcohols had a different trend than aldehydes. The amount of phenethyl alcohol was the highest in the apple fiber microparticles. For the other microparticles, it was observed that with the increase of disaccharides, a decrease in the adsorption of this volatile occurred. For perillyl alcohol, it was determined that microparticles with the lowest amount of disaccharides had higher content than the apple fiber microparticles. Terpenes were the most abundant volatiles in the microparticles. D-limonene was determined in the highest amount in all microparticles. Equal amounts of this terpene were determined in the apple fiber microparticles and the microparticles with sucrose (fiber:disaccharide ratios 1:0.5 and 1:1), while the third sucrose microparticles had the lowest amount of D-limonene. Trehalose microparticles showed different behavior, i.e., the highest amount of this terpene was determined in the microparticles prepared with a fiber:disaccharide ratio of 1:2. At lower amounts of trehalose, absorption of this volatile decreased. Citronellal and linalool had similar behavior. Microparticles with disaccharides (fiber:disaccharide ratios 1:1 and 1:2) had lower contents of those terpenes compared to apple fiber microparticles. A different effect was observed for microparticles with fiber:disaccharide ratio of 1:0.5. Microparticles with sucrose had an equal amount as apple fiber microparticles, while trehalose microparticles had the highest content of these terpenes. For guaiacol, it was determined that all trehalose microparticles had a higher amount of this volatile than the one with just apple fiber, while for sucrose microparticles, this trend was observed only when its lowest amount was used for the formulation. Menthol and β-damascenone had similar behavior. The highest amount of these terpenes was determined in the apple fiber microparticles. For disaccharide microparticles, a decreased adsorption of those terpenes occurred with the increase in disaccharides. Both microparticles with the fiber:disaccharide ratio of 1:0.5 had higher amounts of eugenol and citral than the apple fiber microparticles. For nerol, the highest content was determined in the apple fiber microparticles, α-ionone was the highest in microparticles with disaccharides (fiber:disaccharide ratio of 1:0.5), while microparticles with the fiber:disaccharide ratios of 1:1 and 1:2 had equal or lower amounts of this terpene than apple fiber, respectively. For γ-ionone, it was determined that only microparticles with trehalose (fiber:disaccharide ratio of 1:0.5) had higher content than apple fiber, while other microparticles were lower. Similarly, with the increase in disaccharides, a decrease of this terpene occurred. All disaccharide microparticles had lower β-ionone content than the apple fiber microparticles, but a positive effect of trehalose was noted.

Table 5. Volatile compounds (µg/kg) defined in blackberry juice (BJ), apple fiber (AF), and formulated microparticles ©, their odor threshold, and the flavor descriptor.

Volatiles	BJ	AF	C	RT	RI	OT	Descriptor
Hexanal	-	52.64 ± 1.56	+	5.13	800	20	green
Heptanal	0.72 ± 0.00	6.29 ± 0.10	+	10.76	897	3	green
D-limonene	7.37 ± 0.13	64.90 ± 1.15	+	19.41	1018	10	citrus
Citronellal	1.85 ± 0.02	24.40 ± 0.46	+	21.13	1051	25	floral
Linalool oxide	62.27 ± 1.54	-	-	22.19	1068	100	floral
Guaiacol	104.17 ± 1.80	-	+	23.17	1080	20	green
Linalool	23.39 ± 0.42	80.92 ± 4.45	+	23.96	1096	6	floral
Phenethyl alcohol	47.39 ± 0.48	-	+	24.6	1103	1000	floral
Menthol	13.86 ± 0.28	-	+	27.75	1167	920	minty
Nerol	20.25 ± 0.61	-	+	30.63	1222	290	citrus
Citral	17.02 ± 0.36	4.92 ± 0.21	+	32.64	1265	30	citrus
Perillyl alcohol	29.19 ± 1.98	3.71 ± 0.13	+	33.9284	1290	1660	green
Eugenol	31.19 ± 0.01	-	+	36.5930	1355	30	spicy
β-damascenone	11.90 ± 0.12	5.89 ± 0.24	+	37.6815	1381	10	floral
α-ionone	13.10 ± 0.44	1.64 ± 0.09	+	38.95	1420	0.6	floral
γ-ionone	4.18 ± 0.43	4.24 ± 0.19	+	40.18	1473	0.07	floral
β-ionone	2.58 ± 0.11	5.46 ± 0.43	+	40.34	1480	0.1	floral

"-"—not detected; "+"—detected; RT—retention time (min); RI—retention index; OT—odor threshold (µg/kg), http://www.leffingwell.com (accessed on 23 May 2022).

Table 6. Volatile compounds (μg/kg) in formulated microparticles.

Volatiles	AF_B	S_0.5	S_1	S_2	T_0.5	T_1	T_2
Hexanal	41.75 ± 0.23 [f]	69.16 ± 3.28 [e]	125.91 ± 0.19 [d]	138.37 ± 0.65 [b]	70.55 ± 0.71 [e]	136.82 ± 0.44 [c]	143.68 ± 0.47 [a]
Heptanal	0.81 ± 0.00 [e]	5.80 ± 0.01 [d]	6.87 ± 0.06 [c]	7.74 ± 0.09 [b]	7.74 ± 0.04 [b]	7.57 ± 0.12 [b]	9.04 ± 0.63 [a]
D-limonene	115.62 ± 2.11 [b]	117.08 ± 0.45 [b]	115.37 ± 0.75 [b]	88.22 ± 0.47 [e]	96.25 ± 1.24 [d]	119.33 ± 1.33 [b]	154.68 ± 0.95 [a]
Citronellal	54.67 ± 0.66 [b]	54.61 ± 0.66 [b]	44.95 ± 0.10 [c]	40.47 ± 1.07 [d]	67.03 ± 0.47 [a]	35.74 ± 0.60 [e]	35.19 ± 0.55 [e]
Guaiacol	14.99 ± 0.22 [d]	16.51 ± 0.14 [c]	14.45 ± 0.05 [e]	12.53 ± 0.00 [f]	21.73 ± 0.31 [a]	18.14 ± 0.07 [b]	16.94 ± 0.06 [c]
Linalool	66.58 ± 0.68 [b]	66.94 ± 1.57 [b]	51.80 ± 0.03 [d]	49.86 ± 0.06 [e]	71.76 ± 0.48 [a]	62.66 ± 0.21 [c]	51.26 ± 0.90 [d]
Phenethyl alcohol	22.19 ± 0.09 [a]	19.98 ± 0.05 [b]	11.73 ± 0.07 [f]	9.71 ± 0.08 [h]	17.99 ± 0.11 [c]	13.92 ± 0.06 [d]	12.94 ± 0.11 [e]
Menthol	7.48 ± 0.09 [a]	6.78 ± 0.01 [c]	6.26 ± 0.02 [d]	4.62 ± 0.15 [f]	7.22 ± 0.03 [b]	6.22 ± 0.03 [d]	5.85 ± 0.05 [e]
Nerol	10.21 ± 0.23 [a]	10.01 ± 0.13 [a]	6.38 ± 0.09 [d]	3.40 ± 0.05 [f]	8.79 ± 0.27 [b]	6.75 ± 0.02 [c]	6.12 ± 0.04 [e]
Citral	13.53 ± 0.07 [c]	13.97 ± 0.13 [b]	9.76 ± 0.10 [e]	9.43 ± 0.10 [e]	16.92 ± 0.46 [a]	13.00 ± 0.17 [d]	13.49 ± 0.06 [c]
Perillyl alcohol	10.20 ± 0.25 [b]	10.51 ± 0.00 [b]	9.01 ± 0.05 [c]	8.08 ± 0.11 [d]	11.13 ± 0.15 [a]	9.06 ± 0.03 [c]	8.86 ± 0.16 [c]
Eugenol	3.19 ± 0.16 [c]	3.56 ± 0.13 [b]	2.15 ± 0.02 [e]	2.75 ± 0.01 [d]	4.19 ± 0.30 [a]	2.85 ± 0.03 [d]	3.17 ± 0.19 [c]
β-damascenone	8.29 ± 0.07 [a]	6.68 ± 0.11 [c]	5.69 ± 0.36 [e]	4.70 ± 0.24 [f]	7.58 ± 0.05 [b]	6.02 ± 0.08 [d]	5.62 ± 0.05 [e]
α-ionone	2.61 ± 0.01 [b]	2.81 ± 0.00 [a]	2.65 ± 0.05 [b]	1.73 ± 0.03 [d]	2.83 ± 0.01 [a]	2.63 ± 0.01 [b]	2.27 ± 0.01 [c]
γ-ionone	9.62 ± 0.14 [b]	7.99 ± 0.19 [c]	6.26 ± 0.08 [d]	4.58 ± 0.21 [e]	10.30 ± 0.06 [a]	8.18 ± 0.02 [c]	6.55 ± 0.04 [d]
β-ionone	10.68 ± 0.13 [a]	5.55 ± 0.04 [d]	5.12 ± 0.04 [e]	5.09 ± 0.10 [e]	7.33 ± 0.12 [b]	7.02 ± 0.03 [c]	7.34 ± 0.04 [b]

AF_B—apple fiber/blackberry microparticles (control sample); S—sucrose; T—trehalose; 0.5—ratio of apple fibers:disaccharide (1:0.5); 1—ratio of apple fibers:disaccharide (1:1); 2—ratio of apple fibers:disaccharide (1:2). Values in the same row marked (a–f) with different superscripts are statistically different.

The flavor profile of any product depends on the amount of volatiles, but also on their odor thresholds. To further elucidate the contribution of each volatile to the overall flavor, odor activity values (OAVs) were calculated. OAVs were calculated as the ratio of the volatile concentration to the threshold [22]. A summary of OAVs of all the samples based on published odor thresholds is presented in Table 7. In blackberry juice, linalool, guaiacol, eugenol, β-damascenone, α-ionone, γ-ionone, and β-ionone had OAV values over 1, which indicates their importance to the flavor of blackberry juice. Especially high OAVs (over 10) were calculated for γ-ionone, β-ionone, and α-ionone (29.71, 25.80, and 21.83, respectively). For apple fiber, OAVs between 1 and 10 were calculated for hexanal, heptanal, D-limonene, and α-ionone. Volatiles with OAVs over 10 were linalool, γ-ionone, and β-ionone (13.49, 30.57, and 54.60, respectively). Microparticles contained flavors from both sources. Comparing all microparticles, OAVs between 1 and 10 were calculated for hexanal, citronellal, and α-ionone, while D-limonene and linalool had OAVs slightly over 10, and γ-ionone and β-ionone had very high OAVs. Heptanal was the only volatile that was not important for the flavor of the apple fiber/blackberry microparticles, but it was important for other microparticles with added disaccharides (OAVs were between 1 and 3). Comparing microparticles, especially high differences between the apple fiber/blackberry microparticles and the microparticles with disaccharides were detected for γ-ionone and β-ionone. Even though several volatiles (guaiacol, eugenol, and β-damascenone) had high OAVs in blackberry juice, they were not important for the flavor of the microparticles.

Volatiles of all the samples were grouped according to their flavor descriptors, and the contributions of each flavor note to the overall flavor profile were calculated. Results were used for PCA analysis (Figure 2). PC1 accounted for 52.57% of the total variance, and PC2 for 35.98%. For all samples, dominant flavor notes were floral, green, and citrus. From the biplot, it can be observed that apple fiber had quite a different flavor profile; it was missing spicy and minty flavor notes but had a pronounced floral note, followed by green and citrus ones (approximately equal contributions). Blackberry juice was characterized mostly by a floral flavor note, followed by a green note; the citrus note had the lowest contribution to the overall flavor. The microparticles from blackberry juice and apple fiber were different since their flavor was a mixture of both volatiles that originated from blackberry juice and apple fiber. Apple fiber/blackberry microparticles was characterized mostly by floral and citrus flavor notes. Both microparticles formulated by the addition of sucrose and trehalose (fiber:disaccharide ratio of 1:0.5) were similar in terms of their flavor profile in order to the control sample, i.e., they were characterized mostly by floral and citrus flavor notes. With the increase in added disaccharides, the flavor profile changed, and for those microparticles, green flavor notes prevailed. Interestingly, both microparticles were similar when the fiber:disaccharide ratio was 1:1. The green note was the dominant one, while floral and citrus notes had an equal contribution to the overall flavor. A change in the fiber:disaccharide ratio to 1:2 created a difference between the microparticles with trehalose and sucrose. Green and citrus flavor notes were the dominant ones for microparticles with trehalose, while in the microparticles with sucrose, only the green flavor note was dominant; the other two flavor notes were in equal ratio.

Table 7. Odor activity values (OAVs) of blackberry juice (BJ), apple fiber (AF), and formulated microparticles.

Volatiles	BJ	AF	Samples						
			AF_B	S_0.5	S_1	S_2	T_0.5	T_1	T_2
Hexanal	0.00	2.63	2.09	3.46	6.30	6.92	3.53	6.84	7.18
Heptanal	0.24	2.10	0.27	1.93	2.29	2.58	2.58	2.52	3.01
D-limonene	0.74	6.49	11.56	11.71	11.54	8.82	9.63	11.93	15.47
Citronellal	0.07	0.98	2.19	2.18	1.80	1.62	2.68	1.43	1.41
Linalool oxide	0.62	0.00	0.00	0.00	0.00	0.00	0.00	0.00	0.00
Guaiacol	5.21	0.00	0.75	0.83	0.72	0.63	1.09	0.91	0.85
Linalool	3.90	13.49	11.10	11.16	8.63	8.31	11.96	10.44	8.54
Phenethyl alcohol	0.05	0.00	0.02	0.02	0.01	0.01	0.02	0.01	0.01
Menthol	0.02	0.00	0.01	0.01	0.01	0.01	0.01	0.01	0.01
Nerol	0.07	0.00	0.04	0.03	0.02	0.01	0.03	0.02	0.02
Citral	0.57	0.16	0.45	0.47	0.33	0.31	0.56	0.43	0.45
Perillyl alcohol	0.02	0.00	0.01	0.01	0.01	0.00	0.01	0.01	0.01
Eugenol	1.04	0.00	0.11	0.12	0.07	0.09	0.14	0.10	0.11
β-damascenone	1.19	0.59	0.83	0.67	0.57	0.47	0.76	0.60	0.56
α-ionone	21.83	2.73	4.35	4.68	4.42	2.88	4.72	4.38	3.78
γ-ionone	29.71	30.57	137.43	114.14	89.43	65.43	147.14	116.86	93.57
β-ionone	25.80	54.60	106.80	55.50	51.20	50.90	73.30	70.20	73.40

AF_B—apple fiber/blackberry microparticles; S—sucrose; T—trehalose; 0.5—ratio of apple fibers: disaccharide (1:0.5); 1—ratio of apple fibers:disaccharide (1:1); 2—ratio of apple fibers:disaccharide (1:2).

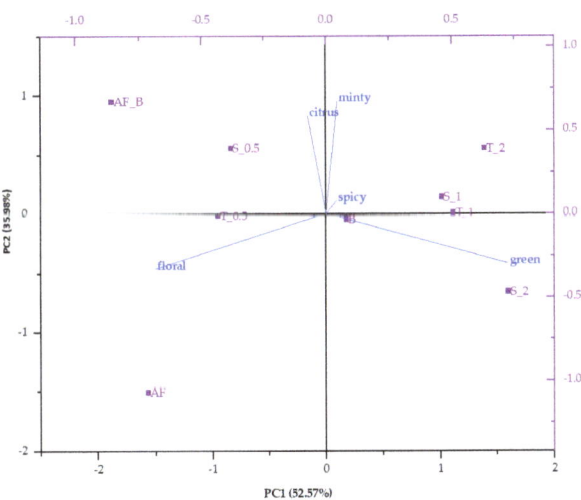

Figure 2. Principal component analysis (PCA) biplot of volatile compounds in blackberry juice, apple fiber, and formulated microparticles. B—blackberry juice; AF—apple fiber; AF_B—apple fibers/blackberry microparticles; S—sucrose; T—trehalose; 0.5—ratio of apple fibers:disaccharide (1:0.5); 1—ratio of apple fibers:disaccharide (1:1); 2—ratio of apple fibers:disaccharide (1:2).

3.3. IR Spectra

The FTIR-ATR was used to observe changes in the IR spectra of apple fibers upon adsorption of blackberry compounds and disaccharides (Figures 3 and 4). Apple fiber/blackberry microparticles had a band at 1013.8 cm^{-1}, while on apple fiber/disaccharide and apple fiber/disaccharide/blackberry microparticles, this band was wider, with the additional band at 991 cm^{-1}, which is a consequence of the binding of disaccharides onto the apple fiber. The band at 1013.8 cm^{-1} is assigned to ring stretching vibrations mixed strongly with CH in-plane bending, while the band at 991 cm^{-1} is assigned to C-O [43]. The band at 1150 cm^{-1} (assigned to C-O stretching mode of the carbohydrates) disappeared due to the binding of sucrose, while the binding of trehalose caused a shift of this band to 1140 cm^{-1}. Additional change in the IR spectra was observed upon the binding of sucrose in the region 950–750 cm^{-1}. Changes in the IR spectra were also observed upon adsorption of blackberry compounds, in both the apple fiber/blackberry and apple fiber/disaccharide/blackberry microparticles. These changes included the shift of a band at 1740 cm^{-1} (which is assigned to C=O) to 1730 cm^{-1}, and its intensity in comparison to 1620 cm^{-1} was higher than in apple fiber/disaccharide microparticles. Additional change upon adsorption of the blackberry compounds was observed in the trehalose microparticles, i.e., the band at 810 cm^{-1} (assigned to ring CH deformation) shifted to 820 cm^{-1} [50].

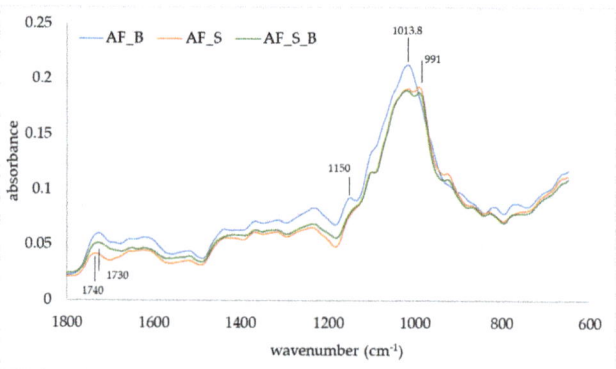

Figure 3. IR spectra of apple fiber/blackberry (AF_B), apple fiber/sucrose (AF_S), and apple fiber/sucrose/blackberry (AF_S_B) microparticles.

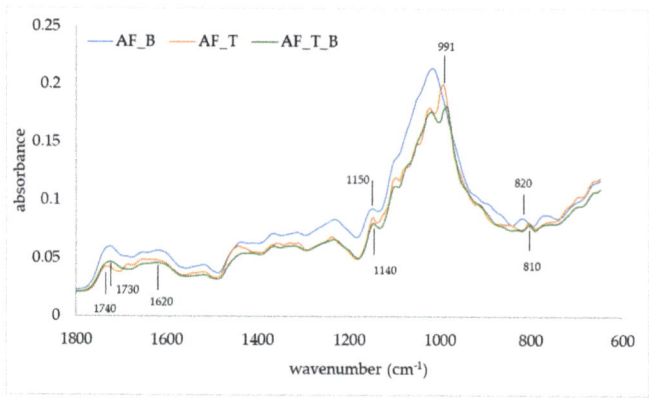

Figure 4. IR spectra of apple fiber/blackberry (AF_B), apple fiber/trehalose (AF_T), and apple fiber/trehalose/blackberry (AF_T_B) microparticles.

4. Discussion

Previous studies showed that the efficiency of the encapsulation of phenolics by carriers depended on the type, amount, and properties of phenolics as well as the carriers [19,51–53]. It was determined that raspberry phenolics effectively bound to cellulose, and the binding depended on the amount of cellulose and the time required for complexation [19]. In the study on the selective binding of ferulic acid, cyanidin-3-glucoside, and catechin onto cellulose-based composites and apple cell walls, it was observed that cellulose was the dominant binding component for catechin, while xyloglucan and arabinoxylan did not contribute to the adsorption of this compound in the presence of cellulose. The other two compounds bound to selected materials with different affinities. Both electrostatic forces and plant cell wall microstructure were very important for binding; thus, the negatively-charged pectin on cell walls exhibited the most extensive binding to positively charged cyanidin-3-glucoside, while its binding with negatively-charged ferulic acid was the least effective [53]. It was determined that the interaction between different phenolics (such as catechin, ferulic acid, chlorogenic acid, gallic acid, and cyanidin-3-glucoside) and cellulose occurred spontaneously, within 1 min, and rapidly increased over 30 min [54]. In addition, the authors determined that chlorogenic acid had different behavior as compared to other investigated phenolics. While all other phenolics bound similarly on a molar basis, the binding of chlorogenic acid was lower [54]. It was postulated by Liu et al. [55] that the initial binding of phenolics and cellulose occurred due to the adsorption of phenolics onto the binding sites of the cellulose surface, which resulted from the presence of labile hydroxyl groups; after that, non-covalent binding, i.e., hydrogen bonding and hydrophobic interactions could occur [55]. This could also be the mechanism of interaction between blackberry juice phenolics and apple fiber and apple fiber/disaccharides. Very important factors for non-covalent binding were also phenolic rings, i.e., their number and their conformational flexibility [56]. In our study, we had mixtures of phenolics, and in addition to apple fiber (consisting of cellulose, hemicellulose, and pectin), we also had disaccharides, which played an important part in the binding of phenolics to apple fiber. It is possible that when interactions between apple fiber and disaccharides were formed, there were no remaining binding sites for phenolics since it was observed that the apple fiber/blackberry microparticles had the highest phenolic content and antioxidant activities. On the other hand, there were differences between the sucrose and trehalose containing microparticles and their amounts, which also proves that the type of disaccharides and their properties can affect the binding of phenolics. The unique structure of trehalose enables trehalose to interact with both hydrophilic and hydrophobic molecules [57]; thus, this property could be the reason for the differences between the sucrose and trehalose microparticles. The differences caused by disaccharide types in fruit samples were proven in several studies [33,35–37,39,58], and in most cases, trehalose had a more positive effect on phenolics and antioxidant activity.

As was already mentioned in the introduction, trehalose had a positive influence on volatiles in different freeze-dried fruit products [28–32,34], and we also observed that trend in our type of products. As it was observed for phenolic compounds as well as for volatiles, we can conclude that the type, amount, and properties of both volatiles and carriers played important roles in the retention of volatile compounds. Additionally, competition between components of binding sites has to be accounted for since it has been proven that this could be a significant factor in the retention of volatiles, especially when they are added as a mixture of compounds [31,34,59], as is the case in this study. Even though trehalose and sucrose are chemical isomers, from our results on phenolics and volatiles, they had different influences on those compounds since they differ in their behavior. The formation of stable intramolecular complexes between trehalose and unsaturated compounds [57,60,61] is also a possible explanation for the positive effect of trehalose on phenolics and volatiles. Volatiles are characterized by their diffusion. Sugars have different diffusion coefficients in water and can change the diffusion coefficient of water [62]; therefore, through the modification of the diffusion coefficient of volatiles, they can influence their retention in the fruit product matrix. We can thus conclude that disaccharides had an influence on water dynamics during complexation, resulting in the effect of sugars on volatile and phenolic compounds.

5. Conclusions

Based on the results of this study, it can be concluded that apple fibers are good encapsulating material for blackberry juice phenolic and volatile compounds. The enhancement of the adsorption of those compounds depended on the type and amount of disaccharides. The addition of sucrose and trehalose decreased the adsorption of phenolics onto the apple fiber. When comparing the influence of applied disaccharides, it was observed that trehalose had a higher positive effect on adsorption than sucrose. Generally, the adsorption of volatiles was higher with the application of disaccharides in comparison only to apple fibers. Comparing sucrose and trehalose, once again, trehalose had a more positive effect than sucrose. These results show that the formulation of encapsulates, i.e., the selection of carriers, had an important role in the final quality of the microparticles. These microparticles can be used for the enrichment of conventional food products with phenolics and volatiles, but also for the development of novel foods, especially functional foods.

Author Contributions: Conceptualization, M.K. and J.Š.; methodology, A.P., M.K., and I.B.; formal analysis, M.N., I.B., and I.I.; investigation, M.N., I.B., and I.I.; data curation, M.N., I.B., and I.I.; writing—original draft preparation, M.K.; writing—review and editing, A.P. and J.Š.; supervision, A.P. and M.K.; project administration, M.K.; funding acquisition, M.K. All authors have read and agreed to the published version of the manuscript.

Funding: This work was part of the PZS-2019-02-1595 project and was fully supported by the "Research Cooperability" Program of the Croatian Science Foundation, funded by the European Union through the European Social Fund under the Operational Programme Efficient Human Resources 2014–2020.

Institutional Review Board Statement: Not applicable.

Informed Consent Statement: Not applicable.

Data Availability Statement: The data sets generated and analyzed in the current study are available from the corresponding author upon reasonable request.

Acknowledgments: The authors would like to thank Hayashibara Company (Nagase group, Japan) for donating the trehalose used in this study.

Conflicts of Interest: The authors declare no conflict of interest.

References

1. Thyavihalli Girijappa, Y.G.; Mavinkere Rangappa, S.; Parameswaranpillai, J.; Siengchin, S. Natural fibers as sustainable and renewable resource for development of eco-friendly composites: A comprehensive review. *Front. Mater.* **2019**, *6*, 226. [CrossRef]
2. Sobczak, P.; Nadulski, R.; Kobus, Z.; Zawiślak, K. Technology for apple pomace utilization within a sustainable development policy framework. *Sustainability* **2022**, *14*, 5470. [CrossRef]
3. Sagar, N.A.; Pareek, S.; Sharma, S.; Yahia, E.M.; Lobo, M.G. Fruit and vegetable waste: Bioactive compounds, their extraction, and possible utilization. *Compr. Rev. Food Sci. Food Saf.* **2018**, *17*, 512–531. [CrossRef]
4. Hussain, S.; Jõudu, I.; Bhat, R. Dietary fiber from underutilized plant resources—A positive approach for valorization of fruit and vegetable wastes. *Sustainability* **2020**, *12*, 5401. [CrossRef]
5. Sendra, E.; Kuri, V.; Fernandez-Lopez, J.; Sayas-Barbera, E.; Navarro, C.; Perez-Alvarez, J.A. Viscoelastic properties of orange fiber enriched yogurt as a function of fiber dose, size and thermal treatment. *LWT Food Sci. Technol.* **2010**, *43*, 708–714. [CrossRef]
6. Bonarius, G.A.; Vieira, J.B.; van der Goot, A.J.; Bodnar, I. Rheological behaviour of fibre-rich plant materials in fat-based food systems. *Food Hydrocoll* **2014**, *40*, 254–261. [CrossRef]
7. Fu, J.-T.; Chang, Y.-H.; Shiau, S.-Y. Rheological, antioxidative and sensory properties of dough and Mantou (steamed bread) enriched with lemon fiber. *LWT Food Sci. Technol.* **2015**, *61*, 56–62. [CrossRef]
8. Su, D.; Zhu, X.; Wang, Y.; Lia, D.; Wang, L. Effect of high-pressure homogenization on rheological properties of citrus fiber. *LWT Food Sci. Technol.* **2020**, *127*, 109366. [CrossRef]

9. Da Silva, L.C.; Viganó, J.; de Souza Mesquita, L.M.; Baião Dias, A.L.; de Souza, M.C.; Sanches, V.L.; Chaves, J.O.; Pizani, R.S.; Contieri, L.S.; Rostagno, M.A. Recent advances and trends in extraction techniques to recover polyphenols compounds from apple by-products. *Food Chem. X* **2021**, *12*, 100133. [CrossRef]
10. Bermúdez-Oria, A.; Rodríguez-Gutiérrez, G.; Fernández-Prior, Á.; Vioque, B.; Fernández-Bolaños, J. Strawberry dietary fiber functionalized with phenolic antioxidants from olives. Interactions between polysaccharides and phenolic compounds. *Food Chem.* **2018**, *280*, 310–320. [CrossRef]
11. Bermúdez-Oria, A.; Rodríguez-Gutiérrez, G.; Rubio-Senent, F.; Lama-Muñoz, A.; Fernández-Bolaños, J. Complexation of hydroxytyrosol and 3,4,-dihydroxyphenylglycol with pectin and their potential use for colon targeting. *Carbohydr. Polym.* **2017**, *163*, 292–300. [CrossRef]
12. Bermúdez-Oria, A.; Rodríguez-Gutiérrez, G.; Rodríguez-Juan, E.; González-Benjumea, A.; Fernández-Bolaños, J. Molecular interactions between 3,4-dihydroxyphenylglycol and pectin and antioxidant capacity of this complex in vitro. *Carbohydr. Polym.* **2018**, *197*, 260–268. [CrossRef]
13. Padayachee, A.; Netzel, G.; Netzel, M.; Day, L.; Zabaras, D.; Mikkelsen, D.; Gidley, M. Binding of polyphenols to plant cell wall analogues—Part 1: Anthocyanins. *Food Chem.* **2012**, *134*, 155–161. [CrossRef]
14. Padayachee, A.; Netzel, G.; Netzel, M.; Day, L.; Zabaras, D.; Mikkelsen, D.; Gidley, M. Binding of polyphenols to plant cell wall analogues—Part 2: Phenolic acids. *Food Chem.* **2012**, *135*, 2292–2297. [CrossRef]
15. Renard, C.M.; Baron, A.; Guyot, S.; Drilleau, J.-F. Interactions between apple cell walls and native apple polyphenols: Quantification and some consequences. *Int. J. Biol. Macromol.* **2001**, *29*, 115–125. [CrossRef]
16. Sun-Waterhouse, D.; Melton, L.D.; O'Connor, C.J.; Kilmartin, P.A.; Smith, B.G. Effect of apple cell walls and their extracts on the activity of dietary antioxidants. *J. Agric. Food Chem.* **2007**, *56*, 289–295. [CrossRef]
17. Sun-Waterhouse, D.; Smith, B.G.; O'Connor, C.J.; Melton, D.L. Effect of raw and cooked onion dietary fiber on the antioxidant activity of ascorbic acid and quercetin. *Food Chem.* **2008**, *11*, 580–585. [CrossRef]
18. Vukoja, J.; Pichler, A.; Ivić, I.; Šimunović, J.; Kopjar, M. Cellulose as a delivery system of raspberry juice volatiles and their stability. *Molecules* **2020**, *25*, 2624. [CrossRef]
19. Vukoja, J.; Buljeta, I.; Pichler, A.; Šimunović, J.; Kopjar, M. Formulation and stability of cellulose-based delivery systems of raspberry phenolics. *Processes* **2021**, *9*, 90. [CrossRef]
20. Kaume, L.; Howard, L.R.; Devareddy, L. The blackberry fruit: A review on its composition and chemistry, metabolism and bioavailability, and health benefits. *J. Agric. Food Chem.* **2012**, *60*, 5716–5727. [CrossRef]
21. Kopjar, M.; Piližota, V. Blackberry Juice. In *Handbook of Functional Beverages and Human Health*; Shahidi, F., Alaalvar, C., Eds.; CRC Press: Boca Raton, FL, USA; Taylor & Francis Group: Abingdon, UK, 2016; pp. 135–145.
22. Qian, M.C.; Wang, Y. Seasonal variation of volatile composition and odor activity value of 'Marion' (Rubus spp. hyb) and 'Thornless Evergreen' (*R. laciniatus* L.) blackberries. *J. Food Sci.* **2005**, *70*, C13–C20. [CrossRef]
23. Du, X.; Finn, C.E.; Qian, M.C. Volatile composition and odour-activity value of thornless 'Black Diamond' and 'Marion' blackberries. *Food Chem.* **2010**, *119*, 117–1134. [CrossRef]
24. Saifullah, M.; Islam Shishir, M.R.; Ferdowsi, R.; Tanver Rahman, M.R.; Van Vuong, Q. Micro and nano encapsulation, retention and controlled release of flavor and aroma compounds: A critical review. *Trends Food Sci. Technol.* **2019**, *86*, 230–251. [CrossRef]
25. Espinosa-Andrews, H.; Morales-Hernández, N.; García-Márquez, E.; Rodríguez-Rodríguez, R. Development of fish oil microcapsules by spray drying using mesquite gum and chitosan as wall materials: Physicochemical properties, microstructure, and lipid hydroperoxide concentration. *Int. J. Polym. Mater. Polym. Biomater.* **2022**, 1–10. [CrossRef]
26. Papoutsis, K.; Golding, J.; Vuong, Q.; Pristijono, P.; Stathopoulos, C.; Scarlett, C.; Bowyer, M. Encapsulation of citrus by-product extracts by spray-drying and freeze-drying using combinations of maltodextrin with soybean protein and ι-carrageenan. *Foods* **2018**, *7*, 115. [CrossRef]
27. Wu, G.; Hui, X.; Stipkovits, L.; Rachman, A.; Tu, J.; Brennan, M.A.; Brennan, C.S. Whey protein-blackcurrant concentrate particles obtained by spray-drying and freeze-drying for delivering structural and health benefits of cookies. *Innov. Food Sci. Emerg. Technol.* **2021**, *68*, 102606. [CrossRef]
28. Komes, D.; Lovrić, T.; Kovačević Ganić, K.; Gracin, L. Study of trehalose addition on aroma retention in dehydrated strawberry puree. *Food Technol. Biotechnol.* **2003**, *41*, 111–119.
29. Komes, D.; Lovrić, T.; Kovačević Ganić, K.; Kljusurić, J.G.; Banović, M. Trehalose improves flavour retention in dehydrated apricot puree. *Int. J. Food Sci. Technol.* **2005**, *40*, 425–435. [CrossRef]
30. Komes, D.; Lovrić, T.; Kovačević Ganić, K. Aroma of dehydrated pear products. *LWT Food Sci. Technol.* **2007**, *40*, 1578–1586. [CrossRef]
31. Kopjar, M.; Piližota, V.; Hribar, J.; Simčič, M.; Zlatić, E.; Tiban, N.N. Influence of trehalose addition and storage conditions on the quality of strawberry cream filling. *J. Food Eng.* **2008**, *87*, 341–350. [CrossRef]
32. Galmarini, M.V.; van Baren, C.; Zamora, M.C.; Chirife, J.; Di Leo Lira, P.; Bandoni, A. Impact of trehalose, sucrose and/or maltodextrin addition on aroma retention in freeze dried strawberry puree. *Int. J. Food Sci. Technol.* **2011**, *46*, 1337–1345. [CrossRef]
33. Kopjar, M.; Jakšić, K.; Piližota, V. Influence of sugars and chlorogenic acid addition on anthocyanin content, antioxidant activity and color of blackberry juice during storage. *J. Food Process. Preserv.* **2012**, *36*, 545–552. [CrossRef]

34. Kopjar, M.; Hribar, J.; Simčič, M.; Zlatić, E.; Tomaž, P.; Piližota, V. Effect of trehalose addition on volatiles responsible for strawberry aroma. *Nat. Prod. Commun.* **2013**, *8*, 1767–1770. [CrossRef] [PubMed]
35. Kopjar, M.; Pichler, A.; Turi, J.; Piližota, V. Influence of trehalose addition on antioxidant activity, colour and texture of orange jelly during storage. *Int. J. Food Sci. Technol.* **2016**, *51*, 2640–2646. [CrossRef]
36. Lončarić, A.; Pichler, A.; Trtinjak, I.; Piližota, V.; Kopjar, M. Phenolics and antioxidant activity of freeze-dried sour cherry puree with addition of disaccharides. *LWT Food Sci. Technol.* **2016**, *73*, 391–396. [CrossRef]
37. Zlatić, E.; Pichler, A.; Kopjar, M. Disaccharides: Influence on volatiles and phenolics of sour cherry juice. *Molecules* **2017**, *22*, 1939. [CrossRef]
38. Zlatić, E.; Pichler, A.; Lončarić, A.; Vidrih, R.; Požrl, T.; Hribar, J.; Piližota, V.; Kopjar, M. Volatile compounds of freeze-dried sour cherry puree affected by addition of sugars. *Int. J. Food Prop.* **2017**, *20*, S449–S456. [CrossRef]
39. Vukoja, J.; Buljeta, I.; Ivić, I.; Šimunović, J.; Pichler, A.; Kopjar, M. Disaccharide type affected phenolic and volatile compounds of citrus fiber-blackberry cream fillings. *Foods* **2021**, *10*, 243. [CrossRef]
40. Van Can, J.G.P.; Van Loon, L.J.C.; Brouns, F.; Blaak, E.E. Reduced glycaemic and insulinaemic responses following treha-lose and isomaltulose ingestion: Implications for postprandial substrate use in impaired glucose-tolerant subjects. *Br. J. Nutr.* **2012**, *108*, 1210–1217. [CrossRef]
41. Neta, T.; Takada, K.; Hirasawa, M. Low-cariogenicity of trehalose as a substrate. *J. Dent.* **2000**, *28*, 571–576. [CrossRef]
42. Buljeta, I.; Nosić, M.; Pichler, A.; Ivić, I.; Šimunović, J.; Kopjar, M. Apple fibers as carriers of blackberry juice polyphenols: Development of natural functional food additives. *Molecules* **2022**, *27*, 3029. [CrossRef] [PubMed]
43. Singleton, V.L.; Rossi, J.A. Colorimetry of total phenolics with phosphomolybdic-phosphotonutric acid reagents. *Am. J. Enol. Vitic.* **1965**, *16*, 144–158.
44. Prior, R.L.; Fan, E.; Ji, H.; Howell, A.; Nio, C.; Payne, M.J.; Reed, J. Multi-laboratory validation of a standard method for quantifying proanthocyanidins in cranberry powders. *J. Sci. Food Agric.* **2010**, *90*, 1473–1478. [CrossRef]
45. Arnao, M.B.; Cano, A.; Acosta, M. The hydrophilic and lipophilic contribution to total antioxidant activity. *Food Chem.* **2001**, *73*, 239–244. [CrossRef]
46. Brand-Williams, W.; Cuvelier, M.E.; Berset, C. Use of a free radical method to evaluate antioxidant activity. *LWT* **1995**, *28*, 25–30. [CrossRef]
47. Apak, R.; Guculu, K.G.; Ozyurek, M.; Karademir, S.E. Novel total antioxidant capacity index for dietary polyphenols and vitamins C and E, using their cupric iron reducing capability in the presence of neocuproine: CUPRAC method. *J. Agric. Food Chem.* **2004**, *52*, 7970–7981. [CrossRef]
48. Benzie, I.F.; Strain, J.J. The ferric reducing ability of plasma (FRAP) as a measure of "antioxidant power": The FRAP assay. *Anal. Biochem.* **1996**, *239*, 70–79. [CrossRef]
49. Ivić, I.; Kopjar, M.; Jakobek, L.; Jukić, V.; Korbar, S.; Marić, B.; Mesić, J.; Pichler, A. Influence of processing parameters on phenolic compounds and color of cabernet sauvignon red wine concentrates obtained by reverse osmosis and nanofiltration. *Processes* **2021**, *9*, 89. [CrossRef]
50. Movasaghi, Z.; Rehman, S.; Rehman, I. Fourier transform infrared (FTIR) spectroscopy of biological tissues. *Appl. Spectrosc. Rev.* **2008**, *43*, 134–179. [CrossRef]
51. Liu, C.; Ge, S.; Yang, J.; Xu, Y.; Zhao, M.; Xiong, L.; Sun, Q. Adsorption mechanism of polyphenols onto starch nanoparticles and enhanced antioxidant activity under adverse conditions. *J. Funct. Foods* **2016**, *26*, 632–644. [CrossRef]
52. Da Rosa, C.G.; Borges, C.D.; Zambiazi, R.C.; Rutz, J.K.; da Luz, S.R.; Krumreich, F.D.; Benvenutti, E.V.; Nunes, M.R. Encapsulation of the phenolic compounds of the blackberry (*Rubus fruticosus*). *LWT Food Sci. Technol* **2014**, *58*, 527–533. [CrossRef]
53. Phan, A.D.T.; Flanagan, B.M.; D'Arcy, B.R.; Gidley, M.J. Binding selectivity of dietary polyphenols to different plant cell wall components: Quantification and mechanism. *Food Chem.* **2017**, *233*, 216–227. [CrossRef] [PubMed]
54. Phan, A.D.T.; Netzel, G.; Wang, D.; Flanagan, B.M.; D'Arcy, B.R.; Gidley, M.J. Binding of dietary polyphenols to cellulose: Structural and nutritional aspects. *Food Chem.* **2015**, *171*, 388–396. [CrossRef] [PubMed]
55. Liu, D.; Martinez-Sanz, M.; Lopez-Sanchez, P.; Gilbert, E.P.; Gidley, M.J. Adsorption behavior of polyphenols on cellulose is affected by processing history. *Food Hydrocoll.* **2017**, *63*, 496–507. [CrossRef]
56. Cartalade, D.; Vernhet, A. Polar interactions in flavan-3-ol adsorption on solid surfaces. *J. Agric. Food Chem.* **2006**, *54*, 3086–3094. [CrossRef]
57. Sakakura, K.; Okabe, A.; Oku, K.; Sakurai, M. Experimental and theoretiacal study on the intermolecular complex formation between trehalose and benzene compounds in aqueous solution. *J. Phys. Chem.* **2011**, *115*, 9823–9830. [CrossRef]
58. Pichler, A.; Pozderović, A.; Moslavac, T.; Popović, K. Influence of sugars, modified starches and hydrocolloids addition on colour and thermal properties of raspberry cream fillings. *Pol. J. Food Nutr. Sci.* **2017**, *67*, 49–58. [CrossRef]
59. Van Ruth, S.M.; King, C. Effect of starch and amylopectin concentrations on volatile flavour release from aqueous model food systems. *Flavour. Fragr. J.* **2003**, *18*, 407–416. [CrossRef]
60. Oku, K.; Watanabe, H.; Kubota, M.; Fukuda, S.; Kurimoto, M.; Tujisaka, Y.; Komori, M.; Inoue, Y.; Sakurai, M. NMR and quantum chemical study on the OH...pi and CH...O interactions between trehalose and unsaturated fatty acids: Implication for the mechanism of antioxidant function of trehalose. *J. Am. Chem. Soc.* **2003**, *125*, 12739–12748. [CrossRef]

61. Oku, K.; Kurose, M.; Kubota, M.; Fukuda, S.; Kurimoto, M.; Tujisaka, Y.; Okabe, A.; Sakurai, M. Combined NMR and quantum chemical studies on the interaction between trehalose and dienes relevant to the antioxidant function of trehalose. *J. Phys. Chem. B* **2005**, *109*, 3032–3040. [CrossRef]
62. Engelsena, S.B.; Monteiro, C.; de Penhoat, C.H.; Pérez, S. The diluted aqueous solvation of carbohydrates as inferred from molecular dynamics simulations and NMR spectroscopy. *Biophys. Chem.* **2001**, *93*, 103–127. [CrossRef]

Article

Starch-Based Hydrogel Nanoparticles Loaded with Polyphenolic Compounds of Moringa Oleifera Leaf Extract Have Hepatoprotective Activity in Bisphenol A-Induced Animal Models

Hend Mohamed Hasanin Abou El-Naga [1,†], Samah A. El-Hashash [1], Ensaf Mokhtar Yasen [1], Stefano Leporatti [2] and Nemany A. N. Hanafy [3,*,†]

1. Nutrition and Food Science Department, Faculty of Home Economics, Al-Azhar University, Nawag, Tanta P.O. Box 31732, Egypt; hendaboelnaga47@gmail.com (H.M.H.A.E.-N.); samahel-hashash@azhar.edu.eg (S.A.E.-H.); ensafyasin.2066@azhar.edu.eg (E.M.Y.)
2. Cnr Nanotec-Istituto di Nanotecnologia, Via Monteroni, 73100 Lecce, Italy; stefano.leporatti@nanotec.cnr.it
3. Nanomedicine Group, Institute of Nanoscience and Nanotechnology, Kafrelsheikh University, Kafr El Sheikh 33516, Egypt
* Correspondence: nemany.hanafy@nano.kfs.edu.eg
† These authors contributed equally to this work.

Citation: Abou El-Naga, H.M.H.; El-Hashash, S.A.; Yasen, E.M.; Leporatti, S.; Hanafy, N.A.N. Starch-Based Hydrogel Nanoparticles Loaded with Polyphenolic Compounds of Moringa Oleifera Leaf Extract Have Hepatoprotective Activity in Bisphenol A-Induced Animal Models. *Polymers* 2022, 14, 2846. https://doi.org/10.3390/polym14142846

Academic Editors: Lorenzo Antonio Picos Corrales, Angel Licea-Claverie and Grégorio Crini

Received: 20 June 2022
Accepted: 11 July 2022
Published: 13 July 2022

Publisher's Note: MDPI stays neutral with regard to jurisdictional claims in published maps and institutional affiliations.

Copyright: © 2022 by the authors. Licensee MDPI, Basel, Switzerland. This article is an open access article distributed under the terms and conditions of the Creative Commons Attribution (CC BY) license (https://creativecommons.org/licenses/by/4.0/).

Abstract: Bisphenol A (BPA) is an xenoestrogenic chemical used extensively in the fabrication of baby bottles, reusable plastic water bottles and polycarbonate plastic containers. The current study aims to investigate the hepatoprotective activity of *Moringa oleifera* Lam leaf extract (MOLE) and hydrogel NPs made of starch-MOLE-Bovine Serum Albumin (BSA) against Bisphenol A-induced liver toxicity in male rats. Fabrication and characterization of hydrogel NPs formed of starch-MOLE-BSA were investigated using FTIR, TEM, zeta potential, UV-visible spectroscopy and fluorescence spectrophotometer. The potential efficacy of hydrogel NPs was studied. Compared to the results of control, the level of liver function, oxidative stress markers and lipid profile status were remodulated in the groups treated with MOLE and hydrogel NPs (Encap. MOLE). Meanwhile, the administration of MOLE and Encap MOLE significantly increased antioxidant activity and decreased the level of apoptotic pathways. Heme oxygenase (HO)-1 and growth arrest -DNA damage-inducible gene 45b (Gadd45b) were also regulated in the groups treated with MOLE and Encap. MOLE compared to the group which received BPA alone. In the present study, MOLE and hydrogel NPs led to remarkable alterations in histological changes during BPA administration. Overall, MOLE has a potential antioxidant activity which can be used in the treatment of liver disorders.

Keywords: starch; Bisphenol A; Moringa leaf extract; encapsulating Moringa leaf extract

1. Introduction

Liver damage is one of the main problems associated with the exposure to harmful materials [1]. The main function of the liver is to filtrate blood from the digestive tract, before passing it to the rest of the body. Thus, the liver can detoxify chemicals and metabolize drugs [2]. For this reason, if the liver is exposed to chemical materials, this can cause physiological disorder [2]. Bisphenol A (BPA) is widely used as a coating layer inside canned foods and beverages, baby bottles, packaged baby formula, pre-packed foodstuffs and containers that were made for food storage in the home. Additionally, BPA can also be found in dental prosthetics and sales receipts that use thermal paper [3]. BPA can leak out of epoxy resin and further diffuse into canned foods, ultimately entering the human body [4]. Furthermore, BPA can interfere with estrogen levels (normal function) within the human body [5]. Therefore, BPA impacts the function of androgen receptors, thyroid hormone receptors and other endocrine system signaling pathways [6]. Previous literature has shown that a small amount of bisphenol can have a significant effect on biological systems because of its ability to change the pathological condition of hormonal signaling pathways [7,8]. This condition can increase psychological and metabolic disorders in children, reproductive disorders in adults and neoplasms resulting from a weakened immune system after their exposure to BPA [9].

Currently, a lot of research has focused on natural compounds extracted from medicinal plants which could assist in the treatment of pathological conditions [10]. *Moringa oleifera* Lam (MOL) contains many bioactive and antioxidants materials which may overcome oxidative stress and degenerating diseases [11,12]. *Moringa oleifera* Lam leaves have many biological activities that can be used in the prevention of cardiovascular diseases [13], such as immune boosting agent and hypotension [14], cholesterol lowering, diuretic and antiulcer properties [15] and antioxidant activity [16]. These properties are mostly useful for protecting the liver against hepatotoxin-induced toxicity [17]. However, the bioactive compounds are highly susceptible to degradation, and this may decrease the antioxidant activity present in MOL. To prevent these limitations, the utilization of the micro/nanoencapsulation technique is necessary [18]. MOL extracts are highly unstable and susceptible to oxidation; they also show limited solubility in water and low bioavailability [19].

The application of nanotechnology in the food industry has grown in the last few decades, causing the use of micro/nanoparticles which are consumer safe [20]. Therefore, the current study aimed to investigate the protective effect of *Moringa oleifera* Lam leaf extract (MOLE) and encapsulating *Moringa oleifera* Lam leaf extract (Encap. MOLE) against BPA toxicity in an experimental animal model.

Starch-based hydrogel NPs are considered a novel approach in the field of drug delivery systems due to the fact that the application of hydrogels in the pharmaceutical industry provides high drug stability, high loading capacity, large chemical interaction surface, drug protection and controlled drug release. Additionally, hydrogel increases mucoadhesive properties leading to improved penetration and cellular uptake.

Starch is a natural biopolymer which can be used as a gelling, thickening or stabilizing mechanism in various industrial and pharmaceutical applications. Amylase and amylopectin are the two main components of starch that form granule assembly. Hanafy succeeded in insulating starch from corn flour by using alkaline hydrolysis and used it to encapsulate anthocyanin inside its moieties. In the current study, an alkaline gelatinization method was used, where the alkaline-treated starch could form a network, generating hydrogels [21]. Then, Moringa oleifera Lam leaf extract was incorporated into the moieties of starch, which were coated with bovine serum albumin (BSA). The nanosized structure was assembled in hydrogel NPs [22].

BSA is a water-soluble molecule which can interact directly with any other organic or inorganic material by different types of chemical interactions. Owing to its charged amino acids, the electrostatic adsorption of negatively or positively charged molecules can be obtained. In this case, substantial amounts of drug can be incorporated within the particle, due to different albumin-binding sites [23].

In the current study, starch-MOLE-BSA was used to form micellar assembly and to investigate the hepatoprotective activities of *Moringa oleifera* Lam leaf extract. For this reason, the parameters were used as follows: The levels of liver function for serum alanine aminotransferase (ALT), aminotransferase (AST), alkaline phosphatase (ALP), total protein, albumin, lipid profiles (total cholesterol, triglyceride, phospholipid), antioxidant markers (liver glutathione (GSH), superoxide dismutase enzyme (SOD)), apoptotic markers (Caspase-3 and Bax), cell homeostasis genes (HO-1 and GADD45B) and histopathology (hematoxylin and eosin (H+E) and Masson trichrome stain).

2. Materials and Methods

2.1. Plant Materials and Chemicals

Moringa (*Moringa oleifera* Lam) leaves were obtained from the Ministry of Agriculture, Kafr El Sheikh Governorate, Egypt. BSA and Bisphenol A (CAS 80-05-7 Sigma–Aldrich Co., St. Louis, MO, USA) were purchased from Sigma Company, Egypt. Trizol Reagent (Invitrogen, Carlsbad, CA, USA) and (3,3′,5,5′-Tetramethylbenzidine) (TMB) were purchased from Sigma company. The kits of biochemical tests were purchased from Gamma Trade Company for biochemicals, Cairo, Egypt. All other required chemicals were obtained from Elgomhouria Company for trading Drugs, Chemicals and Medical Appliances, Cairo, Egypt.

2.2. Preparation of MOLE

Fresh MOLs were washed under running water, then they were shade dried at room temperature. After that, they were crushed into homogenized powder in the blender and stored at room temperature in a closed brown glass container in the dark until used. The extraction of *Moringa oleifera* Lam Leaves was prepared by mixing 10 g of crushed *Moringa oleifera* Lam Leaves in 45 mL of ethanol (70%), which was stirred for 1 h at 70 °C. Upon completing the extraction, 5 mL acetone was added and the stirring continued for an additional 30 min. The solution was further centrifuged and the supernatant was then collected [24].

2.3. Fabrication of Encapsulated MOLE

Starch (0.5 g) was suspended into 50 mL distilled water. Then, a few drops of NaOH (1N) was added dropwise, forming the gelatinized solution under mechanical stirring power. Then, it was centrifuged at 5000 rpm for 5 min and the starch supernatant was removed and kept at room temperature [25].

The extracted starch (20 mL) was completed to 80 mL with distilled water and then 1 mL MOLE was added. The mixture was then stirred for 20 min and 50 mg/50 mL of bovine serum albumin (BSA) solution was added. The stirring then continued for a further 30 min. After that, nanoparticles were dialyzed in the dialysis bags (Molecular weight cutoff of 12 to 14 kDa). Then, the sample was kept at −20 °C for lyophilization (Labconco, Freezone 1 L) at 5 mm Hg at −50 °C for 72 h. The lyophilized powder was stored at −20 °C until analysis [26].

2.4. Extraction and Identification of Phenolic Compound Using HPLC

Phenolic compounds were identified and measured by high-performance liquid chromatography (HPLC) according to a previously reported method [27].

2.5. Characterization Techniques of Nanoparticles

2.5.1. Measurement of Zeta Potential

Dynamic light scattering of the prepared nanoparticle formulations as well as their charges were investigated using a Zeta-sizer (Brookhaven). In total, 1 mg of NPs was diluted into 10 mL distilled water at pH 7.4 for DLS and zeta potential measurements. For the analysis, an average of five successful runs were carried out at 25 °C.

2.5.2. Transmission Electron Microscopy (TEM)

The Encap. MOLE nanoparticles (Encap. MOLE NPs) were acquired by using a TEM (JEOL 2100, Tokyo, Japan). The diluted samples were dropped upon a carbon-coated copper grid and the excess was drawn of. The samples were left to dry for 5 min and then images were acquired.

2.5.3. Absorbance and Fluorescence Spectrophotometers

UV-vis spectroscopy was used to assess the wavelength and to identify the specific absorption for MOLE, BSA and Encap. For MOLE, 1 mL of the samples was added to 3 mL of distilled water, which was then measured using the UV-Vis spectrophotometer (Jasco

V-770 UV Visible Absorbance Spectrophotometer) in the 200–800 nm wavelength range. Additionally, the fluorescence spectrophotometer was used for the same procedure. The obtained results were analyzed utilizing Origin 8.

2.5.4. Fourier Transform Infrared Spectroscopy (FTIR)

FTIR experiments were carried out using JASCO Fourier Transform Infrared Spectrometer (Japan, model no. AUP1200343) to detect the surface molecular structures in the range of 400–4000 cm^{-1}. Few dried samples of starch, BSA, MOLE and Encap. MOLE were grounded with KBr into homogenous powder and pressed into a suitable tablet with good thickness. For all the tests, at least three scans were recorded for different regions on the samples and the representative spectra were analyzed [28].

2.6. Animals and Ethical Approval

A total of 36 adult Sprague Dawley male albino rats weighing 200 ± 5 g were obtained from Helwan Farm, an animal colony, VI Org., Cairo, Egypt. They were kept in polypropylene cages at a room temperature of 22 ± 1 °C, relative humidity of 50 ± 20% and under a 12 h light/dark cycle. Water was supplied *ad libitum*. They were left to adjust to the laboratory conditions for one week before beginning the experiment. This study was approved by the Ethics Committee of Kafrelsheikh University [25].

2.7. Experimental Diet

Pelleted food was purchased to feed the rats from the Agricultural Development Company, 6-October City, Giza Governorate, Egypt. The food consisted of sunflower oil (15%), concentrate mixture 45% (10%), yellow corn (49%), soybean meal 44% (11%), wheat bran (10%), molasses (3%), common salt (0.5%), ground limestone (0.2%), dicalcium phosphate (0.1%), lysine (0.2%), dl-methionine (0.7%) and mineral-vitamin premix (0.3%).

2.8. Experimental Design and Sampling

2.8.1. Concentration of BPA

The dose of BPA (50 mg/kg/daily) was chosen according to many previous studies [29,30].

2.8.2. Experimental Design

Male rats were randomly assigned [31–33] to 6 groups of 6 rats for 4 weeks as follows: group I: received normal saline and kept as an untreated group (control group); group II: given BPA (50 mg/kg/day); group III: given MOLE (50 mg/kg/day); group IV: given MOLE (50 mg/kg/day) + BPA (50 mg/kg/day); group V: given Encap. MOLE (50 mg/kg/day); and group VI: given Encap. MOLE (50 mg/kg/day) + BPA (50 mg/kg/day). BPA, MOLE and Encap. MOLE were given orally. Rats in groups four and six (IV and VI) received MOLE and Encap. MOLE two hours after BPA administration.

Following this, animals were fasted overnight, then anesthetized by intraperitoneal injection of 70 mg/kg pentobarbital sodium. Blood samples were collected from the hepatic portal vein of the rats and placed into dry, clean centrifuge tubes. Sera were carefully separated by the centrifugation of blood samples (3000 rpm for 10 min) at room temperature, then placed into dry, clean Eppendorf tubes and kept frozen at −20 °C for biochemical determinations. Livers were carefully dissected, washed in ice-cold saline (0.9 g/100 mL), and dried using filter paper. After that, a specimen from each liver was immersed in buffered neutral formalin solution (10%) for histopathological examination, while other specimens were stored at −80 °C for other biochemical and molecular investigations [25–27].

2.9. Homogenization of Liver Tissue

To prepare the liver tissue homogenate, 1 g of liver tissue was removed and cut into small pieces, then it was homogenized using 4710 Ultrasonics Homogenizer (Cole-Parmer Instrument Co., Salisbury, NC, USA) in (1.15 g/100 mL) KCl (ice cold solution) in the presence of 50 mmol/L potassium phosphate-buffered solution (pH 7.4). The homogenized tissues were further centrifuged at 4000 rpm at 4 °C for 5 min. After that, supernatants were used in the experiments [34].

2.10. Biochemical Indices in Liver Tissue Homogenate and Sera

2.10.1. Antioxidant Enzymes

In the liver tissue homogenate, the activities of reduced glutathione (GSH) and superoxide dismutase (SOD) were measured following referenced methods [35,36].

2.10.2. Lipid Profile

Total cholesterol (TC) and triglycerides (TG) were determined in the homogenate of the liver tissue according to the methods described by Richmond [37] and Jacobs and VanDenmark [38], respectively. Additionally, the concentration of phospholipids (PhLs) was calculated in liver tissue homogenate according to the method of Ray et al. [39].

2.10.3. Liver Function Biomarkers

In sera, the activities of liver enzymes, including aminotransferases (ALT and AST) and alkaline phosphatase (ALP), were determined following the methods of Reitman and Frankel [40] and Kind and King [41], respectively. In addition, albumin and total protein (TP) were determined by using the following referenced methods [42,43].

2.11. ELISA for Caspase-3 and Bax Detection

Enzyme-linked immunosorbent assay (ELISA) was used to measure the level of Caspase-3 and Bax [44]. Briefly, the homogenate of liver tissue and standards (100 µL) were added separately into the wells. Then, they were incubated for 2 h at 37 °C. The unreacted materials were washed and then 100 µL of biotin-conjugated antibody was added as a specific detector for Caspase-3 and Bax. After cleaning, 100 µL of avidin conjugated horseradish peroxidase (HRP) was added to the wells and then the samples and standards were incubated for 1 h at 37 °C, followed by the addition of 90 µL of TMB substrate solution. After that, incubation for 20 min was carried out at 37 °C to obtain proportional color to the amount of Caspase-3 and Bax. Then, the reaction was finished by adding stop reaction, then the microplate was tapped gently for thorough mixing and 450 nm was used. The activity of caspase-3 and Bax is expressed as ng/mL.

2.12. RNA Extraction

Total ribonucleic acid (RNA) was extracted by using a specific reagent called Trizol reagent following the standard's protocol [45]. Complementary deoxyribonucleic acid (cDNA) was synthesized using a cDNA synthesis kit based on the standard's protocol. The cDNA thermocycler was left at 37 °C for 30 min. qPCR procedure was run under three conditions, 95 °C for 5 min, 45 cycles at 95 °C for 30 s and 60 °C for 1 min. mRNA expression level was normalized into endogenous control (GAPDH). Then, the calculation was carried out as the relative differences between the control and treatment groups. Primers and probes for the qPCR were designed using Allele ID 6. All primers are listed in Table 1.

Table 1. Forward and reverse primers of the selected genes.

Gene Name	Forward Primer (5'-3')	Reverse Primer (5'-3')
GAPDH	CTACATGGCCTCCAAGGAGTAAG	TGGAATTGTGAGGGAGATGCTC
GADD45B	GAAGATGCAGGCGGTGACTG	CCTCCTCTTCTTCGTCTATGGC
HO-1	ACAGCATGTCCCAGGATTTGTC	GGAGGCCATCACCAGCTTAAAG

2.13. Histopathological Examination

Liver specimens were immersed in 10% phosphate-buffered neutral formalin (dehydrated, cleared in xylene), then the specimens were processed into paraffin blocks and cut off at 5 µm thickness. Sections were stained by normal histology routes using hematoxylin and eosin [46] or using Masson trichrome stain for collagen fibers [47]. Images were acquired using an inverted light microscope.

2.14. Biostatistics

The results were expressed as mean ± standard division of mean (SEM). Data were analyzed by SPSS version 20 using one-way analysis of variance (ANOVA), followed by Duncan's test for comparison between different treatment groups. The data are shown as * $p < 0.05$ ** $p < 0.01$ and *** $p < 0.001$. The data are representative of at least three independent experiments.

3. Results

3.1. HPLC Identification and Quantification

The composition of polyphenolic compounds extracted from MOLE and Encap. MOLE was determined by high-performance liquid chromatography (HPLC) and the results are listed in Table 2 and Figure 1.

Table 2. Identification and concentration of polyphenolic compounds in MOLE and Encap. MOLE.

Polyphenol Compounds	MOLE (µg/mL)	Encap. MOLE (µg/mL)
Gallic acid	2.5	1.44
Chlorogenic acid	9.55	5.54
Methyl gallat	0.5	0.25
Syringic acid	0.33	0.59
Pyro catechol	4.83	1.07
Rutin	3.3	2.76
Ellagic acid	26	4.09
Coumaric acid	6.87	1.08
Ferulic acid	2.97	1.49
Naringenin	4.82	3.08
Taxifolin	0.69	0.26
Cinnamic acid	0.03	0.04

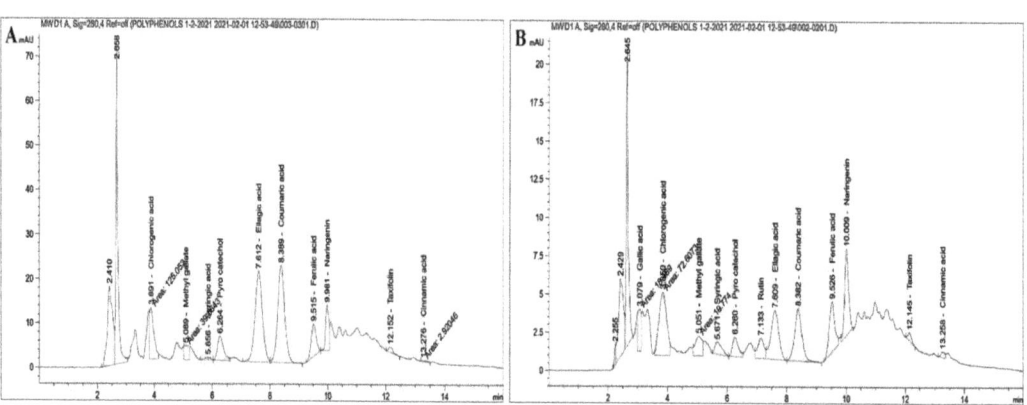

Figure 1. Quantification and identification of polyphenol and flavonoids content isolated from MOL (A) and were then encapsulated (B).

It was found that the major polyphenolic compounds found in ethanolic extract of MOLE were ellagic acid (26 µg/mL). Chlorogenic acid was the second compound found (9.55 µg/mL), followed by coumaric acid, pyrocatechol and naringenin. Another seven compounds were found in small concentrations, ranging between 3.3 µg/mL (rutin) and 0.03 µg/mL (cinnamic acid) [48,49]. On the other hand, the major polyphenolic compounds in Encap. MOLE were found to be chlorogenic acid (5.54 µg/mL), followed by ellagic acid, naringenin, rutin, ferulic acid and gallic acid. Another six compounds were found in Encap. MOLE, but in very small concentrations, ranging between 1.08 µg/mL (coumaric acid) and 0.04 µg/mL (cinnamic acid).

3.2. Characterization

Starch is a natural biodegradable polymer which contains amylose and amylopectin as the main units of its components [50]. These two units are assembled in the shape of granules, the size ranging from 1 to 100 µm [51,52]. In the current study, TEM images showed spherical nanoparticles with diameters ranging between 40 and 75 nm. Their assembly confirms the successful formation of starch-MOLE-BSA NPs (Figure 2A). For instance, the two main components of starch may contribute to the formation of starch backbone, leading to the final structural shape of NPs. The UV visible spectrophotometer of the extract showed the absorption peak at 269 nm [53] and the characteristic absorbance peaks of MOLE were previously shown at 210 nm and 265 nm ($\pi \rightarrow \pi^*$ transition of the aromatic conjugated ring), and at 330 nm ($n \rightarrow \pi^*$ transition of hydroxyl groups (non-bonding electron) within the phenolic ring), while encapsulated MOLE peaked at 276 nm. This shift may be due to its interaction with starch components. Meanwhile, pure BSA had an absorption peak at 277 nm due to the weak absorption of tryptophan (Trp), aromatic amino acids phenylalanine (Phe), and tyrosine (Tyr), (Figure 2C). [54]. Additionally, MOLE and Encap. MOLE exhibited fluorescence intensity at (339–442 nm) and (351–444 nm), respectively. Zeta potential of Encap. MOLE showed potential surface charge at (21 mV) with good distribution (Figure 3). This result shows that they are capable of being stable drug carriers in humans.

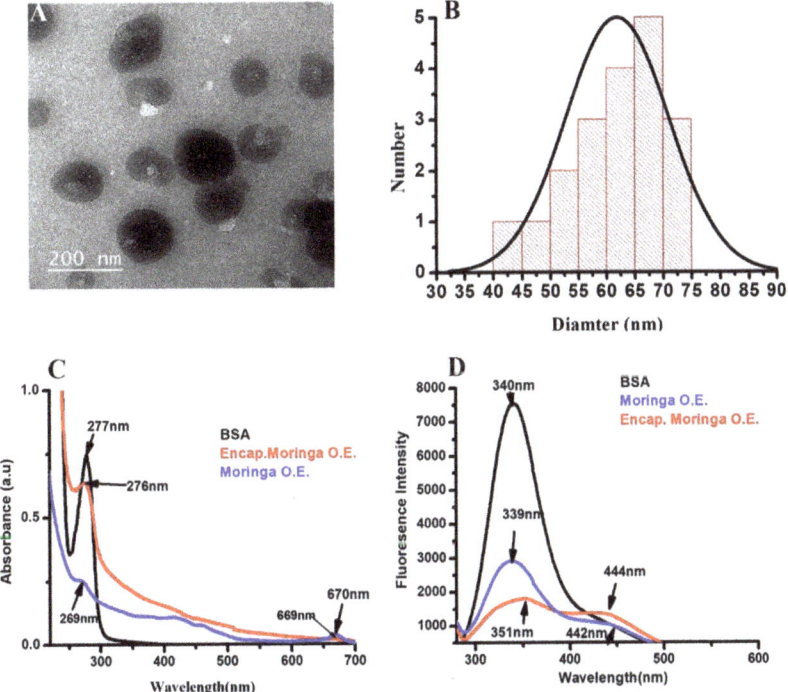

Figure 2. Characterization of MOLE NPs. (**A**) TEM image. (**B**) Quantification of NP diameter by using image J and Origin 8 program. (**C**) UV-Visible spectrophotometer for MOLE, BSA and Encap. MOLE. (**D**) Fluorescence spectrophotometer for fluorescence intensity of BSA, MOLE and Encap. MOLE.

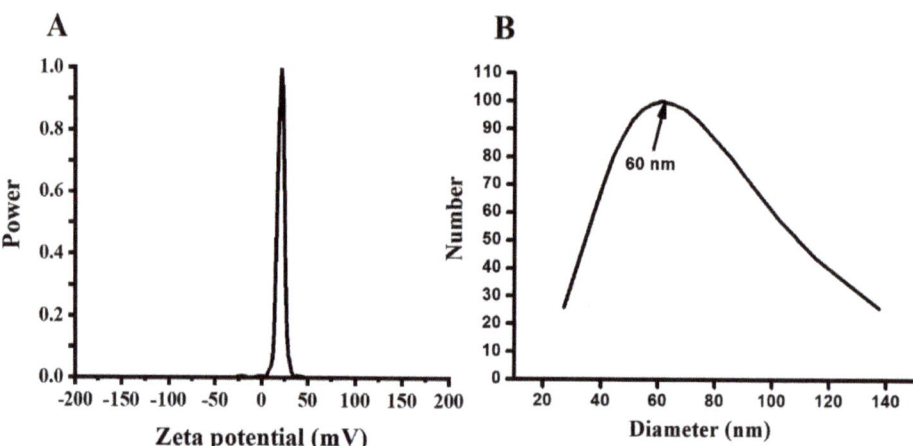

Figure 3. Zeta potential measurement of Encap. MOLE NPs (**A**). Nanosizer of Encap. MOLE NPs (**B**).

In Figure 4, the FTIR spectrum of starch showed a band at 3421 cm^{-1}, which is associated with the stretching O-H vibration. The 2922 cm^{-1} band was related to C-H stretches due to the presence of the ring methane hydrogen atoms. Bands between 1652 to 1000 cm^{-1} were attributed to hydrogen bonds of O-H groups stretching vibration, O-H bending vibrations and C-O stretching vibrations [55].

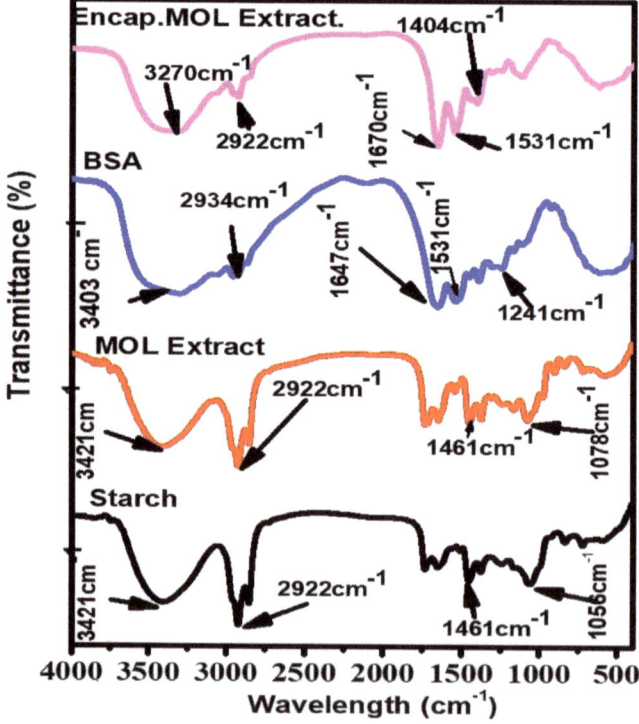

Figure 4. FTIR spectra of starch, MOLE, BSA and Encap. MOLE.

The FTIR spectrum of MOLE showed a broad band at 2922 cm^{-1}, indicating the presence of vibration stretching of the aromatic (C-H) group, while a band located at 3435 cm^{-1} belonged to (O-H) stretching vibration that was associated with phenols and alcohols. A weak band at 1461 cm^{-1} was attributed to the -OH bond. The results obtained in the present study are in agreement with [56].

In the spectrum of BSA, a broad band located at 3356 cm^{-1} and 2934 cm^{-1} can be attributed to the stretching vibration of the -NH stretch and -CH, respectively. Bands at 1647 cm^{-1} and 1531 cm^{-1} responded to C=O stretching and -N-H bending of amide I and II band [57]. Meanwhile, Encap. MOLE observed the main peaks of MOLE at 3270 cm^{-1}, 2922 cm^{-1} and 1461 cm^{-1}.

3.3. In Vivo Studies

3.3.1. Liver Functions

Animals exposed to the oral administration of BPA showed significant elevation in the levels of serum ALT, AST and ALP ($p < 0.05$) compared to the untreated group (control) [58], while a reduction in the levels of serum albumin and total protein ($p < 0.05$) were detected [59]. In contrast, hepatic function parameters were remodulated in the BPA groups treated with MOLE and Encap. MOLE separately (Figure 5) [60].

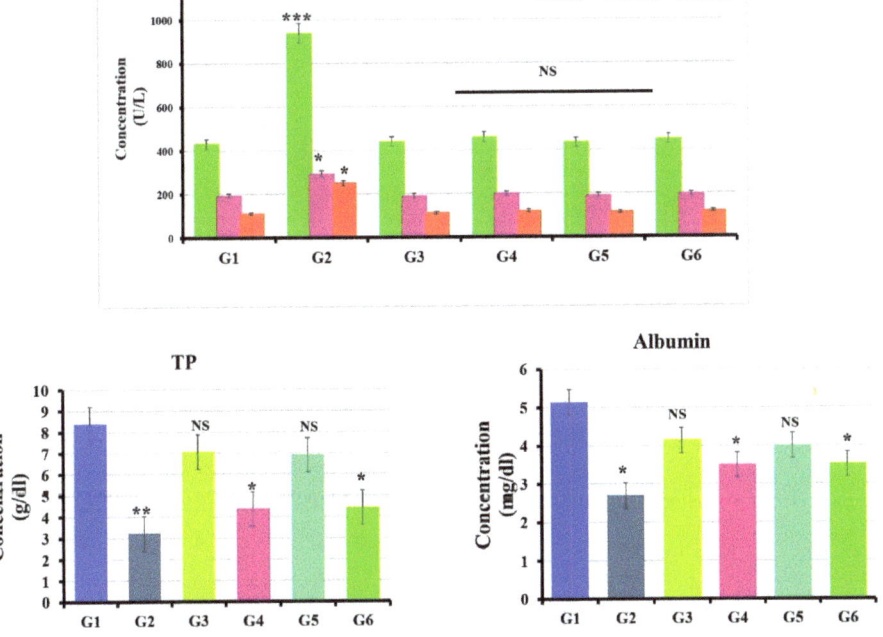

Figure 5. Effect of Moringa and nano-Moringa leaf extracts on liver function in Bisphenol A—Exposed versus normal rats. G1: control. G2: BPA. G3: MOLE. G4: MOLE-BPA. G5: Encap. MOLE. G6: Encap. MOLE-BPA. The data are shown as * $p < 0.05$ ** $p < 0.01$ and *** $p < 0.001$.

3.3.2. Oxidative Stress Markers

In the current study, animals that were given BPA orally showed a significant decrease in the levels of GSH and SOD ($p < 0.05$) compared to the untreated group (control group) [61]. On the contrary, the levels of GSH and SOD were significantly increased in groups treated separately with MOLE and Encap. MOLE simultaneously with their exposure to BPA (Figure 6).

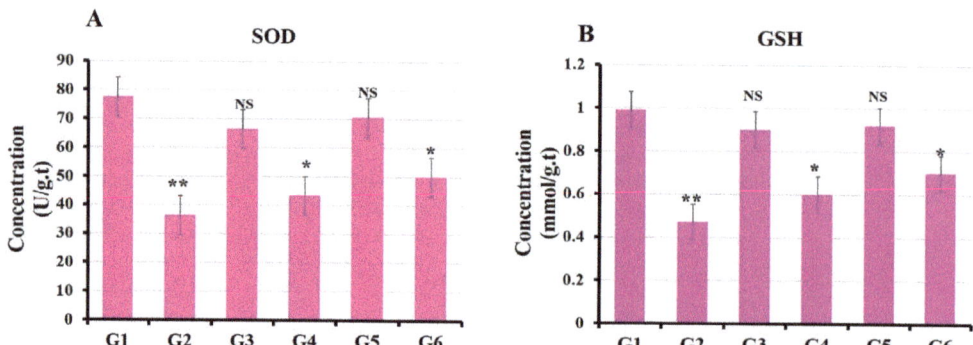

Figure 6. Effect of both Moringa and nano-Moringa leaf extracts on antioxidant enzyme activities in the liver tissue homogenates of Bisphenol A—Exposed versus normal rats. G1: control. G2: BPA. G3: MOLE. G4: MOLE-BPA. G5: Encap. MOLE. G6: Encap. MOLE-BPA. The data are shown as * $p < 0.05$ and ** $p < 0.01$.

3.3.3. Evaluations of Lipid Profile Status

In comparison to the untreated group (control group), the levels of total cholesterol (TC) and triglyceride (TG) increased significantly in the group treated with BPA [62], while an improvement in the alteration of lipid profiles was observed in the groups treated separately with MOLE and Encap. MOLE simultaneously with their exposure to BPA (Figure 7) [63].

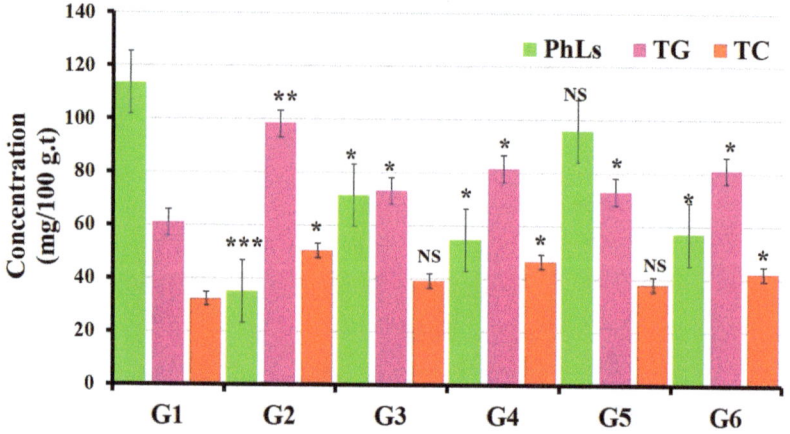

Figure 7. Effect of both Moringa and nano-Moringa leaf extracts on lipid profile in liver tissue homogenates of Bisphenol A—Exposed versus normal rats. G1: control. G2: BPA. G3: MOLE. G4: MOLE-BPA. G5: Encap. MOLE. G6: Encap. MOLE-BPA. The data are shown as * $p < 0.05$ ** $p < 0.01$ and *** $p < 0.001$.

3.4. ELISA Kits for Caspase-3 and Bax Detection

During apoptosis, the focal adhesion kinase, actin and poly (ADP-ribose) polymerase (PARP) were cleaved by Caspase-3 [64]. Meanwhile, endonuclease that called CAD was also activated by Caspase-3, leading to the fragmentation of DNA. Caspase-3 is regulated by bcl-2 and bcl-xL, which work to maintain mitochondrial membrane integrity and prevent the cleavage of Caspase-3 from its proenzyme state [65]. Meanwhile, the mitochondrial outer membrane permeability can be controlled by Bax, which allows the release of proapoptotic molecules (e.g., cytochrome c) [66].

The levels of Caspase-3 and Bax in the liver tissues were significantly increased in the animal model exposed to BPA (50 mg/kg) for month by (6.3 ± 0.1 ng/mL $p < 0.01$) and (4.2 ± 0.1 ng/mL $p < 0.001$), respectively, compared to the control (0.57 ± 0.01) and (0.38 ± 0.01), respectively (Figure 8) [67]. On the other hand, groups (G4 and G6) treated separately with MOLE and Encap. MOLE during their exposure to BPA showed significant inhibition in the level of Caspase-3 and Bax by (2.3 ± 0.2 ng/mL) and (0.9 ± 0.1 ng/mL) and (1.96 ± 0.02 ng/mL) and (0.7 ± 0.04 ng/mL), respectively [68]. In contrast, the levels of Caspase-3 and Bax increased slightly in groups (G3 and G5) treated separately with MOLE and Encap. MOLE alone (0.8 ± 0.1 ng/mL) [69].

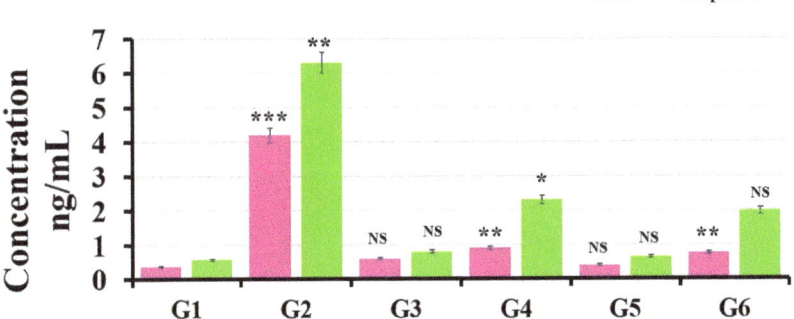

Figure 8. Effect of both MOLE and Encap. MOLE on the levels of Caspase-3 and Bax in liver tissues of BPA—Exposed versus normal rats. G1: control. G2: BPA. G3: MOLE. G4: MOLE-BPA. G5: Encap. MOLE. G6: Encap. MOLE-BPA. The data are shown as * $p < 0.05$ ** $p < 0.01$ and *** $p < 0.001$.

3.5. The Findings of Real-Time PCR

In the current study, the cyto-protective enzyme (Heme oxygenase-1 (HO-1)) was studied. This enzyme degrades heme into carbon monoxide, free iron and biliverdin, turning into bilirubin. This mechanism is important in the regulation of oxidative stress, apoptosis and inflammation. In the normal state, the expression of HO-1 is normal, while it may be increased in the pathology state due to its role in the regulation of cell homeostasis [70].

The growth arrest and DNA damage-inducible gene 45b (Gadd45b) mediates DNA damage repair, cell cycle arrest and apoptosis in response to cell injury [71]. High expression of GADD45 is used as an indicator for a variety of diseases, such as tumors [72] and nephropathy [73].

The expression levels of HO-1 and Gadd45b increased significantly in groups (G4 and G6) treated separately with MOLE and Encap. MOLE during the course of BPA exposure compared to the untreated group (control group) ($p < 0.05$) (Figure 9) [74]. Meanwhile, MOLE and Encap. MOLE alone have the ability to maintain the homeostasis of HO-1 and Gadd45b in animal groups (G3 and G5).

3.6. Histopathology Results

In the present study, the control group showed normal liver structure. The hepatic cords were radiating from the central vein and formed anastomosing plates that were separated by blood sinusoids and the hepatocytes are located with eosinophilic cytoplasm, central rounded and vesicular nuclei. In contrast, multiple histopathological degenerative changes in hepatic tissues were shown as a result of BPA-induced cytotoxic effect on male albino rats, including vascular dilatation and congestion, Kupffer cell proliferation, inflammatory cell infiltration and nuclear degenerative changes. These findings were consistent with the previous literature [75,76], while groups (G4 and G6) treated with MOLE and Encap. MOLE, respectively, during their exposure to BPA showed an improved histological architecture (Figures 10 and 11) [77]. However, there was slight observation of vascular dilatation and blood congestion in the group (G4) treated with MOLE alone compared to G6 which was treated with Encap. MOLE. This indicates the potential therapeutic effect of *Moringa oleifera* Lam leaf extract incorporated inside hydrogel NPs. Meanwhile, an eosinophilic cytoplasm appears in both groups which could be due to the presence of isothiocyanate as a *Moringa oleifera* Lam leaf extract. For instance, isothiocyanate can produce glycosides which cause necrotic cells. This indication was clearly observed in groups (G3 and G5) that were treated separately with MOLE and Encap. MOLE, showing an eosinophilic structure and pyknotic stages [78].

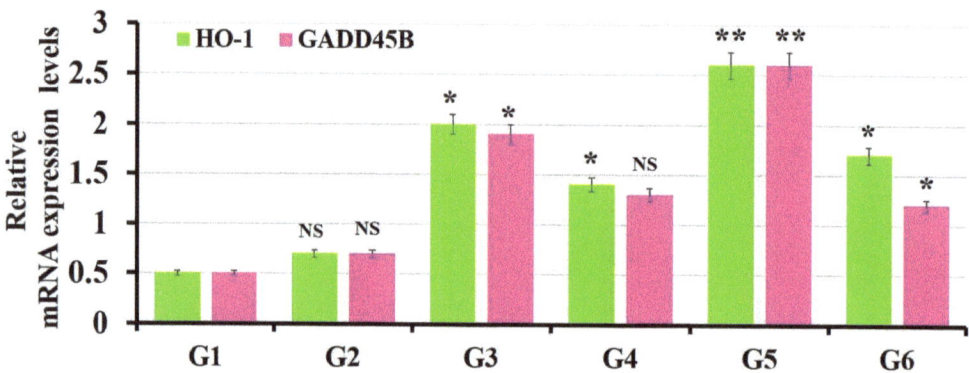

Figure 9. Effect of both MOLE and Encap. MOLE on the expression levels of HO-1 and Gadd45b in BPA—Exposed versus normal rats. G1: control. G2: BPA. G3: MOLE. G4: MOLE-BPA. G5: Encap. MOLE. G6: Encap. MOLE-BPA. The data are shown as * $p < 0.05$ ** $p < 0.01$ and *** $p < 0.001$.

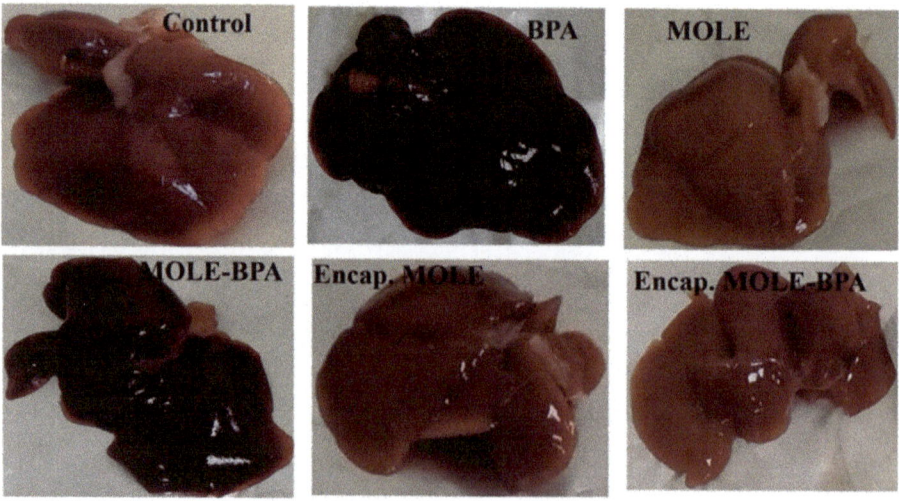

Figure 10. Photomicrograph of individual livers showing the morphological appearance.

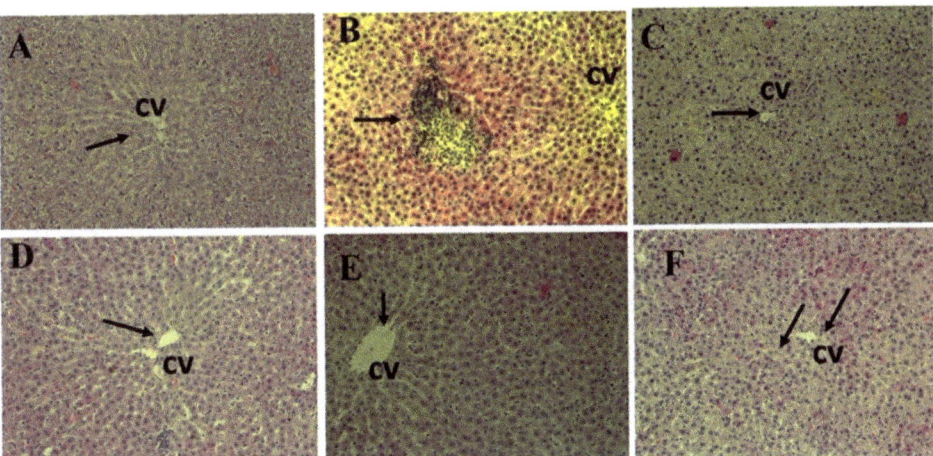

Figure 11. Histopathological examination of liver sections stained with H&E (200X). (**A**) Control; (**B**) BPA; (**C**) MOLE; (**D**) MOLE-BPA; (**E**) Encap. MOLE; (**F**) Encap. MOLE-BPA.

In Figure 12, Masson trichrome stain of liver sections in the control groups revealed that collagen fibers (blue stain) were distributed normally around the central vein (CV) and portal tract (PT). However, there was an marked increase in the density of collagen fibers around the portal tract (PT) (Figure 12A–C) in the group treated with BPA [79], while encapsulation MOLE improved the distribution and density of collagen fibers during the oral administration of BPA [80]. On the other hand, collagen fibers were distributed in the portal vein area of the group treated with MOLE during the course of BPA.

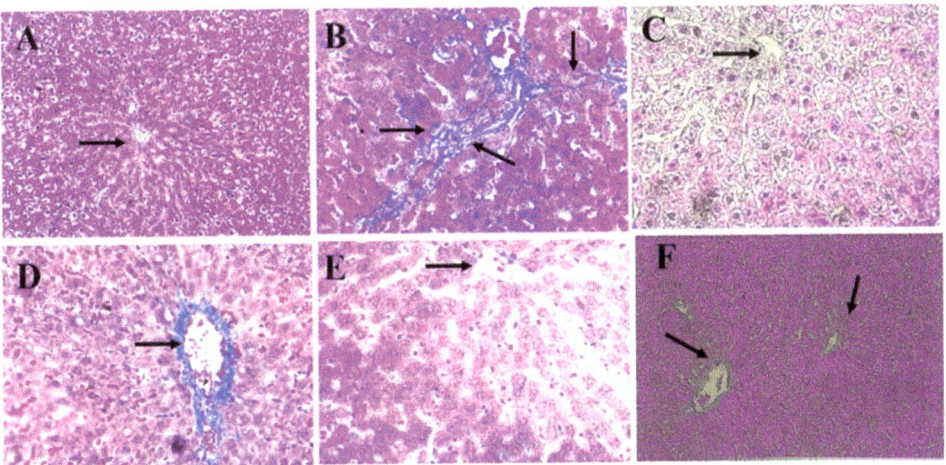

Figure 12. Masson Trichrome staining of liver sections (200X). (**A**) Control; (**B**) BPA; (**C**) MOLE; (**D**) MOLE-BPA; (**E**) Encap. MOLE; (**F**) Encap. MOLE-BPA.

In the current study, the pathological profile of liver damage scores was calculated according to a previous publication [27]. Histopathological examination was carried out and vascular dilatation and congestion, Kupffer cell proliferation, inflammatory cell infiltration, nuclear degenerative changes and collagen fibers were evaluated. The grading scale to score pathologic findings was as follows: 0 = no injury; 1 = slight injury; 2 = moderate injury; 3 = severe injury; and 4 = very severe injury (Figure 13).

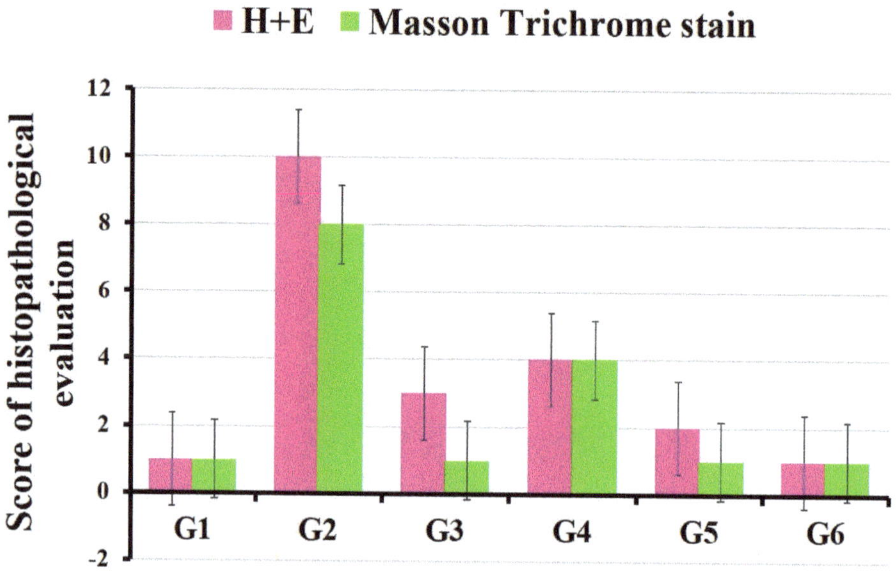

Figure 13. Scores of histopathological evaluation of different animal groups. G1: control. G2: BPA. G3: MOLE. G4: MOLE-BPA. G5: Encap. MOLE. G6: Encap. MOLE-BPA.

4. Discussion

The exposure of humans to BPA is becoming ubiquitous and continues due to its presence in components of polycarbonate plastic, dental sealant resin, flame retardants and liners for food packaging. Unfortunately, BPA is an endocrine-disrupting chemical, causing injury in the brain, liver, kidney, epididymal sperm in rodents and other organs. Xenobiotic chemical such as BPA was metabolized in the liver as the main organ for chemical detoxification. Thereby, inside the liver microsomes, BPA is glucuronidated and mediated by UGT2B1. The resultant was excreted mainly into the bile in a male rat and nonpregnant female rat. In this metabolic condition, reactive oxygen species formed [81].

For the first time, in the current study, MOLE was encapsulated inside hydrogel NPs and used to evaluate the hepatoprotective activity alongside the administration of BPA. Indeed, MOL contains antioxidants (Table 2) that can remodulate histopathological evidence induced by BPA. The physiological disorder administrated in the hepatic enzymes (ALP, ALT, AST) and in the levels of albumin and total protein was significantly remodulated in groups treated with MOLE and Encap. MOLE (Figure 5). This result indicates the hepatoprotective effect of MOLE and Encap. MOLE in eliminating free radicals that were generated by the metabolization of BPA [82,83].

The antioxidant enzymes such as SOD and GSH were significantly decreased in animals exposed to 50 mg/kg BPA, indicating that BPA strongly lowered the hepatic antioxidant status, while MOLE and Encap. MOLE improved the enzyme antioxidant activity (Figure 6). Meanwhile, the activation of Caspase-3 and Bax was significantly demonstrated in the group exposed to BPA, while MOLE and Encap. MOLE inhibited Caspase-3 and Bax levels significantly (Figure 8).

It is well known that the expression of HO-1 and GADD45B can affect oxidative stress. To provide better understanding for this hypothesis, the expression of HO-1 as a cyto-protective enzyme to maintain cell homeostasis and GADD45B as an indicator for blocking cell cycle survival, apoptosis and DNA repair were studied. In the current study, the expression of HO-1 increased significantly in groups (G4 and G6) treated separately with MOLE and Encap MOLE during the course of BPA compared to the control group ($p < 0.05$). Additionally, its expression was upregulated in groups (G3 and G5) treated with

MOLE ($p < 0.05$) and Encap. MOLE ($p < 0.001$), respectively, compared to the untreated group (control group). These data were in agreement with [84], revealing the ability of MOLE to increase the expression of HO-1 (Figure 9).

Moreover, the expression of GADD45B reveals the rate of cell damage at the gene level. In the current study, a significant regulation in the level of GADD45B was obtained in groups treated with MOLE ($p < 0.05$) and Encap. MOLE ($p < 0.001$). Meanwhile, its expression was significantly reduced in the group treated with Encap. MOLE ($p < 0.05$) over the course of BPA and was reduced non-significantly in the group treated with MOLE compared to control. Nevertheless, the expression of HO-1 and GADD45B was maintained in the group that received BPA, as BPA was suspended in sesame oil, which is in agreement with [85].

The histopathology results revealed normal radial arrangement of hepatocytes along the central vein. However, hepatocytes were disordered in the group treated with BPA for a month with obvious identification of inflammatory cell infiltration. Additionally, serious eosinophilic structure was clearly shown in the cytoplasm of hepatocytes. Conversely, MOLE and Encap. MOLE significantly improved the histopathological architecture of the liver structure providing no inflammation. Moreover, collagen fibers accumulated along the portal area of the group treated with BPA, while collagen fibers were maintained in the group treated with MOLE alongside BPA and they were significantly reduced in the group treated with Encap. MOLE alongside BPA (Figures 10–12).

It can be summarized that MOLE contains many antioxidant and bioactive materials that could use to protect and prevent hepatotoxicity produced by the exposure to environmentally toxic chemicals such as BPA. Encapsulation of MOLE saves its bioactive materials from temperature, humidity and enzymatic degradation. Additionally, it improves their adhesion in the small intestine, improving their adsorption.

5. Conclusions

M. oleifera leaf extract rich in bioactive compounds such as phenolic compounds, minerals, protein and fibers that have antioxidant capacity. However, its sensitivity to pH, temperature and other physiological enzyme degradation limits their use in biomedical applications. In the current study, phenolic compounds inserted into a starch system may alter functional properties of starch, such as gelatinization, rheological properties, gelling and retrogradation, which can improve the nutritional quality of food. Indeed, the non-covalent interactions between starch and phenolics result in either the formation of V-type amylose inclusion complex or the non-inclusive complex with much weaker binding forces. This hydrogen bridge greatly affects the hydrodynamic radius of amylose, resulting in the enhancement of phenolic compound bioavailability and control of starch digestion.

Author Contributions: H M H A F.-N.: methodology, data curation, investigation, writing original draft preparation and ideation of using encapsulating MOLE NPs in nutrition. S.A.E.-H.: supervision, revision and biostatistics analysis. E.M.Y.: supervision and revision. S.L.: revision and supporting. N.A.N.H.: methodology, data curation, investigation, format analysis, review, editing and supervision. All authors have read and agreed to the published version of the manuscript.

Funding: The current work was self-funded by Nemany A. N. Hanafy and Hend Mohamed Abou El-Naga and it was not received any external funds.

Institutional Review Board Statement: The animal study protocol was approved by Ethics Committee controlled by Kafrelsheikh University.

Informed Consent Statement: Not applicable.

Data Availability Statement: Data available in a publicly accessible repository.

Acknowledgments: Hend Mohamed Abou El-Naga would like to thank Yacien Mohammed Alsodany of the Plant Environment and Flora Department and Assistant Lecturer Mohammed Mahmoud Elkhalafy of the Plant Environment and Flora Department of Botany and Microbiology, Faculty of Science, Kafrelsheikh University, Kafr ELSheikh, Egypt, for their help in identifying MOL. Hend M. Abou El-Naga would like to thank, Ensaf Mokhtar Yasen, Samah Ahmed El-Hashash and Nemany A. N. Hanafy for their great supervision. Hend, M. Abou El-Naga would like to thank her family (her mother, sisters and brother) for their encouragement and support.

Conflicts of Interest: The authors declare that there are no competing interest or personal relationships that could have appeared to influence the work reported in this paper.

References

1. Cichoż-Lach, H.; Michalak, A. Oxidative stress as a crucial factor in liver diseases. *World J. Gastroenterol.* **2014**, *20*, 8082–8091. [CrossRef]
2. Chiang, J. Liver Physiology: Metabolism and Detoxification. *Pathobiol. Hum. Dis.* **2014**, 1770–1782. [CrossRef]
3. Wahlang, B.; Jin, J.; Beier, J.I.; Hardesty, J.E.; Daly, E.F.; Schnegelberger, R.D.; Falkner, K.C.; Prough, R.A.; A Kirpich, I.; Cave, M.C. Mechanisms of Environmental Contributions to Fatty Liver Disease. *Curr. Environ. Health Rep.* **2019**, *6*, 80–89. [CrossRef] [PubMed]
4. Thoene, M.; Rytel, L.; Nowicka, N.; Wojtkiewicz, J. The state of bisphenol research in the lesser developed countries of the EU: A mini-review. *Toxicol. Res.* **2018**, *7*, 371–380. [CrossRef] [PubMed]
5. Goodson, A.; Robin, H.; Summerfield, W.; Cooper, I. Migration of bisphenol A from can coatings—Effects of damage, storage conditions and heating. *Food Addit. Contam.* **2004**, *21*, 1015–1026. [CrossRef]
6. Acconcia, F.; Pallottini, V.; Marino, M. Molecular Mechanisms of Action of BPA. *Dose Response* **2015**, *13*, 1559325815610582. [CrossRef]
7. Cantonwine, D.E.; Hauser, R.; Meeker, J.D. Bisphenol A and human reproductive health. *Expert Rev. Obstet. Gynecol.* **2013**, *8*, 329–335. [CrossRef]
8. Inadera, H. Neurological Effects of Bisphenol A and its Analogues. *Int. J. Med Sci.* **2015**, *12*, 926–936. [CrossRef]
9. Thoene, M.; Rytel, L.; Dzika, E.; Włodarczyk, A.; Kruminis-Kaszkiel, E.; Konrad, P.; Wojtkiewicz, J. Bisphenol A Causes Liver Damage and Selectively Alters the Neurochemical Coding of Intrahepatic Parasympathetic Nerves in Juvenile Porcine Models under Physiological Conditions. *Int. J. Mol. Sci.* **2017**, *18*, 2726. [CrossRef]
10. Greenwell, M.; Rahman, P.K. Medicinal Plants: Their Use in Anticancer Treatment. *Int. J. Pharm. Sci. Res.* **2015**, *6*, 4103–4112. [CrossRef]
11. Mbikay, M. Therapeutic Potential of Moringa oleifera Leaves in Chronic Hyperglycemia and Dyslipidemia: A Review. *Front. Pharmacol.* **2012**, *3*, 24. [CrossRef] [PubMed]
12. Vergara-Jimenez, M.; Almatrafi, M.M.; Fernandez, M.L. Bioactive Components in *Moringa oleifera* Leaves Protect against Chronic Disease. *Antioxidants* **2017**, *6*, 91. [CrossRef] [PubMed]
13. Salama, A.A.A.; Fayed, A.M.; Attia, T.A.; Elbatrna, S.A.; Ismaiel, E.I.; Hassan, A. Protective Effects of *Moringa oleifera* extract on Isoniazid and Rifampicin Induced Hepatotoxicity in Rats: Involvement of Adiponectin and Tumor Necrosis Factor-α. *Egypt. J. Vet. Sci.* **2018**, *49*, 25–34.
14. Alia, F.; Putri, M.; Anggraeni, N.; Yamsunarno, M.R.A.A. The Potency of *Moringa oleifera* Lam. as Protective Agent in Cardiac Damage and Vascular Dysfunction. *Front. Pharmacol.* **2022**, *12*, 724439. [CrossRef]
15. Anwar, F.; Latif, S.; Ashraf, M.; Gilani, A.H. *Moringa oleifera*: A food plant with multiple medicinal uses. *Phytother. Res.* **2007**, *21*, 17–25. [CrossRef] [PubMed]
16. Santos, A.F.S.; Argolo, A.C.C.; Paiva, P.M.G.; Coelho, L.C.B.B. Antioxidant Activity of *Moringa oleifera* Tissue Extracts. *Phytother. Res.* **2012**, *26*, 1366–1370. [CrossRef] [PubMed]
17. Fakurazi, S.; Sharifudin, S.A.; Arulselvan, P. *Moringa oleifera* Hydroethanolic Extracts Effectively Alleviate Acetaminophen-Induced Hepatotoxicity in Experimental Rats through Their Antioxidant Nature. *Molecules* **2012**, *17*, 8334–8350. [CrossRef]
18. Osamede Airouyuwa, J.; Kaewmanee, T. Microencapsulation of *Moringa oleifera* leaf extracts with vegetable protein as wall materials. *Food Sci. Technol. Int.* **2019**, *25*, 533–543. [CrossRef]
19. Munin, A.; Edwards-Lévy, F. Encapsulation of Natural Polyphenolic Compounds; a Review. *Pharmaceutics* **2011**, *3*, 793–829. [CrossRef]
20. Singh, T.; Shukla, S.; Kumar, P.; Wahla, V.; Bajpai, V.K.; Rather, I.A. Application of Nanotechnology in Food Science: Perception and Overview. *Front. Microbiol.* **2017**, *8*, 1501, Erratum in *Front. Microbiol.* **2017**, *8*, 2517. [CrossRef]
21. Labelle, M.; Ispas-Szabo, P.; Mateescu, M.A. Structure-Functions Relationship of Modified Starches for Pharmaceutical and Biomedical Applications. *Starch Stärke* **2020**, *72*, 2000002. [CrossRef]
22. Hanafy, N.A.N.; El-Kemary, M.; Leporatti, S. Micelles Structure Development as a Strategy to Improve Smart Cancer Therapy. *Cancers* **2018**, *10*, 238. [CrossRef] [PubMed]

23. Hanafy, N.A.N.; Quarta, A.; Di Corato, R.; Dini, L.; Nobile, C.; Tasco, V.; Carallo, S.; Cascione, M.; Malfettone, A.; Soukupova, J.; et al. Hybrid polymeric-protein nano-carriers (HPPNC) for targeted delivery of TGFβ inhibitors to hepatocellular carcinoma cells. *J. Mater. Sci. Mater. Med.* **2017**, *28*, 120. [CrossRef] [PubMed]
24. Mabrouk Zayed, M.M.; Sahyon, H.A.; Hanafy, N.A.N.; El-Kemary, M.A. The Effect of Encapsulated Apigenin Nanoparticles on HePG-2 Cells through Regulation of P53. *Pharmaceutics* **2022**, *14*, 1160. [CrossRef] [PubMed]
25. Nayak, G.; Honguntikar, S.D.; Kalthur, S.G.; D'Souza, A.S.; Mutalik, S.; Setty, M.M.; Kalyankumar, R.; Krishnamurthy, H.; Kalthur, G.; Adiga, S.K. Ethanolic extract of *Moringa oleifera* Lam. leaves protect the pre-pubertal spermatogonial cells from cyclophosphamide-induced damage. *J. Ethnopharmacol.* **2016**, *182*, 101–109. [CrossRef] [PubMed]
26. Hanafy, N.A.N. Starch based hydrogel NPs loaded by anthocyanins might treat glycogen storage at cardiomyopathy in animal fibrotic model. *Int. J. Biol. Macromol.* **2021**, *183*, 171–181. [CrossRef]
27. Hanafy, N.A.N.; Leporatti, S.; El-Kemary, M.A. Extraction of chlorophyll and carotenoids loaded into chitosan as potential targeted therapy and bio imaging agents for breast carcinoma. *Int. J. Biol. Macromol.* **2021**, *182*, 1150–1160. [CrossRef]
28. Hanafy, N.A.N.; El-Kemary, M.A. Silymarin/curcumin loaded albumin nanoparticles coated by chitosan as muco-inhalable delivery system observing anti-inflammatory and anti COVID-19 characterizations in oleic acid triggered lung injury and in vitro COVID-19 experiment. *Int. J. Biol. Macromol.* **2022**, *198*, 101–110. [CrossRef]
29. Simon, S.; Joseph, J.; George, D. Optimization of extraction parameters of bioactive components from *Moringa oleifera* leaves using Taguchi method. *Biomass Convers. Biorefinery* **2022**, 1–10. [CrossRef]
30. El-Hashash, S.A.; El-Sakhawy, M.A.; El-Nahass, E.E.; Abdelaziz, M.A.; Abdelbasset, W.K.; Elwan, M.M. Prevention of Hepatorenal Insufficiency Associated with Lead Exposure by *Hibiscus sabdariffa* L. Beverages Using In Vivo Assay. *BioMed Res. Int.* **2022**, *2022*, 7990129. [CrossRef]
31. Hass, U.; Christiansen, S.; Boberg, J.; Rasmussen, M.G.; Mandrup, K.; Axelstad, M. Low-dose effect of developmental bisphenol A exposure on sperm count and behaviour in rats. *Andrology* **2016**, *4*, 594–607. [CrossRef]
32. Vandenberg, L.N.; Ehrlich, S.; Belcher, S.M.; Ben-Jonathan, N.; Dolinoy, D.C.; Hugo, E.R.; Hunt, P.A.; Newbold, R.R.; Rubin, B.S.; Saili, K.S.; et al. Low dose effects of bisphenol A: An integrated review of in vitro, laboratory animal, and epidemiology studies. *Endocr. Disruptors* **2013**, *1*, e26490. [CrossRef]
33. Ahmed, M.S.W.; Moselhy, A.W.; Nabil, T.M. Bisphenol A Toxicity in Adult Male Rats: Hematological, Biochemical and Histopathological Approach. *Glob. Vet.* **2015**, *14*, 228–238.
34. Sahu, C.; Singla, S.; Jena, G. Studies on male gonadal toxicity of bisphenol A in diabetic rats: An example of exacerbation effect. *J. Biochem. Mol. Toxicol.* **2022**, *36*, e22996. [CrossRef] [PubMed]
35. Doshi, T.; D'Souza, C.; Dighe, V.; Vanage, G. Effect of neonatal exposure on male rats to bisphenol a on the expression of DNA methylation machinery in the postimplantation embryo. *J. Biochem. Mol. Toxicol.* **2012**, *26*, 337–343. [CrossRef]
36. Ellman, G.L. Tissue sulfhydryl groups. *Arch. Biochem. Biophys.* **1959**, *82*, 70–77. [CrossRef]
37. Beauchamp, C.; Fridovich, I. Superoxide dismutase: Improved assays and an assay applicable to acrylamide gels. *Anal. Biochem.* **1971**, *44*, 276–287. [CrossRef]
38. Richmond, N. Enzymatic colorimetric test for cholesterol determination. *Clin. Chem.* **1973**, *19*, 1350–1356. [CrossRef]
39. Jacobs, N.J.; VanDenmark, P.J. Enzymatic colorimetric determination of triglycerides. *Arch. Biochem. Biophys.* **1960**, *88*, 250–255. [CrossRef]
40. Ray, T.K.; Skipski, V.P.; Barclay, M.; Essner, E.; Archibald, F.M. Lipid Composition of Rat Liver Plasma Membranes. *J. Biol. Chem.* **1969**, *244*, 5528–5536. [CrossRef]
41. Reitman, S.; Frankel, S. A colorimetric method for determination of oxaloacetic transaminase and serum glutamic pyruvic transaminase. *Am. J. Clin. Pathol.* **1957**, *28*, 56–60. [CrossRef] [PubMed]
42. Kind, P.R.N.; King, E.J. Estimation of Plasma Phosphatase by Determination of Hydrolysed Phenol with Amino-antipyrine. *J. Clin. Pathol.* **1954**, *7*, 322–326. [CrossRef] [PubMed]
43. Gornall, A.G.; Bardawill, C.J.; David, M.M. Determination of serum proteins by means of the biuret reaction. *J. Biol. Chem.* **1949**, *177*, 751–766. [CrossRef]
44. Doumas, B.T.; Watson, W.A.; Biggs, H.G. Albumin standards and the measurement of serum albumin with bromcresol green. *Clin. Chim. Acta* **1971**, *31*, 87–96. [CrossRef]
45. Somade, O.T.; Ajayi, B.O.; Olunaike, O.E.; Jimoh, L.A. Hepatic oxidative stress, up-regulation of pro-inflammatory cytokines, apoptotic and oncogenic markers following 2-methoxyethanol administrations in rats. *Biochem. Biophys. Rep.* **2020**, *24*, 100806. [CrossRef]
46. Kazemi, S.; Mousavi, S.N.; Aghapour, F.; Rezaee, B.; Sadeghi, F.; Moghadamnia, A.A. Induction Effect of Bisphenol A on Gene Expression Involving Hepatic Oxidative Stress in Rat. *Oxidative Med. Cell. Longev.* **2016**, *2016*, 6298515. [CrossRef]
47. Safer, A.M.; Afzal, M.; Hanafy, N.; Sosamma, O.; Mousa, S.A. Curative propensity of green tea extract towards hepatic fibrosis induced by CCl4: A histopathological study Corrigendum in /etm/10/2/835. *Exp. Ther. Med.* **2012**, *3*, 781–786. [CrossRef]
48. Safer, A.; Afzal, M.; Hanafy, N.; Mousa, S. Green tea extract therapy diminishes hepatic fibrosis mediated by dual exposure to carbon tetrachloride and ethanol: A histopathological study Corrigendum in /etm/10/3/1239. *Exp. Ther. Med.* **2015**, *9*, 787–794. [CrossRef]

49. Oboh, G.; Ademiluyi, A.O.; Ademosun, A.O.; Olasehinde, T.A.; Oyeleye, S.I.; Boligon, A.A.; Athayde, M.L. Phenolic Extract from *Moringa oleifera* Leaves Inhibits Key Enzymes Linked to Erectile Dysfunction and Oxidative Stress in Rats' Penile Tissues. *Biochem. Res. Int.* **2015**, *2015*, 175950. [CrossRef]
50. Fattah, M.E.A.; Sobhy, H.M.; Reda, A.; Abdelrazek, H.M.A. Hepatoprotective effect of *Moringa oleifera* leaves aquatic extract against lead acetate–induced liver injury in male Wistar rats. *Environ. Sci. Pollut. Res.* **2020**, *27*, 43028–43043. [CrossRef]
51. Domene-López, D.; García-Quesada, J.C.; Martin-Gullon, I.; Montalbán, M.G. Influence of Starch Composition and Molecular Weight on Physicochemical Properties of Biodegradable Films. *Polymers* **2019**, *11*, 1084. [CrossRef] [PubMed]
52. Spinozzi, F.; Ferrero, C.; Perez, S. The architecture of starch blocklets follows phyllotaxic rules. *Sci. Rep.* **2020**, *10*, 20093. [CrossRef] [PubMed]
53. Sit, N.; Deka, S.C.; Misra, S. Optimization of starch isolation from taro using combination of enzymes and comparison of properties of starches isolated by enzymatic and conventional methods. *J. Food Sci. Technol.* **2014**, *52*, 4324–4332. [CrossRef] [PubMed]
54. Diniz, P.H.G.D.; Barbosa, M.F.; De Melo Milanez, K.D.T.; Pistonesi, M.F.; de Araújo, M.C.U. Using UV–Vis spectroscopy for simultaneous geographical and varietal classification of tea infusions simulating a home-made tea cup. *Food Chem.* **2016**, *192*, 374–379 . [CrossRef] [PubMed]
55. Bronze-Uhle, E.; Costa, B.C.; Ximenes, V.F.; Lisboa-Filho, P.N. Synthetic nanoparticles of bovine serum albumin with entrapped salicylic acid. *Nanotechnol. Sci. Appl.* **2016**, *10*, 11–21. [CrossRef] [PubMed]
56. Ramezani, H.; Behzad, T.; Bagheri, R. Synergistic effect of graphene oxide nanoplatelets and cellulose nanofibers on mechanical, thermal, and barrier properties of thermoplastic starch. *Polym. Adv. Technol.* **2020**, *31*, 553–565. [CrossRef]
57. El-Houssiny, A.S.; Fouad, E.A.; Hegazi, A.G. A Comparative Antimicrobial Activity Study of *Moringa oleifera* Extracts Encapsulated within ALg Nanoparticles. *Nanosci. Nanotechnol.-Asia* **2021**, *11*, 144–152. [CrossRef]
58. Yadav, P.; Yadav, A.B. Preparation and characterization of BSA as a model protein loaded chitosan nanoparticles for the development of protein-/peptide-based drug delivery system. *Future J. Pharm. Sci.* **2021**, *7*, 200. [CrossRef]
59. Hassan, Z.K.; Elobeid, M.A.; Virk, P.; Omer, S.A.; ElAmin, M.; Daghestani, M.H.; AlOlayan, E.M. Bisphenol A Induces Hepatotoxicity through Oxidative Stress in Rat Model. *Oxidative Med. Cell. Longev.* **2012**, *2012*, 194829. [CrossRef]
60. Olukole, S.G.; Ola-Davies, E.O.; Lanipekun, D.O.; Oke, B.O. Chronic exposure of adult male Wistar rats to bisphenol A causes testicular oxidative stress: Role of gallic acid. *Endocr. Regul.* **2020**, *54*, 14–21. [CrossRef]
61. Fakurazi, S.; Hairuszah, I.; Nanthini, U. Moringa oleifera Lam prevents acetaminophen induced liver injury through restoration of glutathione level. *Food Chem. Toxicol.* **2008**, *46*, 2611–2615. [CrossRef] [PubMed]
62. Moghaddam, H.S.; Samarghandian, S.; Farkhondeh, T. Effect of bisphenol A on blood glucose, lipid profile and oxidative stress indices in adult male mice. *Toxicol. Mech. Methods* **2015**, *25*, 507–513. [CrossRef] [PubMed]
63. Wang, B.; Wang, S.; Zhao, Z.; Chen, Y.; Xu, Y.; Li, M.; Xu, M.; Wang, W.; Ning, G.; Bi, Y.; et al. Bisphenol A exposure in relation to altered lipid profile and dyslipidemia among Chinese adults: A repeated measures study. *Environ. Res.* **2020**, *184*, 109382. [CrossRef] [PubMed]
64. Mehta, K.; Balaraman, R.; Amin, A.H.; Bafna, P.A.; Gulati, O.D. Effect of fruits of Moringa oleifera on the lipid profile of normal and hypercholesterolaemic rabbits. *J. Ethnopharmacol.* **2003**, *86*, 191–195. [CrossRef]
65. Mallat, Z.; Tedgui, A. Apoptosis in the vasculature: Mechanisms and functional importance. *Br. J. Pharmacol.* **2000**, *130*, 947–962. [CrossRef]
66. Elmore, S. Apoptosis: A review of programmed cell death. *Toxicol. Pathol.* **2007**, *35*, 495–516. [CrossRef]
67. Garrido, C.; Galluzzi, L.; Brunet, M.; Puig, P.E.; Didelot, C.; Kroemer, G. Mechanisms of cytochrome c release from mitochondria. *Cell Death Differ.* **2006**, *13*, 1423–1433. [CrossRef]
68. Xia, W.; Jiang, Y.; Li, Y.; Wan, Y.; Liu, J.; Ma, Y.; Mao, Z.; Chang, H.; Li, G.; Xu, B.; et al. Early-Life Exposure to Bisphenol A Induces Liver Injury in Rats Involvement of Mitochondria-Mediated Apoptosis. *PLoS ONE* **2014**, *9*, e90443. [CrossRef]
69. Akter, T.; Rahman, M.A.; Moni, A.; Apu, M.A.I.; Fariha, A.; Hannan, M.A.; Uddin, M.J. Prospects for Protective Potential of *Moringa oleifera* against Kidney Diseases. *Plants* **2021**, *10*, 2818. [CrossRef]
70. Cirmi, S.; Ferlazzo, N.; Gugliandolo, A.; Musumeci, L.; Mazzon, E.; Bramanti, A.; Navarra, M. Moringin from *Moringa oleifera* Seeds Inhibits Growth, Arrests Cell-Cycle, and Induces Apoptosis of SH-SY5Y Human Neuroblastoma Cells through the Modulation of NF-κB and Apoptotic Related Factors. *Int. J. Mol. Sci.* **2019**, *20*, 1930. [CrossRef]
71. Liu, B.; Qian, J.-M. Cytoprotective role of heme oxygenase-1 in liver ischemia reperfusion injury. *Int. J. Clin. Exp. Med.* **2015**, *8*, 19867–19873. [PubMed]
72. Rodríguez-Jiménez, P.; Fernández-Messina, L.; Ovejero-Benito, M.C.; Chicharro, P.; Vera-Tomé, P.; Vara, A.; Cibrian, D.; Martínez-Fleta, P.; Jiménez-Fernández, M.; Sánchez-García, L.; et al. Growth arrest and DNA damage-inducible proteins (GADD45) in psoriasis. *Sci. Rep.* **2021**, *11*, 14579. [CrossRef]
73. Verzella, D.; Bennett, J.; Fischietti, M.; Thotakura, A.K.; Recordati, C.; Pasqualini, F.; Capece, D.; Vecchiotti, D.; D'Andrea, D.; Di Francesco, B.; et al. GADD45β Loss Ablates Innate Immunosuppression in Cancer. *Cancer Res.* **2018**, *78*, 1275–1292. [CrossRef] [PubMed]
74. Wu, Y.-S.; Liang, S.; Li, D.-Y.; Wen, J.-H.; Tang, J.-X.; Liu, H.-F. Cell Cycle Dysregulation and Renal Fibrosis. *Front. Cell Dev. Biol.* **2021**, *9*, 714320. [CrossRef] [PubMed]

75. Cheng, W.N.; Jeong, C.H.; Seo, H.G.; Han, S.G. Moringa Extract Attenuates Inflammatory Responses and Increases Gene Expression of Casein in Bovine Mammary Epithelial Cells. *Animals* **2019**, *9*, 391. [CrossRef] [PubMed]
76. Kamel, A.H.; Foaud, M.A.; Moussa, H.M. The adverse effects of bisphenol A on male albino rats. *J. Basic Appl. Zool.* **2018**, *79*, 6. [CrossRef]
77. Poormoosavi, S.M.; Najafzadehvarzi, H.; Behmanesh, M.A.; Amirgholami, R. Protective effects of Asparagus officinalis extract against Bisphenol A-induced toxicity in Wistar rats. *Toxicol. Rep.* **2018**, *5*, 427–433. [CrossRef]
78. Gasmalbari, E.; El-Kamali, H.H.; Abbadi, O.S. Biochemical and Haematological Effects and Histopathological Changes caused by *Moringa oleifera* on Albino Rats. *Chin. J. Med. Res.* **2020**, *3*, 84–88. [CrossRef]
79. Elswefy, S.E.-S.; Abdallah, F.R.; Atteia, H.H.; Wahba, A.S.; Hasan, R.A. Inflammation, oxidative stress and apoptosis cascade implications in bisphenol A-induced liver fibrosis in male rats. *Int. J. Exp. Pathol.* **2016**, *97*, 369–379. [CrossRef]
80. Wilujeng, L.K.; Safitri, F.N.; Supriono, S.; Kalim, H.; Poeranto, S. The effect of *Moringa oleifera* (Lam) leaves ethanol extracts as anti-inflammatory and anti-fibrotic through TNF-α and p38-MAPK expression: In Vivo model of liver fibrosis approach. In Proceedings of the International Conference on Life Sciences and Technology (ICoLiST 2020), Java, Indonesia, 29 September 2020; Volume 2353, p. 030046. [CrossRef]
81. Darbandi, M.; Darbandi, S.; Agarwal, A.; Sengupta, P.; Durairajanayagam, D.; Henkel, R.; Sadeghi, M.R. Reactive oxygen species and male reproductive hormones. *Reprod. Biol. Endocrinol.* **2018**, *16*, 87. [CrossRef]
82. Sharma, V.; Paliwal, R.; Janmeda, P.; Sharma, S. Chemopreventive efficacy of *Moringa oleifera* pods against 7,12-dimethylbenz[a]ant hracene induced hepatic carcinogenesis in mice. *Asian Pac. J. Cancer Prev.* **2012**, *13*, 2563–2569. [CrossRef] [PubMed]
83. Albrahim, T.; Binobead, M.A. Roles of *Moringa oleifera* Leaf Extract in Improving the Impact of High Dietary Intake of Monosodium Glutamate-Induced Liver Toxicity, Oxidative Stress, Genotoxicity, DNA Damage, and PCNA Alterations in Male Rats. *Oxidative Med. Cell. Longev.* **2018**, *2018*, 4501097. [CrossRef]
84. Duranti, G.; Maldini, M.; Crognale, D.; Horner, K.; Dimauro, I.; Sabatini, S.; Ceci, R. *Moringa oleifera* Leaf Extract Upregulates Nrf2/HO-1 Expression and Ameliorates Redox Status in C2C12 Skeletal Muscle Cells. *Molecules* **2021**, *26*, 5041. [CrossRef] [PubMed]
85. Woo, M.; Han, S.; Song, Y.O. Sesame Oil Attenuates Renal Oxidative Stress Induced by a High Fat Diet. *Prev. Nutr. Food Sci.* **2019**, *24*, 114–120. [CrossRef] [PubMed]

Article

Development of Turmeric Oil—Loaded Chitosan/Alginate Nanocapsules for Cytotoxicity Enhancement against Breast Cancer

Htet Htet Moe San [1,2], Khent Primo Alcantara [1,2], Bryan Paul I. Bulatao [1,2], Waraluck Chaichompoo [2], Nonthaneth Nalinratana [2,3], Apichart Suksamrarn [4], Opa Vajragupta [2,5], Pranee Rojsitthisak [2,6] and Pornchai Rojsitthisak [2,7,*]

1. Pharmaceutical Sciences and Technology Program, Faculty of Pharmaceutical Sciences, Chulalongkorn University, Bangkok 10330, Thailand; htethtetmoesan17.hhms@gmail.com (H.H.M.S.); khentalcantara@gmail.com (K.P.A.); bibulatao@up.edu.ph (B.P.I.B.)
2. Center of Excellence in Natural Products for Ageing and Chronic Diseases, Chulalongkorn University, Bangkok 10330, Thailand; waraluck_kik@hotmail.com (W.C.); nonthaneth.n@pharm.chula.ac.th (N.N.); opa.v@chula.ac.th (O.V.); pranee.l@chula.ac.th (P.R.)
3. Department of Pharmacology and Physiology, Faculty of Pharmaceutical Sciences, Chulalongkorn University, Bangkok 10330, Thailand
4. Department of Chemistry and Center of Excellence for Innovation in Chemistry, Faculty of Science, Ramkhamhaeng University, Bangkok 10240, Thailand; s_apichart@ru.ac.th
5. Molecular Probes for Imaging Research Network, Faculty of Pharmaceutical Sciences, Chulalongkorn University, Bangkok 10330, Thailand
6. Metallurgy and Materials Science Research Institute, Chulalongkorn University, Bangkok 10330, Thailand
7. Department of Food and Pharmaceutical Chemistry, Faculty of Pharmaceutical Sciences, Chulalongkorn University, Bangkok 10330, Thailand
* Correspondence: pornchai.r@chula.ac.th; Tel.: +66-841111704

Abstract: Turmeric oil (TO) exhibits various biological activities with limited therapeutic applications due to its instability, volatility, and poor water solubility. Here, we encapsulated TO in chitosan/alginate nanocapsules (CS/Alg-NCs) using o/w emulsification to enhance its physicochemical characteristics, using poloxamer 407 as a non-ionic surfactant. TO-loaded CS/Alg-NCs (TO-CS/Alg-NCs) were prepared with satisfactory features, encapsulation efficiency, release characteristics, and cytotoxicity against breast cancer cells. The average size of the fabricated TO-CS/Alg-NCs was around 200 nm; their distribution was homogenous, and their shapes were spherical, with smooth surfaces. The TO-CS/Alg-NCs showed a high encapsulation efficiency, of 70%, with a sustained release of TO at approximately 50% after 12 h at pH 7.4 and 5.5. The TO-CS/Alg-NCs demonstrated enhanced cytotoxicity against two breast cancer cells, MDA-MB-231 and MCF-7, compared to the unencapsulated TO, suggesting that CS/Alg-NCs are potential nanocarriers for TO and can serve as prospective candidates for in vivo anticancer activity evaluation.

Keywords: *ar*-turmerone; polymeric nanoparticles; anticancer activity; release study; biodegradable polymers

1. Introduction

Breast cancer (BC) is the most common type of cancer to be diagnosed and the primary cause of cancer death among women, according to the Global Cancer Observatory 2020, with an estimated incidence and mortality rate of 24.5% and 15.5%, respectively [1]. Generally, BC can be categorized as estrogen-receptor-positive (ER+) or -negative (ER−). Other types, based on biomarkers such as progesterone receptor (PR) and human epidermal growth factor receptor 2 (HER2), are further sub-categorized as luminal A and B, basal-like, and HER2+ [2,3]. Basal-like BC or triple-negative BC (TNBC) is a unique type due to the absence of the biomarkers ER, PR, and HER2 [4]. The development of an effective treatment strategy for breast cancer remains very complex due to its multifaceted behavior against protein expression. Different types of BC respond differently to treatments, making BA treatment almost intractable. Current therapy for BC involves a multimodal strategy combining surgery, chemotherapy, radiotherapy, adjuvant therapy, and hormonal therapy [5–7]. However, long- or short-term use could result in an economic and psychological burden on patients and, worse, a high chance of multidrug resistance and detrimental side effects [8]. Thus, the survival rate of patients with BC is still unsatisfactory. Currently, researchers are leaning toward finding alternative forms of treatment for BC, whether in the form of therapeutics, adjuvant treatments, chemopreventive agents, or effective targeting and delivery systems [9,10]. For many years, phytochemicals have been viewed as novel approaches to the targeting and killing of cancer cells while mitigating the harmful side effects of conventional therapies [8].

Turmeric (*Curcuma longa* L.), which belongs to the family Zingiberaceae, has been used as a traditional home remedy, dye, and food additive in Southeast Asia. One of the major components of turmeric is turmeric oil (TO) which mainly contains *ar*-turmerone. TO has been widely used in pharmaceutical applications due to its broad range of biological activities, particularly its antioxidant [11] and anticancer properties [12]. Previous *ar*-turmerone studies on breast cancer showed the inhibition of enzymatic activity and the expression of matrix metallopeptidase 9 (MMP-9) and cyclooxygenase-2 (COX-2) through the nuclear factor kappa-light-chain-enhancer of activated-B-cells (NF-κB) pathway [13]. Furthermore, this compound was proven to stimulate peripheral blood mononuclear cell (PBMC) proliferation and cytokine production [14]. However, despite numerous reports on promising anti-cancer and immunomodulatory activities, TO possesses various disadvantages, such as instability, volatility, and highly lipophilic properties, limiting its therapeutic applications [15,16].

Recently, nanoparticles (NPs) have been the primary source of interest in therapeutic formulations for amplifying stability, bioavailability, and delivery to the target site [17]. Alginate (Alg) and chitosan (CS) are interesting in pharmaceutical applications due to their non-immunogenicity, biocompatibility, biodegradability, sustained release into the bloodstream or cancerous tissue, and enhanced drug-encapsulating efficiency [18,19]. Various preparation techniques have been developed concerning the production methods of chitosan/alginate nanoparticles (CS/Alg-NPs), including sonication [20], electrostatic gelation [21], the self-assembly of polysaccharides [22], the extrusion of polymer dispersions [23], electrospraying [24], and microfluidic methods [25]. The ionotropic gelation method, based on electrostatic interaction, is one of the most frequently utilized formulation methods [26]. The ionotropic gelation method produces CS/Alg-NPs through pre-gelation and polyelectrolyte complexation phases. While the pre-gelation phase occurs via the ionic cross-linking of divalent cations with Alg, the polyelectrolyte complexation phase occurs via electrostatic interactions between the negatively charged carboxylic acid groups of Alg and the positively charged amino groups of CS [27,28]. CS/Alg-NPs have been reported as useful nanocarriers for the encapsulation of chemotherapeutic compounds. Alternatively, emulsification solvent diffusion/evaporation methods are developed for enhancing the solubility of encapsulated compounds in CS/Alg-NPs through oil/water (o/w) emulsion, using stabilizers such as poloxamers [29]. Kumar et al. [29] and Das et al. [30] revealed the positive effect of poloxamer on the encapsulation efficiency of hydrophobic curcumin in CS/Alg-NPs. Sorasitthiyanukarn et al. [31,32] successfully fabricated CS/Alg nanocarriers using o/w emulsification and ionotropic gelation methods by encapsulating curcumin diethyl diglutarate [31] and curcumin diglutaric acid [32] by improving bioavailability and enhancing anticancer activity.

In 2008, Lertsutthiwong et al. [33] reported an approach to overcome these restrictions by encapsulating TO in a biopolymer network, forming Alg nanocapsules (Alg-NCs). However, the TO-loaded Alg-NCs (TO-Alg-NCs) displayed low stability at room temperature and poor drug loading capacity. To overcome these limitations, CS, a natural polysaccharide consisting of β-(1→4) glycosidic linked D-glucosamine and N-acetyl-D-glucosamine, can be used to coat TO-Alg-NCs to obtain TO-loaded chitosan/alginate nanocapsules (TO-CS/Alg-NCs) for physicochemical property improvement [15,16,34]. However, the information on the biological activities of TO-CS/Alg-NCs is limited, and the CS/Alg-NC system for the encapsulation of TO needs to be developed. Therefore, this study was undertaken to establish TO-CS/Alg-NCs with improved physicochemical characteristics and cytotoxicity against two invasive breast carcinoma cell lines, hormone-dependent MCF-7 (ER+ and PR+) and basal-like MDA-MB-231 (TNBC), both of which are invasive breast carcinoma cells.

2. Materials and Methods

Chemicals. TO was purchased from Thai–China Flavours and Fragrances Industry (Nonthaburi, Thailand). The amount of *ar*-turmerone in TO was found to be about 12%. *Ar*-turmerone was provided by the Department of Chemistry and Center of Excellence for Innovation in Chemistry, Faculty of Science, Ramkhamhaeng University (Bangkok, Thailand). CS (MW = 63 kDa, 91.74% DD) was supplied by Marine Bio-Resources (Samut Sakorn, Thailand). Sodium Alg (medium viscosity) and poloxamer 407 were purchased from Sigma-Chemicals (St. Louis, MO, USA). Acetonitrile was purchased from RCI Labscan (Bangkok, Thailand). Absolute ethanol, glacial acetic acid, calcium chloride, and other chemicals were purchased from Carlo Erba reagents (Val de Reuil, France).

Cell Culture. Human breast cancer cells (MCF-7 and MDA-MB-231) and HEK293 were cultured in Dulbecco's modified Eagle's medium (DMEM) supplemented with 10% fetal bovine serum and 100 units/mL penicillin/streptomycin (Gibco™ Thermo Fisher Scientific Inc., Waltham, MA, USA) in humidified atmosphere of 5% CO_2 at 37 °C.

2.1. Preparation of TO-CS/Alg-NCs

TO-CS/Alg-NCs were prepared by o/w emulsification of TO in the aqueous solution of Alg followed by ionotropic gelation with calcium chloride and coating with a CS solution using the method previously described by Lertsutthiwong et al. [15], with slight modifications. Briefly, 1% (v/v) ethanolic TO solution was added dropwise using a syringe pump (NE 100, New Era, Pump System Inc., New York, USA), at a speed of 20 mL/h, into the aqueous Alg solution (20 mL, 0.6 mg/mL) containing poloxamer 407 (0.65% (w/v)), and continuously stirred at 1000 rpm for 30 min using a magnetic stirrer (Onilab LLC Scientific Inc., MS-H380-Pro, Riverside, CA, USA). The o/w emulsion was then sonicated for 15 min, and calcium chloride solution (4 mL, 0.67 mg/mL) was added and continuously mixed for another 30 min. Subsequently, the CS solution (0.1 mg/mL) was added dropwise into the mixture, followed by continuous mixing for 30 min. The TO-CS/Alg-NC suspension was equilibrated overnight in the dark before characterization.

2.2. Physicochemical Characterization

The particle size, polydispersity index (PDI), and zeta potential of obtained TO-CS/Alg-NCs were characterized using a Nano-ZS Zetasizer (Malvern Instruments Ltd., Worcertershire, UK). The particle size and PDI were determined by dynamic light scattering and zeta potential was measured by electrophoretic mobility of the NCs [15]. The morphology of the obtained TO-CS/Alg-NCs was visualized using a transmission electron microscope (JEM-2100, JEOL, Tokyo, Japan). The functional groups and interaction between TO and excipients were analyzed using a Fourier transform infrared spectrometer (FT-IR, PerkinElmer Inc., Boston, MA, USA) at a range of 400–4000 cm^{-1} with a resolution of 2 cm^{-1} and 64 scans per spectra.

The encapsulation efficiency (EE) and loading capacity (LC) were determined via the indirect method and quantified using ultra-high-performance liquid chromatography (UHPLC, Agilent 1290 Infinity II LC System, CA, USA) according to the reported method, with some modifications [15]. The TO-CS/Alg-NC suspension was ultracentrifuged (Ultracentrifuge, Hitachi CP 100NX, Ibaraki, Japan) at 4 °C and 45,000 rpm for 1 h. The settled NCs were lyophilized (Lyophilizer, FreeZone, Labconco, MO, USA) for 24 h, and the unencapsulated TO in the supernatant was determined by UHPLC. Briefly, the collected supernatant was diluted with ethanol and filtered through a 0.45-micrometer syringe filter before injection into an Intersil® ODS-3 column (4.6 mm × 150 mm, i.d., 5 µm) (GL Sciences Inc., Tokyo, Japan) maintained at 33 °C. The mobile phase was a mixture of water and acetonitrile (25:75) in an isocratic elution. The injection volume was set at 20 µL with a 0.5 mL/min flow rate. A diode array detector was used to detect the analyte at a wavelength of 254 nm. The chromatographic analysis data running time was 30 min per sample with standard *ar*-turmerone eluted at a retention time of 12.8 min. The quantity of TO in the NCs was computed as the difference between the total amount of TO initially added into the formulation (TO$_{formulation}$) and the amount of TO present in the supernatant (TO$_{supernatant}$). The EE and LC were evaluated using Equations (1) and (2).

$$EE\ (\%) = \frac{(TO\ formulation\ -\ TO\ supernatant)}{TO\ formulation} \times 100 \qquad (1)$$

$$LC\ (\%) = \frac{(TO\ formulation\ -\ TO\ supernatant)}{Dry\ mass\ of\ NCs} \times 100 \qquad (2)$$

2.3. In Vitro Release and Kinetics Studies

The release study of TO from CS/Alg-NCs was performed using a dialysis diffusion method based on a previous report [35], with modifications. Phosphate-buffered saline (PBS) solution (1 mg/mL potassium dihydrogen phosphate, 2 mg/mL dipotassium hydrogen phosphate, 8.5 mg/mL sodium chloride in deionized water, pH 7.4) and sodium acetate buffer (50 mg/mL sodium acetate in 1% acetic acid, with the pH adjusted with 4.2 g/L sodium hydroxide to pH 5.5) were used, with 40 % (*v/v*) ethanol in each medium. A dialysis bag (SnakeSkin™, 10,000 Da MWCO, 33 mm diameter; Thermo Scientific, Illinois, USA) with a molecular weight cut-off at 12,000–14,000 Da (Cellu-Sep® T4, TX, USA) was first soaked in the respective media for 24 h before the experiment. TO-CS/Alg-NC suspension (20 mL) was added into the dialysis bag and sealed with clips on both ends. The dialysis bag was immersed in the release medium (500 mL) and maintained at 37 °C under continuous agitation at 100 rpm. Sampling times were set between 0 and 24 h, wherein 5 mL of medium were withdrawn at specific time points. The withdrawn samples were replaced with an equal volume of fresh medium to maintain sink conditions throughout the experiment. The concentration of TO in the medium was quantified using UHPLC and calculated against the calibration curve. The cumulative TO released (%) was computed based on Equation (3):

$$CR\ (\%) = \frac{V_e \sum_{i=1}^{n-1} C_{n-1} + V_o C_n}{m} \times 100 \qquad (3)$$

where CR is the cumulative amount of TO released (%), Ve is the sampling volume (5 mL), Vo is the total volume of release medium (500 mL), Cn is the concentration of TO at a particular time point (mg/mL), and m is the total amount of TO in TO-CS/Alg-NCs (mg).

The mechanisms involved in the release of TO from CS/Alg-NCs at pH 5.5 and 7.4 were analyzed using non-linear regression by fitting the release data to different kinetic models using the add-in DDsolver software in Microsoft Excel [36]. The kinetic constant (k) was derived using zero-order kinetics, first-order kinetics, Korsmeyer–Peppas' power law equation, and Hixson–Crowell's cube root-of-time equation. The release exponent (*n*) was also determined using Korsmeyer–Peppas' power law equation. The goodness-of-fit of the release of TO was evaluated by comparing the coefficient of determination (r^2) of the different models [37]. For the Korsmeyer–Peppas model, the release exponent (*n*) can be categorized into values for *n* < 0.43 corresponding to a spherical matrix and a drug release mechanism with Fickian diffusion, with 0.43 < *n* < 0.85 indicating anomalous transport from spheres and *n* > 0.85 suggesting drug release from spheres by polymer swelling [38].

2.4. In Vitro Biological Assay

The cell viability assay was adapted from previous work with modifications [39]. Briefly, the cytotoxicities of the unencapsulated TO, TO-CS/Alg-NCs, and the nanocarrier (CS/Alg-NCs) were evaluated using MDA-MB-231, MCF-7, and HEK293 cell lines. The cells were seeded at a density of 3×10^4 cells per 100 µL into each well of 96-well culture plates and incubated for 24 h. Next, the cells were treated with five serial concentrations of pure TO and TO-CS/Alg-NCs in a serum-free medium and incubated at 37 °C for 24 h. The cytotoxicity of the CS/Alg-NCs was also evaluated at a concentration range of 10 to 60% (v/v). After 24 h of treatment, the culture medium was removed and 100 µL of an MTT reagent (0.5 mg/mL in serum-free medium) was added to each well and further incubated at 37 °C. After 4 h, the MTT medium was removed, and the insoluble formazan crystals were dissolved by adding dimethyl sulfoxide (DMSO). After complete dissolution, the absorbance was measured at 570 nm using a microplate reader (CLARIOstar, BMG Labtech, Ortenau, Baden-Württemberg, Germany). The percentage of cell viability was calculated using Equation (4):

$$\text{Cell viability (\%)} = \left(OD_{Sample}\right) / \left(OD_{Control}\right) \times 100 \quad (4)$$

2.5. Statistical Analysis

All experiments were performed in triplicate and data were expressed as mean ± standard deviation (SD). The half-maximal inhibitory concentrations (IC_{50}) for TO and TO-CS/Alg-NCs were determined through a non-linear regression-curve-fit analysis. A two-way ANOVA was then used to analyze the cell viability data and IC_{50} values. Tukey's multiple comparisons test was used as the post hoc test. All statistical analyses were performed using GraphPad® Prism software version 9.3.0 (San Diego, CA, USA), with $p < 0.05$ considered statistically significant.

3. Results and Discussion

3.1. Preparation and Characterization of TO-CS/Alg-NCs

The TO-CS/Alg-NCs were prepared by o/w emulsification followed by inotropic gelation based on the previous method, with some modifications [33]. Concerning the procedure of the NC preparation, the NC suspension was used for the subsequent experiments without washing. Although potential impurities, including the organic solvent and surfactant, might have been carried over to the next experiments, these impurities were minimal due to the usage of organic solvent and surfactant at very low concentrations. In addition, the prepared nanosuspension was ultracentrifuged at 45,000 rpm at −4 °C for 1 h to separate the unencapsulated TO and surfactant before lyophilization. The NC preparation was equilibrated overnight to complete the crosslinking process and allow the NCs to form with a uniform size [32]. The o/w emulsification proceeded through the dropwise addition of the ethanolic TO to the aqueous Alg solution containing poloxamer as an emulsifier. Poloxamer is an amphiphilic block copolymer, which consists of poly(ethylene oxide)-poly(propylene oxide)-poly(ethylene oxide) triblock copolymer (PEO-PPO-PEO) [40]. The hydrophobic block in the middle interacted with the TO, while the hydrophilic block on both ends interacted with the aqueous phase, forming micelles. Next, the polymeric Alg micelles with a TO core underwent ionotropic gelation with calcium chloride to form a rigid egg-box structure of oligopolyguluronic sequences, which were presented in the ionic form of the carboxylate group of Alg crosslinking with calcium ion in mildly acidic pH [41]. Following ionotropic gelation, a polyelectrolyte complex between the protonated amino groups of the CS and the ionized carboxylate groups of the Alg was formed. The strength of polyelectrolyte complexation is mainly affected by the pH [42], and, therefore, the carboxylate groups of Alg are ionized at a pH greater than its pKa of 4.4 [43]. However, the amino groups of CS are protonated at pH less than its pKa of 6.5 [44]. In this study, we adjusted the Alg and CS solution to pH 4.9 and 6.0, respectively, to provide sufficient protonated and ionized groups for electrostatic linkages. The magnitude of the electrostatic interaction between the Alg and the Ca^+ ions and cationic CS polymer could have affected the characteristics of the NCs. High interaction is represented as positive or less negative zeta potential values due to the conservation of the free cationic groups on the surfaces, resulting in the more rigid and compact structure of the NCs (Figure 1) [27]. The results showed an average size of 184.8 ± 14.8 nm and a PDI of 0.192 ± 0.1, indicating the monodispersity of the particles (Figure 2A). The relatively small particle size obtained using poloxamer 407 as a stabilizer was attributed to its higher HLB value (HLB = 18) than Tween 80 (HLB = 15), which was used in a previous study [15]. Surfactants possessing higher HLB values are more suitable for o/w emulsification due to their high water solubility; thus, less aggregation occurs, resulting in more stable and smaller particles [45]. The zeta potential was also determined to predict the stability of the NCs in the aqueous system based on their surface charge. The NCs rendered a negative zeta potential value of about -21.8 ± 1.1 (Figure 2B). Wu et al. [46] suggested that a nanosuspension formulated using poloxamer 407 as a non-ionic surfactant is stable if the zeta potential ranges between -20 and -30 mV. Thus, the use of poloxamer 407 in this study can improve the stability of the NCs.

The morphology of the TO-CS/Alg-NCs was visualized by TEM after diluting the nanosuspension 50× in ultrapure water. The NCs were spherical and had smooth surfaces, with a particle size of approximately 200 nm (Figure 2C,D). It was evident in the image that a thin layer of CS was coated onto the TO-Alg-NCs, as indicated by the arrow shown in (Figure 2D). The EE and LC were determined by the indirect method. The concentration of the unencapsulated *ar*-turmerone in the TO-CS/Alg nanosuspension was determined using the calibration curve of the standard *ar*-turmerone ($y = 86.082x + 10.617$, $R^2 > 0.9999$). The EE and LC of the TO-CS/Alg-NCs using poloxamer 407 were $70.3 \pm 1.3\%$ and $3.4 \pm 1.3\%$, respectively. The LC was similar to that in the previous report using Tween 80 as a surfactant [15]. However, the EE of the TO-CS/Alg-NCs using poloxamer 407 (HLB = 18) was higher than that of using Tween 80 (HLB = 15). Ranjith and Wijewardene [47] suggested that a high HLB value renders more water-soluble stabilizers; hence, hydrophobic drugs such as TO can be more stable in an aqueous phase, resulting in high encapsulation efficiency.

Further characterization was carried out by FT-IR to investigate possible interactions on the TO-CS/Alg-NCs (Figure 3). In the empty NCs, an IR peak at 1102 cm^{-1} represented –CH–OH in cyclic alcohol and C–O stretching of the CS, while the peak near 1341 cm^{-1} belonged to the C–H bending in poloxamer 407. The peaks at about 1605 cm^{-1} and 1414 cm^{-1} were attributed to the stretching of the –COO- groups in the Alg and the N–H twisting vibration of the essential amine of CS, respectively. These results suggest the interaction between the –COO$^-$ of the Alg and the –NH_3^+ of the CS to form a strong polyelectrolyte complexation and the encapsulation of the TO in the CS/Alg-NCs [48]. In the spectrum of the TO, a peak at 1685 cm^{-1} could be assigned to the C=O group from the active component, *ar*-turmerone. The bands between 3100 cm^{-1} and 2900 cm^{-1} were attributed to the –OH group or the –NH and aliphatic C–H of the TO [49,50]. Additionally, the TO peaks, 1112, 1618, and 1685 cm^{-1}, shifted to 1108, 1576, and 1635 cm^{-1}, respectively in the TO-loaded CS/Alg-NCs, indicating the presence of the C=O and CH–OH functional groups, respectively [49]. Moreover, most of the characteristic absorption bands of the TO in the spectrum of the TO-CS/Alg-NCs broadened with the reduction in the intensity, indicating the successful encapsulation of the TO into the CS/Alg-NCs.

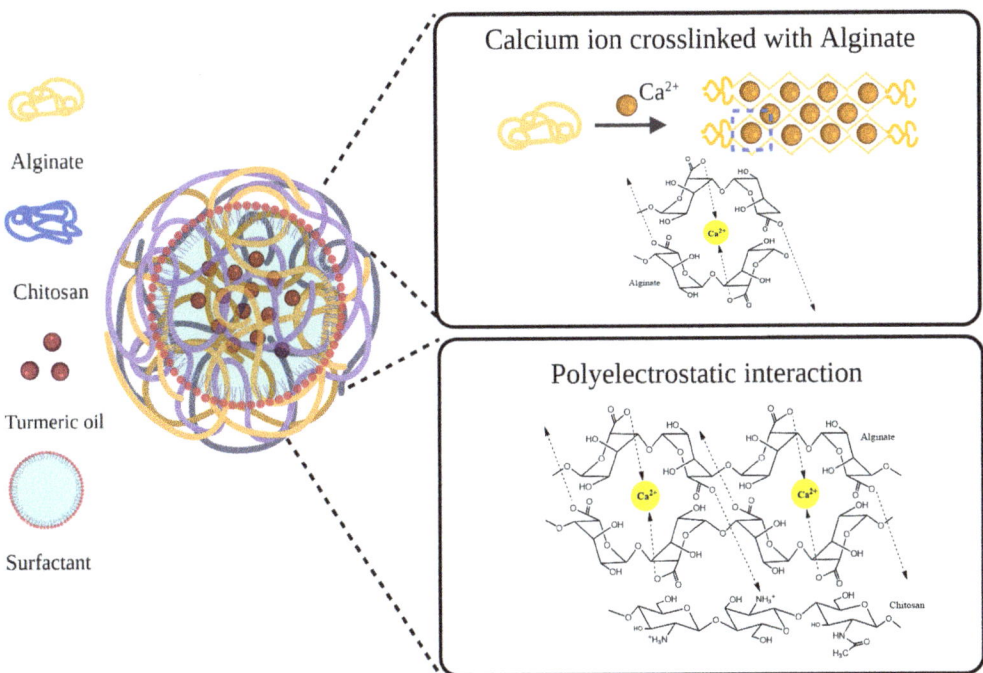

Figure 1. Electrostatic interaction of the anionic Alg with Ca$^+$ ions and cationic CS polymer.

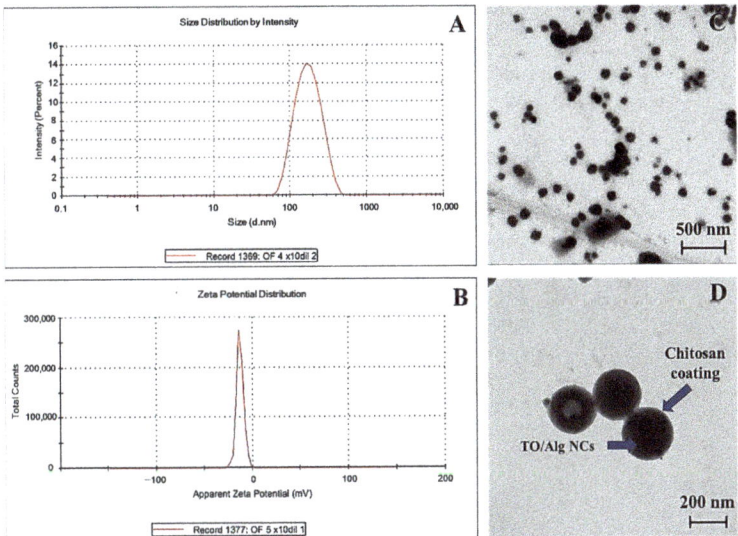

Figure 2. Physical characteristics of TO-CS/Alg-NCs. (**A**): Size distribution by intensity, (**B**): Zeta potential distribution, (**C,D**): TEM images at 50,000 and 100,000× magnification, respectively.

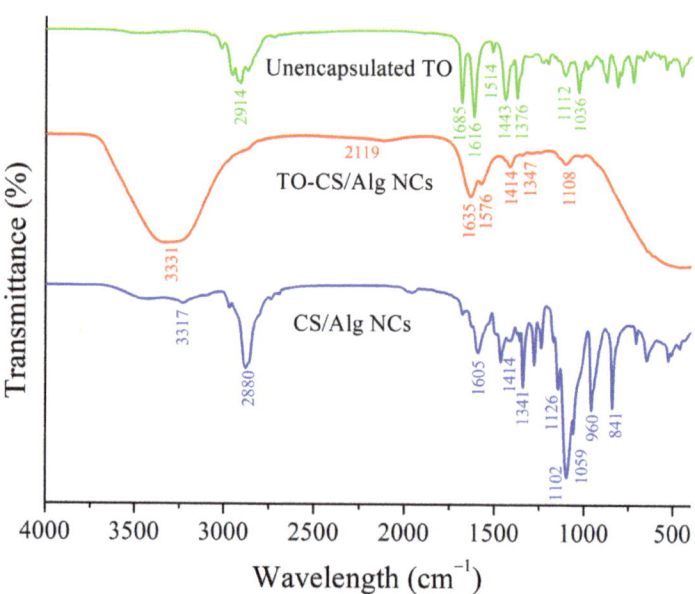

Figure 3. Fourier transform infrared spectra of unencapsulated TO, TO-CS/Alg-NCs, and CS/Alg-NCs.

3.2. In Vitro Release Study of TO

The pH-responsive behavior of the release of the TO from the TO-CS/Alg-NCs was observed in the buffers simulating the blood/normal tissues (pH 7.4) and the acidic endosome environment (pH 5.5) of the cancer cells [51]. Changes in pH changes may be observed as NCs traverse different biological compartments in the human body. For the characterization of the release profile of hydrophobic molecules, the release medium often includes solubility-enhancing agents to better capture the in vivo performance of a drug delivery system [52]. The actual release of the TO from the TO-CS/Alg-NCs was also a function of the mild shear stress due to the constant agitation of the system while incubating at 37 °C. This process mimicked the movement of extracellular fluids around the particles [52]. The NCs inside the dialysis bag contained only the aqueous dispersion of the NCs. The presence of ethanol (40% (v/v)) in the release buffers, contained in the receiver compartment, facilitated a faster molecular diffusion of the TO throughout the aqueous phase [53]. Thus, it served to maintain a sink condition, preventing the back diffusion of the TO from the receiver to the donor (dialysis bag) compartment [54]. In fact, the use of ethanol in release experiments ranges from 10 to 96%, alone or in combination with water or buffers [35,55–60].

Nonetheless, the present study acknowledges some limitations on the conditions of the release experiments that may have had a substantial influence on the quantification of the release of the TO. As is evident in Figure 4A, the slow release profile of the unencapsulated oil in both release media may suggest non-sink conditions, which may have taken the form of the precipitation of the TO inside the dialysis chamber or the affinity of the TO with the dialysis membrane, preventing the unhindered permeation of the TO and its translocation to the receiver compartment [61]. This was evidenced by the apparently large variations in the % TO released in largest number of time points and the absence of a trend showing the immediate and complete release of the TO. However, it can be observed in Figure 4A that the release of the TO from the CS/Alg-NCs provided a biphasic pattern that is typically observed in polymeric NPs. This profile is characterized by a burst release effect followed by a slow release phase [62]. Phase 1 is characterized by the rapid release of TO molecules that are surface-bound or near the water layer. This may be attributed to the diffusion and migration of TO molecules during the fabrication and drying processes, leading to burst release effects as water molecules move to the gel surface, carrying TO molecules via convection and leading to higher TO concentrations at the surface of the carrier. Phase 2, characterized by a slow release, results in a few TO molecules diffusing from the core to maintain a slow and sustained release [63]. This effect can be governed by slow TO diffusion through the polymer matrix or existing pores and is simultaneous with polymer hydrolysis and degradation [62]. Moreover, the diffusion of water would hydrolytically break the bonds of the polyelectrolyte complex [62].

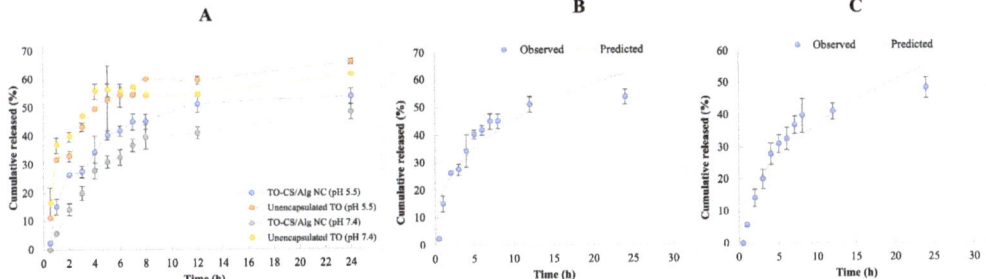

Figure 4. Release study showing (**A**) cumulative release of TO from unencapsulated TO and TO-CS/Alg-NCs in the release media, and nonlinear curve fitting to the experimental data using Korsmeyer–Peppas model in (**B**) pH 5.5 and (**C**) pH 7.4.

The lower standard deviations in the release of the TO from the TO-CS/Alg-NCs in both release media may imply the efficiency of the CS/Alg-NCs at entrapping, releasing, and dispersing the hydrophobic TO within the dialysis chamber. The efficient release from the NCs could maintain the sink conditions by ensuring that the TO concentration in the receiver compartment was low compared to the TO concentration within the dialysis chamber throughout the entire release study. These reasons are based on a number of assumptions, such as the improved dispersion of the hydrophobic TO due to its encapsulation in the CS/Alg-NCs. In fact, the amount of compound in the receiver compartment should not exceed more than 80% of the total amount of compound used in the experiment [54]. This requirement was achieved, since less than 80% of the TO was translocated to the receiver compartment. This result was important to overcome the effect of the receiver compartment in driving osmosis across the dialysis membrane [64].

In addition, the swelling of polymers and the penetration of fluids into the particles may well take a finite amount of time. This argument may implicate the overall release study and potentially underestimate the release profile of the TO when the release experiment was halted before reaching 100% TO. However, the 24-h dynamic dialysis method to determine the release profile was used in conjunction with the duration of the cell experiment, which was also set to 24 h. With reference to Section 3.3, the slow liberation of the TO from the CS/Alg-NCs may have been responsible for maintaining a therapeutic intracellular concentration of TO, resulting in the reasonable cytotoxicity of the TO in the breast cancer cells. This was the case when CS-NPs encapsulated TO in a previous report, with <30% of the TO released within 24 h in both neutral and acidic media [65]. The same study revealed that extending the release study even for up to 20 days only released the TO by 52%. Our previous study utilized CS/Alg NPs to encapsulate TO, but the in vitro release profile and kinetics were not characterized [65].

Based on the zeta potential of the TO-CS/Alg-NCs (−21.8 mV), it was evident that the negative charge was due to the higher charge density of the carboxyl groups of the Alg, resulting in a higher cross-linking with the CS, a stronger polyelectrolyte membrane, and smaller pores. This was desirable, as an excess of CS results in a positively-charged value for zeta potential, producing increased hypertonicity, which, in turn, increases the osmotic pressure on the polyelectrolyte membrane and leads to a premature bursting effect [37]. A previous study demonstrated that the swelling of the polyelectrolyte complex (PEC) of CS and Alg was higher in acidic than in neutral conditions. At pH 5.5, a stronger PEC membrane is expected, since the CS and Alg are still completely ionized, increasing their counterion charge density [37]. However, the presence of counterions in the buffer solutions results in the charge neutralization of the particles, bringing the apparently negatively charged functional groups of the TO-CS/Alg-NCs close to the isoelectric point [66,67].

Furthermore, the TO release at pH 5.5 was higher than that at pH 7.4. The favored release at pH 5.5 can be explained by the protonation of the NH_2 groups of CS (up to pH 6.8), which were randomly distributed along with the PEC, resulting in the swelling of CS and promoting the release of the TO through the porous CS/Alg-NCs into the release media [63]. By contrast, the TO from the NCs was slowly released at pH 7.4 because of the decreased solubility and shrinkage of the CS, which prevented the release of the TO from the CS/Alg-NCs [32,37]. At higher pH, the solubility of CS decreases. These observed effects are desirable, since TO should be preferentially released within the acidic intracellular environments of cancer cells [63]. These results demonstrate the pH-responsiveness of the TO-CS/Alg-NCs, which can be regarded as useful in the delivery of compounds due to the differential pH that can be observed in cancerous and normal tissues [68]. Therefore, the CS/Alg-NC carrier is suitable to deliver TO by relying on pH, an internal stimulus that is practical, convenient, and non-invasive in the human body [69].

Comparing the r^2 among the release kinetic models in Table 1, the highest value was attained for the Korsmeyer–Peppas model. This implies that the Korsmeyer–Peppas' power-law release kinetics was the best-fit model for describing the release behavior of the TO from the CS/Alg-NCs in both release media (Figure 4B,C). The release exponent (n) of the Korsmeyer–Peppas model can be used as a parameter to analyze the configuration of the nanoparticle and the release mechanism involved. Drug release mechanisms from polymeric NPs can be classified into diffusion through water-filled pores, diffusion through the polymer matrix, osmotic pumping, and erosion [62]. It was apparent that the zero-order release profile did not appropriately represent the release of the TO from the CS/Alg-NCs. The values of n at pH 5.5 and 7.4 were 0.356 and 0.455, respectively. Since the n value at pH 5.5 is less than 0.43, TO-CS/Alg-NCs can correspond to a spherical matrix and a drug release mechanism with Fickian diffusion. In this type of observed behavior, the rate of solvent diffusion is much greater than the process of polymeric chain relaxation, promoting the rapid equilibration of solvent on the surface exposure of the polymeric system. On the other hand, a value between 0.43 and 0.85 can be associated with non-Fickian diffusion, specifically an anomalous transport from the spheres. In this case, the release of TO from the polymeric matrix was governed by the diffusion or swelling of the matrix. Moreover, the velocity of the solvent diffusion and the polymeric relaxation possessed similar magnitudes. These observations may be attributed to the differences in solubility in different media [70]. Additional in vitro release studies would be valuable to assess time points beyond 24 h to better capture more predictive results for in vivo studies.

Table 1. Comparison of release kinetics model of TO from CS/Alg-NCs in different media.

Medium	Zero-Order		First-Order		Korsmeyer–Peppas			Hixson–Crowell	
	k_0	r^2_0	k_1	r^2_1	n	k_k	r^2_k	k_H	r^2_H
pH 5.5	3.705	−0.5653	0.078	0.4225	0.356	20.650	0.8352	0.018	0.1962
pH 7.4	2.813	0.1404	0.0654	0.6207	0.455	11.442	0.8390	0.013	0.4935

3.3. In Vitro Cytotoxicity Assay

The cytotoxic effects of the unencapsulated TO and TO-CS/Alg-NCs (with equivalent TO concentrations ranging from 20 to 120 µg/mL) were evaluated against the MDA-MB-231 and MCF-7 breast cancer cells using the MTT reduction assay (Figure 5A,B). The two breast cancer cell lines were chosen for comparative purposes and to differentiate the responses of the cancer cell lines to the treatments. The MDA-MB-231 and MCF-7 cells are commonly used because they differ in origin, survival, and recurrence rates. The MCF-7 cell line, which originates from human breast adenocarcinoma, is non-metastatic and positive for estrogen receptor (ER+) and progesterone receptor (PR+). The MDA-MB-231 cells belong to the triple-negative breast cancer subtype, which lacks the estrogen receptor (ER), progesterone receptor (PR−), and human epidermal growth factor receptor (HER2−). MDA-MB-231 cells are highly invasive and commonly represent late-stage breast cancer [71,72]. The possible off-target effects of the unencapsulated TO and TO-CS/Alg-NCs were evaluated using HEK293 cells (Figure 5C). The HEK293 is the human embryonic kidney epithelial cell line commonly used to study the toxic effects of nanoparticles [73]. It is an appropriate model as a non-breast-cancer cell line since it demonstrates no tissue-specific gene expression signature, and differentiation markers of several tissues are highly expressed [74].

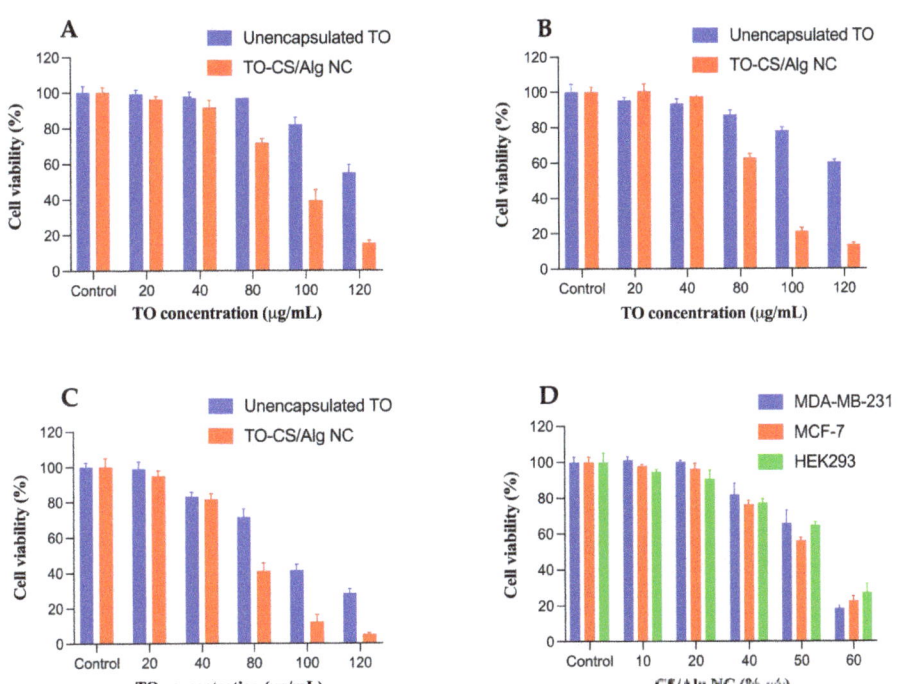

Figure 5. The viability of (**A**) MDA-MB-213, (**B**) MCF-7, and (**C**) HEK293 cells treated with TO and TO-CS/Alg-NCs (equivalent to 20 to 120 µg/mL TO), and (**D**) MDA-MB-231, MCF-7, and HEK293 cells treated with 10 to 60% (v/v) of CS/Alg-NCs.

ANOVA was applied to the cell viability data to determine if there was a significance in the mean difference among the unencapsulated TO and TO-CS/Alg-NCs across the cell lines. The results indicate no significant differences in the cell viability across all the treatments in the cancer cells and HEK293 cell line at 20 µg/mL of TO. Evaluating the safety of the nanocarrier system in a normal cellular environment is an essential tool to ascertain the suitability of the biomaterial for administration to the human body. Therefore, the biocompatibility of the nanocarrier (CS/Alg-NCs) was evaluated using the HEK293 cells (Figure 5D). The present study highlights pronounced toxicity toward HEK293 cells of at least 50% (v/v) in the CS/Alg-NCs, with the HEK293 cells showing the viability of less than 70%. Figure 5D also shows that up to 40% (v/v) of the CS/Alg-NCs were considered non-toxic (>70% viability) toward HEK293 and the two breast cancer cell lines, demonstrating the most relevant concentration of the nanocarrier to be considered for subsequent studies. The TO alone did not induce significant cell death in the MDA-MB-231 and MCF-7 cells, even up to 80 µg/mL. The cytotoxic response was augmented when the TO was encapsulated in the CS/Alg-NCs. At 40% (v/v) of CS/Alg-NCs, equivalent to 80 µg/mL of TO in the TO-CS/Alg-NCs, significant percentages of the MDA-MB-231 (28%) and MCF-7 (40%) were inhibited. Furthermore, both the MDA-MB-231 and the MCF-7 had at least a 60% reduction in cell viability at 100 µg/mL. At 120 µg/mL, at least 80% of the cells were not viable, demonstrating that the CS/Alg-NCs enhanced the toxicity of the TO against the MDA-MB-231 and MCF-7. This demonstrates the TO-concentration-dependent cytotoxicity of the MDA-MB-231 and MCF-7. Overall, the sensitivity of both the MDA-MB-231 and the MCF-7 to the TO-CS/Alg-NCs was prominent between 80 to 120 µg/mL, confirming the cytotoxicity of the TO nanocapsules in different subtypes of breast cancer cells. This implies that the internalization of TO can be enhanced through its encapsulation in CS/Alg-NCs and the endocytosis of TO-CS/Alg-NCs. The factors that may have contributed to the enhanced activity of the TO-CS/Alg-NCs toward the breast cancer cells include their size <200 nm, negative zeta potential, and spherical shape. The negative zeta potential (−21.8 mV) implies that the primary amine groups of the chitosan may have electrostatically interacted with the carboxyl groups of the alginate. On the other hand, this result also indicates that there is an increased association among the hydrophobic acetyl groups of chitosan, promoting stronger hydrophobic interactions with cancer cell membranes [75]. Moreover, the TO-CS/Alg-NCs, which have a negative surface charge, may act as proton sponges, increasing the osmotic pressure within the endosome to release the TO within the cytosol [76]. The significantly higher cytotoxic effect of the TO-CS/Alg-NCs compared to the unencapsulated TO may also be attributed to the sustained release of the TO from the CS/Alg-NCs in the acidic intracellular pH. Concerning the release of the TO from the CS/Alg-NCs, the observed release profile could suggest that a sufficient amount of TO might not have been liberated within the duration of the release study. Nevertheless, the cell viability results imply that the TO encapsulated in the CS/Alg-NCs showed enhanced cytotoxic activity compared to the TO alone, providing evidence that a sufficient concentration of TO was released in the cytosol. The chitosan–alginate polyelectrolyte (PEC) complex was needed to protect the TO and release most of it intracellularly. The PEC may be responsible for the higher mechanical strength of the nanocarrier against possible degradation from chitosanases and lysozymes than either polymer alone [76]. Another possible mechanism underlying the enhanced cytotoxicity of the TO-CS/Alg-NCs includes the presence of the CD44 receptor, a transmembrane glycoprotein, in the MDA-MB-231 and the MCF-7. Although this mechanism may be more relevant for hyaluronic acid, the natural ligand of CD44, its structural similarity to chitosan through the functional group N-acetyl glucosamine could have led to the binding and endocytosis of the nanocapsules. Studies have shown that the level of CD44 in MDA-MB-231 was four times as high as that in MCF-7 cells [73,77]. Considering this plausible mechanism, this study partly demonstrates the prominent differential susceptibility of the luminal subtype MCF-7 and the basal subtype MDA-MB-231 to cytotoxic compounds, confirming the documented resistance of MDA-MB-231 to most of the chemotherapeutic drugs, including TO, in the present study [72].

The IC_{50} values of the TO and TO-CS/Alg-NCs' activity against the MDA-MB-231, MCF-7, and HEK293 cells are presented in Table 2. The IC_{50} values of the TO-CS/Alg-NCs differed significantly from the IC_{50} values of the unencapsulated TO against the MDA-MB-231 and MCF-7, indicating that the TO-CS/Alg-NCs had significantly higher cytotoxicity against the MDA-MB-231 and MCF-7 than the equivalent dose of the unencapsulated TO. The cytotoxic effects of the TO-CS/Alg-NCs observed in both breast cancer and normal cells demonstrate the non-selectivity of the nanocarriers.

Table 2. Mean IC_{50} values of unencapsulated TO and TO-CS/Alg-NCs against MDA-MB-231, MCF-7, and HEK293 cell lines.

Cell Line	IC_{50} (µg/mL)	
	Unencapsulated TO	TO-CS/Alg-NCs
MDA-MB-231	329.53 ± 8.06	99.11 ± 3.40 *
MCF-7	344.60 ± 42.5	82.88 ± 4.40 *,^
HEK293	141.33 ± 11.09	84.30 ± 9.60

ANOVA results: * $p < 0.0001$ (compared to unencapsulated TO), ^ $p = 0.8765$, no significant difference (compared to mean IC_{50} of TO-CS/Alg-NCs in MDA-MB-231).

The potential off-target effects of the TO-CS/Alg-NCs (at an equivalent concentration of at least 80 µg/mL TO) were shown by the reduced viability of HEK293 cells of 41%. Since the nanocarrier CS/Alg-NCs resulted in a >70% viability for HEK293 cells at the equivalent concentration, the significant off-target effects imply that the reduced viability resulted from the encapsulation of the TO in the CS/Alg-NCs. The cell death of HEK293 cells may have come from the wide variety of phytochemicals present in the TO [78]. Based on the cell viability findings of the present study, a possible strategy to overcome this limitation is to optimize the concentration of TO in CS/Alg-NCs to achieve an optimum therapeutic efficacy while reducing the toxic side effects. The cytotoxic effects may also be attributed to the components of the nanocarrier system itself. Nanoparticle-induced apoptosis in both normal and cancer cells can occur via ROS generation triggering the activation of caspase 9, consequently inducing the intrinsic apoptosis pathway through the mitochondria [79]. Therefore, another strategy to overcome these unwanted findings is to conjugate a ligand with CS. The selective toxicity toward breast cancer cells can be increased by targeting the folate receptors, which are highly expressed in breast cancer cells. The nanocarrier can result in even higher cytotoxicity toward breast cancer cells through ligand-receptor-mediated endocytosis. Conjugating folic acid with chitosan has been shown to increase the cytotoxicity of drug-loaded nanocarriers toward breast cancer cells and their biocompatibility with normal cells, as demonstrated in several studies [69,80–82]. Nevertheless, it was clear that the encapsulation of the TO in the CS/Alg-NCs significantly enhanced their cytotoxicity toward the breast cancer cells. Therefore, the use of TO-CS/Alg-NCs can be proposed as a potential alternative therapeutic strategy for breast cancer treatment. Further optimization studies are required to appropriately design the nanoformulation for intravenous administration with minimal off-target effects on normal cells.

4. Conclusions

TO-CS/Alg-NCs containing poloxamer 407 as a non-ionic surfactant were prepared using the o/w emulsification, ionotropic gelation, and freeze-drying method with a CS/Alg mass ratio of 0.03:1, 0.65% poloxamer, and 1% TO. The prepared NCs were spherical in shape, with an average particle size of 200 nm and a zeta potential of −21.8. The EE of the TO was 70. The FTIR analysis confirmed the PEC between the primary amines of CS and the carboxylic groups of Alg. The PEC formation was also confirmed by the significant change in the surface charge of the TO-Alg-NCs following CS coating. No new chemical entity formation was observed, indicating chemical compatibility between the TO and the CS/Alg-NCs. The sustained release of TO of approximately 50% after 12 h in both media suggests that CS/Alg-NCs can be used as sustained and controlled nanocarriers for TO. The cytotoxicity of the TO-CS/Alg-NCs was significantly more potent than that of the unencapsulated TO against both the MDA-MB-231 and the MCF-7. With a computed EE of 70%, we can assume that the unencapsulated or unbound TO was 30%. The cytotoxicity of the TO-CS/Alg Alg-NCs was likely to have been due to both the encapsulated and the unencapsulated TO. In the future, the enhanced cytotoxicity of TO-CS/Alg-NCs against cancer cells can be further improved by the functionalization of polymeric materials with cancer-specific ligands for developing active targeted delivery systems.

Author Contributions: Conceptualization, H.H.M.S., O.V., P.R. (Pranee Rojsitthisak) and P.R. (Pornchai Rojsitthisak); methodology, H.H.M.S. and K.P.A.; experimentation, H.H.M.S., K.P.A., W.C., N.N. and A.S.; data curation, H.H.M.S.; Analysis, K.P.A., B.P.I.B., W.C., N.N. and A.S.; writing—original draft preparation, H.H.M.S., K.P.A., and B.P.I.B.; writing—review and editing, O.V., P.R. (Pranee Rojsitthisak) and P.R. (Pornchai Rojsitthisak); supervision, O.V., P.R. (Pranee Rojsitthisak) and P.R. (Pornchai Rojsitthisak); project administration, P.R. (Pornchai Rojsitthisak); funding acquisition, P.R. (Pranee Rojsitthisak) and P.R. (Pornchai Rojsitthisak). All authors have read and agreed to the published version of the manuscript.

Funding: This work was supported by the Ratchadaphiseksomphot Endowment Fund, Chulalongkorn University (RCU_H_64_035_62, Pranee Rojsitthisak), the Ratchadaphiseksomphot Endowment Fund for Center of Excellence in Natural Products for Ageing and Chronic Diseases (GCE 6503433003-1, Pornchai Rojsitthisak), and the 90th Anniversary of Chulalongkorn University Fund from the Ratchadaphiseksomphot Endowment Fund (H.H.M.S. and Pornchai Rojsitthisak).

Institutional Review Board Statement: Not applicable.

Informed Consent Statement: Not applicable.

Data Availability Statement: All the data are available within the manuscript.

Acknowledgments: We would like to thank the Pharmaceutical Research Instrument Center of the Faculty of Pharmaceutical Sciences at Chulalongkorn University for providing the research facilities. H.H.M.S. was financially supported by the Graduate Scholarship Programme for ASEAN and Non-ASEAN Countries from the Office of Academic Affairs, Chulalongkorn University. W.C. was financially supported by Ratchadapisek Somphot Fund for Postdoctoral Fellowship, Graduate School, Chulalongkorn University.

Conflicts of Interest: The authors declare no competing interests.

References

1. Sung, H.; Ferlay, J.; Siegel, R.L.; Laversanne, M.; Soerjomataram, I.; Jemal, A.; Bray, F. Global cancer statistics 2020: GLOBOCAN estimates of incidence and mortality worldwide for 36 cancers in 185 countries. *CA Cancer J. Clin.* **2021**, *71*, 209–249. [CrossRef] [PubMed]
2. Liedtke, C.; Mazouni, C.; Hess, K.; Andre, F.; Tordai, A.; Mejia, J.; Symmans, W.; Gonzalez-Angulo, A.; Hennessy, B.; Green, M.; et al. Response to neoadjuvant therapy and long-term survival in patients with triple-negative breast cancer. *Am. J. Clin. Oncol.* **2008**, *26*, 1275–1281. [CrossRef] [PubMed]
3. Lehmann, B.D.; Bauer, J.A.; Chen, X.; Sanders, M.E.; Chakravarthy, A.B.; Shyr, Y.; Pietenpol, J.A. Identification of human triple-negative breast cancer subtypes and preclinical models for selection of targeted therapies. *J. Clin. Investig.* **2011**, *121*, 2750–2767. [CrossRef] [PubMed]
4. Barzaman, K.; Karami, J.; Zarei, Z.; Hosseinzadeh, A.; Kazemi, M.H.; Moradi-Kalbolandi, S.; Safari, E.; Farahmand, L. Breast cancer: Biology, biomarkers, and treatments. *Int. Immunopharmacol.* **2020**, *84*, 106535. [CrossRef]
5. Chew, H.K. Adjuvant therapy for breast cancer: Who should get what? *West. J. Med.* **2001**, *174*, 284–287. [CrossRef]
6. de Matteis, A.; Nuzzo, F.; D'Aiuto, G.; Labonia, V.; Landi, G.; Rossi, E.; Mastro, A.A.; Botti, G.; De Maio, E.; Perrone, F. Docetaxel plus epidoxorubicin as neoadjuvant treatment in patients with large operable or locally advanced carcinoma of the breast: A single-center, phase II study. *Cancer* **2002**, *94*, 895–901. [CrossRef]
7. Fisusi, F.A.; Akala, E.O. Drug combinations in breast cancer therapy. *Pharm. Nanotechnol.* **2019**, *7*, 3–23. [CrossRef]
8. Brahmachari, G. Chapter 1—Discovery and development of anti-breast cancer agents from natural products: An overview. In *Discovery and Development of Anti-Breast Cancer Agents from Natural Products*; Brahmachari, G., Ed.; Elsevier Science Publishing: New York, NY, USA, 2021; pp. 1–6.
9. Aung, T.N.; Qu, Z.; Kortschak, R.D.; Adelson, D.L. Understanding the effectiveness of natural compound mixtures in cancer through their molecular mode of action. *Int. J. Mol. Sci.* **2017**, *18*, 656. [CrossRef]
10. Wilczewska, A.Z.; Niemirowicz, K.; Markiewicz, K.H.; Car, H. Nanoparticles as drug delivery systems. *Pharmacol. Rep.* **2012**, *64*, 1020–1037. [CrossRef]
11. Jayaprakasha, G.K.; Jena, B.S.; Negi, P.S.; Sakariah, K.K. Evaluation of antioxidant activities and antimutagenicity of turmeric oil: A byproduct from curcumin production. *J. Biosci.* **2002**, *57*, 828–835. [CrossRef]
12. Aratanechemuge, Y.; Komiya, T.; Moteki, H.; Katsuzaki, H.; Imai, K.; Hibasami, H. Selective induction of apoptosis by ar-turmerone isolated from turmeric (*Curcuma longa* L.) in two human leukemia cell lines, but not in human stomach cancer cell line. *Int. J. Mol. Med.* **2002**, *9*, 481–484. [CrossRef] [PubMed]
13. Park, S.Y.; Kim, Y.H.; Kim, Y.; Lee, S.J. Aromatic-turmerone attenuates invasion and expression of MMP-9 and COX-2 through inhibition of NF-κB activation in TPA-induced breast cancer cells. *J. Cell. Biochem.* **2012**, *113*, 3653–3662. [CrossRef] [PubMed]

14. Yue, G.G.L.; Chan, B.C.L.; Hon, P.M.; Lee, M.Y.H.; Fung, K.P.; Leung, P.C.; Lau, C.B.S. Evaluation of in vitro anti-proliferative and immunomodulatory activities of compounds isolated from *Curcuma longa*. *Food Chem. Toxicol.* **2010**, *48*, 2011–2020. [CrossRef] [PubMed]
15. Lertsutthiwong, P.; Rojsitthisak, P.; Nimmannit, U. Preparation of turmeric oil-loaded chitosan-alginate biopolymeric nanocapsules. *Mater. Sci. Eng. C Biomim. Supramol. Syst.* **2009**, *29*, 856–860. [CrossRef]
16. Lertsutthiwong, P.; Rojsitthisak, P. Chitosan-alginate nanocapsules for encapsulation of turmeric oil. *Pharmazie* **2011**, *66*, 911–915. [PubMed]
17. Begines, B.; Ortiz, T.; Pérez-Aranda, M.; Martínez, G.; Merinero, M.; Argüelles-Arias, F.; Alcudia, A. Polymeric nanoparticles for drug delivery: Recent developments and future prospects. *Nanomaterials* **2020**, *10*, 1403. [CrossRef]
18. Parveen, S.; Sahoo, S.K. Polymeric nanoparticles for cancer therapy. *J. Drug Target* **2008**, *16*, 108–123. [CrossRef]
19. Prabhu, R.H.; Patravale, V.B.; Joshi, M.D. Polymeric nanoparticles for targeted treatment in oncology: Current insights. *Int. J. Nanomed.* **2015**, *10*, 1001–1018.
20. Maity, S.; Mukhopadhyay, P.; Kundu, P.P.; Chakraborti, A.S. Alginate coated chitosan core-shell nanoparticles for efficient oral delivery of naringenin in diabetic animals—An in vitro and in vivo approach. *Carbohydr. Polym.* **2017**, *170*, 124–132. [CrossRef]
21. Yoncheva, K.; Benbassat, N.; Zaharieva, M.M.; Dimitrova, L.; Kroumov, A.; Spassova, I.; Kovacheva, D.; Najdenski, H.M. Improvement of the antimicrobial activity of oregano oil by encapsulation in chitosan—Alginate nanoparticles. *Molecules* **2021**, *26*, 7017. [CrossRef]
22. Giri, T.K. 5-Nanoarchitectured polysaccharide-based drug carrier for ocular therapeutics. In *Nanoarchitectonics for Smart Delivery and Drug Targeting*; Holban, A.M., Grumezescu, A.M., Eds.; William Andrew Publishing: Oxford, UK, 2016; pp. 119–141.
23. Mirtič, J.; Rijavec, T.; Zupančič, Š.; Zvonar Pobirk, A.; Lapanje, A.; Kristl, J. Development of probiotic-loaded microcapsules for local delivery: Physical properties, cell release, and growth. *Eur. J. Pharm. Sci.* **2018**, *121*, 178–187. [CrossRef] [PubMed]
24. Kianersi, S.; Solouk, A.; Saber-Samandari, S.; Keshel, S.H.; Pasbakhsh, P. Alginate nanoparticles as ocular drug delivery carriers. *J. Drug Deliv. Sci. Technol.* **2021**, *66*, 102889. [CrossRef]
25. Yu, L.; Sun, Q.; Hui, Y.; Seth, A.; Petrovsky, N.; Zhao, C.-X. Microfluidic formation of core-shell alginate microparticles for protein encapsulation and controlled release. *J. Colloid Interface Sci.* **2019**, *539*, 497–503. [CrossRef] [PubMed]
26. Niculescu, A.J.; Grumezescu, A.M. Applications of chitosan-alginate-based nanoparticles—An up-to-date review. *Nanomaterials* **2022**, *12*, 186. [CrossRef] [PubMed]
27. Patil, J.S.; Kamalapur, M.V.; Marapur, S.C.; Kadam, D.V. Ionotropic gelation and polyelectrolyte complexation: The novel techniques to design hydrogel particulate sustained, modulated drug delivery system: A review. *Dig. J. Nanomater. Biostruct.* **2010**, *5*, 241–248.
28. Pedroso-Santana, S.; Fleitas-Salazar, N. Ionotropic gelation method in the synthesis of nanoparticles/microparticles for biomedical purposes. *Polym. Int.* **2020**, *69*, 443–447. [CrossRef]
29. Kumar, S.; Dilbaghi, N.; Rani, R.; Bhanjana, G. Nanotechnology as emerging tool for enhancing solubility of poorly water-soluble drugs. *Bio. Nano Sci.* **2012**, *2*, 227–250. [CrossRef]
30. Das, R.K.; Kasoju, N.; Bora, U. Encapsulation of curcumin in alginate-chitosan-pluronic composite nanoparticles for delivery to cancer cells. *Nanomedicine* **2010**, *6*, 153–160. [CrossRef]
31. Sorasitthiyanukarn, F.N.; Bhuket, P.R.N.; Muangnoi, C.; Rojsitthisak, P.; Rojsitthisak, P. Chitosan/alginate nanoparticles as a promising carrier of novel curcumin diethyl diglutarate. *Int. J. Biol. Macromol.* **2019**, *131*, 1125–1136. [CrossRef]
32. Sorasitthiyanukarn, F.N.; Muangnoi, C.; Bhuket, P.R.N.; Rojsitthisak, P.; Rojsitthisak, P. Chitosan/alginate nanoparticles as a promising approach for oral delivery of curcumin diglutaric acid for cancer treatment. *Mater. Sci. Eng. C* **2018**, *93*, 178–190. [CrossRef]
33. Lertsutthiwong, P.; Noomun, K.; Jongaroonngamsang, N.; Rojsitthisak, P.; Nimmannit, U. Preparation of alginate nanocapsules containing turmeric oil. *Carbohydr. Polym.* **2008**, *74*, 209–214. [CrossRef]
34. Prabaharan, M. Review paper: Chitosan derivatives as promising materials for controlled drug delivery. *J. Biomater. Appl.* **2008**, *23*, 5–36. [CrossRef] [PubMed]
35. Yang, B.; Jiang, J.; Jiang, L.; Zheng, P.; Wang, F.; Zhou, Y.; Chen, Z.; Li, M.; Lian, M.; Tang, S.; et al. Chitosan mediated solid lipid nanoparticles for enhanced liver delivery of zedoary turmeric oil in vivo. *Int. J. Biol. Macromol.* **2020**, *149*, 108–115. [CrossRef] [PubMed]
36. Zhang, Y.; Huo, M.; Zhou, J.; Zou, A.; Li, W.; Yao, C.; Xie, S. DDSolver: An add-in program for modeling and comparison of drug dissolution profiles. *AAPS J.* **2010**, *12*, 263–271. [CrossRef]
37. Sankalia, M.G.; Mashru, R.C.; Sankalia, J.M.; Sutariya, V.B. Reversed chitosan-alginate polyelectrolyte complex for stability improvement of alpha-amylase: Optimization and physicochemical characterization. *Eur. J. Pharm. Biopharm.* **2007**, *65*, 215–232. [CrossRef]
38. Picos-Corrales, L.A.; Garcia-Carrasco, M.; Licea-Claverie, A.; Chavez-Santoscoy, R.A.; Serna-Saldívar, S.O. NIPAAm-containing amphiphilic block copolymers with tailored LCST: Aggregation behavior, cytotoxicity and evaluation as carriers of indomethacin, tetracycline and doxorubicin. *J. Macromol. Sci. A* **2019**, *56*, 759–772. [CrossRef]
39. Bhunchu, S.; Muangnoi, C.; Rojsitthisak, P.; Rojsitthisak, P. Curcumin diethyl disuccinate encapsulated in chitosan/alginate nanoparticles for improvement of its in vitro cytotoxicity against MDA-MB-231 human breast cancer cells. *Pharmazie* **2016**, *71*, 691–700.

40. Qiu, Y.; Hamilton, S.K.; Temenoff, J. 4—Improving mechanical properties of injectable polymers and composites. In *Injectable Biomaterials: Science and applications*; Vernon, B., Ed.; Woodhead Publishing: Cambridge, UK, 2011; pp. 61–91.
41. Rajaonarivony, M.; Vauthier, C.; Couarraze, G.; Puisieux, F.; Couvreur, P. Development of a new drug carrier made from alginate. *J. Pharm. Sci.* **1993**, *82*, 912–917. [CrossRef]
42. Gierszewska, M.; Ostrowska-Czubenko, J.; Chrzanowska, E. pH-responsive chitosan/alginate polyelectrolyte complex membranes reinforced by tripolyphosphate. *Eur. Polym. J.* **2018**, *101*, 282–290. [CrossRef]
43. Yerramathi, B.B.; Kola, M.; Annem, B.M.; Aluru, R.; Thirumanyam, M.; Zyryanov, G.V. Structural studies and bioactivity of sodium alginate edible films fabricated through ferulic acid crosslinking mechanism. *J. Food Eng.* **2021**, *301*, 110566. [CrossRef]
44. Pan, C.; Yue, H.; Zhu, L.; Ma, G.H.; Wang, H.L. Prophylactic vaccine delivery systems against epidemic infectious diseases. *Adv. Drug Deliv. Rev.* **2021**, *176*, 113867. [CrossRef] [PubMed]
45. Lin, W.J.; Huang, L. Influence of pluronics on protein-loaded poly (?-caprolactone) microparticles. *J. Microencapsul.* **2001**, *18*, 191–197. [PubMed]
46. Wu, L.; Zhang, J.; Watanabe, W. Physical and chemical stability of drug nanoparticles. *Adv. Drug Deliv. Rev.* **2011**, *63*, 456–469. [CrossRef] [PubMed]
47. Ranjith, H.P.; Wijewardene, U. Lipid emulsifiers and surfactants in dairy and bakery products. In *Modifying Lipids for Use in Food*; Gunstone, F.D., Ed.; Woodhead Publishing Ltd.: Cambridge, UK, 2006; pp. 393–428.
48. Ramli, R.; Soon, C.; Anika, Z.M.R. Synthesis of chitosan/alginate/silver nanoparticles hydrogel scaffold. *MATEC Web Conf.* **2016**, *78*, 1031. [CrossRef]
49. Natrajan, D.; Srinivasan, S.; Sundar, K.; Ravindran, A. Formulation of essential oil-loaded chitosan-alginate nanocapsules. *J. Food Drug Anal.* **2015**, *23*, 560–568. [CrossRef]
50. Araújo, L.A.; Araújo, R.G.; Gomes, F.O.; Lemes, S.R.; Almeida, L.M.; Maia, L.J.; Gonçalves, P.J.; Mrué, F.; Silva-Junior, N.J.; Melo-Reis, P.R. Physicochemical/photophysical characterization and angiogenic properties of *Curcuma longa* essential oil. *An. Acad. Bras. Cienc.* **2016**, *88*, 1889–1897. [CrossRef]
51. Santadkha, T.; Skolpap, W.; Thitapakorn, V. Diffusion modeling and in vitro release kinetics studies of curcumin-loaded superparamagnetic nanomicelles in cancer drug delivery system. *J. Pharm. Sci.* **2021**. [CrossRef]
52. Sorasitthiyanukarn, F.N.; Muangnoi, C.; Rojsitthisak, P.; Rojsitthisak, P. Chitosan oligosaccharide/alginate nanoparticles as an effective carrier for astaxanthin with improving stability, in vitro oral bioaccessibility, and bioavailability. *Food Hydrocoll.* **2022**, *124*, 107246. [CrossRef]
53. Qu, Y.; Harte, F.M.; Elias, R.J.; Coupland, J.N. Effect of ethanol on the solubilization of hydrophobic molecules by sodium caseinate. *Food Hydrocoll.* **2018**, *77*, 454–459. [CrossRef]
54. Zambito, Y.; Pedreschi, E.; Di Colo, G. Is dialysis a reliable method for studying drug release from nanoparticulate systems?—A case study. *Int. J. Pharm.* **2012**, *434*, 28–34. [CrossRef]
55. Barone, A.; Mendes, M.; Cabral, C.; Mare, R.; Paolino, D.; Vitorino, C. Hybrid nanostructured films for topical administration of simvastatin as coadjuvant treatment of melanoma. *J. Pharm. Sci.* **2019**, *108*, 3396–3407. [CrossRef] [PubMed]
56. Hamdi, M.; Nasri, R.; Li, S.; Nasri, M. Design of blue crab chitosan responsive nanoparticles as controlled-release nanocarrier: Physicochemical features, thermal stability and in vitro pH-dependent delivery properties. *Int. J. Biol. Macromol.* **2020**, *145*, 1140–1154. [CrossRef] [PubMed]
57. Kakkar, V.; Kaur, I.P.; Kaur, A.P.; Saini, K.; Singh, K.K. Topical delivery of tetrahydrocurcumin lipid nanoparticles effectively inhibits skin inflammation: In vitro and in vivo study. *Drug Dev. Ind. Pharm.* **2018**, *44*, 1701–1712. [CrossRef] [PubMed]
58. Lin, C.C.; Lin, H.Y.; Chen, H.C.; Yu, M.W.; Lee, M.H. Stability and characterisation of phospholipid-based curcumin-encapsulated microemulsions. *Food Chem.* **2009**, *116*, 923–928. [CrossRef]
59. Rezaee, M.; Askari, G.; EmamDjomeh, Z.; Salami, M. Effect of organic additives on physiochemical properties and anti-oxidant release from chitosan-gelatin composite films to fatty food simulant. *Int. J. Biol. Macromol.* **2018**, *114*, 844–850. [CrossRef] [PubMed]
60. Scomoroscenco, C.; Teodorescu, M.; Raducan, A.; Stan, M.; Voicu, S.N.; Trica, B.; Ninciuleanu, C.M.; Nistor, C.L.; Mihaescu, C.L.; Petcu, C.; et al. Novel gel microemulsion as topical drug delivery system for curcumin in dermatocosmetics. *Pharmaceutics* **2021**, *13*, 505. [CrossRef]
61. Abouelmagd, S.A.; Sun, B.; Chang, A.C.; Ku, Y.J.; Yeo, Y. Release kinetics study of poorly water-soluble drugs from nanoparticles: Are we doing it right? *Mol. Pharm.* **2015**, *12*, 997–1003. [CrossRef]
62. Kamaly, N.; Yameen, B.; Wu, J.; Farokhzad, O.C. Degradable controlled-release polymers and polymeric nanoparticles: Mechanisms of controlling drug release. *Chem. Rev.* **2016**, *116*, 2602–2663. [CrossRef]
63. Campos, J.; Varas-Godoy, M.; Haidar, Z.S. Physicochemical characterization of chitosan-hyaluronan-coated solid lipid nanoparticles for the targeted delivery of paclitaxel: A proof-of-concept study in breast cancer cells. *Nanomedicine* **2017**, *12*, 473–490. [CrossRef]
64. Yu, M.; Yuan, W.; Li, D.; Schwendeman, A.; Schwendeman, S.P. Predicting drug release kinetics from nanocarriers inside dialysis bags. *J. Control Release* **2019**, *315*, 23–30. [CrossRef]
65. Valizadeh, M.; Behnamian, M.; Dezhsetan, S.; Karimirad, R. Controlled release of turmeric oil from chitosan nanoparticles extends shelf life of *Agaricus bisporus* and preserves its postharvest quality. *Food Biosci.* **2021**, *44*, 101401. [CrossRef]

66. Bhattacharjee, S. DLS and zeta potential—What they are and what they are not? *J. Control Release* **2016**, *235*, 337–351. [CrossRef] [PubMed]
67. Sheikhi, A.; Afewerki, S.; Oklu, R.; Gaharwar, A.K.; Khademhosseini, A. Effect of ionic strength on shear-thinning nanoclay-polymer composite hydrogels. *Biomater. Sci.* **2018**, *6*, 2073–2083. [CrossRef] [PubMed]
68. Kanamala, M.; Wilson, W.R.; Yang, M.; Palmer, B.D.; Wu, Z. Mechanisms and biomaterials in pH-responsive tumour targeted drug delivery: A review. *Biomaterials* **2016**, *85*, 152–167. [CrossRef]
69. Di Martino, A.; Trusova, M.E.; Postnikov, P.S.; Sedlarik, V. Folic acid-chitosan-alginate nanocomplexes for multiple delivery of chemotherapeutic agents. *J. Drug Deliv. Sci. Technol.* **2018**, *47*, 67–76. [CrossRef]
70. Bruschi, M.L. Mathematical models of drug release. In *Strategies to Modify the Drug Release from Pharmaceutical Systems*; Bruschi, M.L., Ed.; Woodhead Publishing Ltd.: Kidlington, UK, 2015; pp. 63–86.
71. Abu-Huwaij, R.; Abbas, M.M.; Al-Shalabi, R.; Almasri, F.N. Synthesis of transdermal patches loaded with greenly synthesized zinc oxide nanoparticles and their cytotoxic activity against triple-negative breast cancer. *Appl. Nanosci.* **2021**, *12*, 69–78. [CrossRef]
72. Kavalappa, Y.P.; Gopal, S.S.; Ponesakki, G. Lutein inhibits breast cancer cell growth by suppressing antioxidant and cell survival signals and induces apoptosis. *J. Cell Physiol.* **2021**, *236*, 1798–1809. [CrossRef]
73. Vignesh, K.S.; Renuka, D.P.; Hemananthan, E. In vitro studies to analyze the stability and bioavailability of thymoquinone encapsulated in the developed nanocarrier. *J. Dispers. Sci. Technol.* **2019**, *41*, 243–256.
74. Stepanenko, A.A.; Dmitrenko, V.V. HEK293 in cell biology and cancer research: Phenotype, karyotype, tumorigenicity, and stress-induced genome-phenotype evolution. *Gene* **2015**, *569*, 182–190. [CrossRef]
75. Nilsen-Nygaard, J.; Strand, S.; Vårum, K.; Draget, K.; Nordgård, C. Chitosan: Gels and interfacial properties. *Polymers* **2015**, *7*, 552–579. [CrossRef]
76. Rafiee, A.; Alimohammadian, M.H.; Gazori, T.; Riazi-rad, F.; Fatemi, S.M.R.; Parizadeh, A.; Haririan, I.; Havaskary, M. Comparison of chitosan, alginate and chitosan/alginate nanoparticles with respect to their size, stability, toxicity and transfection. *Asian Pac. J. Trop. Dis.* **2014**, *4*, 372–377. [CrossRef]
77. Chang, G.; Wang, J.; Zhang, H.; Zhang, Y.; Wang, C.; Xu, H.; Zhang, H.; Lin, Y.; Ma, L.; Li, Q.; et al. CD44 targets Na(+)/H(+) exchanger 1 to mediate MDA-MB-231 cells' metastasis via the regulation of ERK1/2. *Br. J. Cancer* **2014**, *110*, 916–927. [CrossRef] [PubMed]
78. Balaji, S.; Chempakam, B. Toxicity prediction of compounds from turmeric (*Curcuma longa* L.). *Food Chem. Toxicol.* **2010**, *48*, 2951–2959. [CrossRef] [PubMed]
79. Liu, N.; Tang, M. Toxic effects and involved molecular pathways of nanoparticles on cells and subcellular organelles. *J. Appl. Toxicol.* **2020**, *40*, 16–36. [CrossRef]
80. Jiang, K.; Chi, T.; Li, T.; Zheng, G.; Fan, L.; Liu, Y.; Chen, X.; Chen, S.; Jia, L.; Shao, J. A smart pH-responsive nano-carrier as a drug delivery system for the targeted delivery of ursolic acid: Suppresses cancer growth and metastasis by modulating P53/MMP-9/PTEN/CD44 mediated multiple signaling pathways. *Nanoscale* **2017**, *9*, 9428–9439. [CrossRef]
81. Mathew, M.E.; Mohan, J.C.; Manzoor, K.; Nair, S.V.; Tamura, H.; Jayakumar, R. Folate conjugated carboxymethyl chitosan–manganese doped zinc sulphide nanoparticles for targeted drug delivery and imaging of cancer cells. *Carbohydr. Polym.* **2010**, *80*, 442–448. [CrossRef]
82. Ramya, A.N.; Joseph, M.M.; Maniganda, S.; Karunakaran, V.; Sreelekha, T.T.; Maiti, K.K. Emergence of gold-mesoporous silica hybrid nanotheranostics: Dox-encoded, folate targeted chemotherapy with modulation of SERS fingerprinting for apoptosis toward tumor eradication. *Small* **2017**, *13*, 1700819. [CrossRef]

Article

Hybrid Microcapsules for Encapsulation and Controlled Release of Rosemary Essential Oil

Doha Berraaouan [1], Kamal Essifi [1], Mohamed Addi [2,*], Christophe Hano [3], Marie-Laure Fauconnier [4] and Abdesselam Tahani [1,*]

[1] Physical Chemistry of Natural Substances and Process Research Team, Laboratory of Applied Chemistry and Environment (LCAE-CPSUNAP), Faculty of Sciences, Université Mohamed Premier, BV Mohammed VI BP 717, Oujda 60000, Morocco

[2] Laboratoire d'Amélioration des Productions Agricoles, Biotechnologie et Environnement (LAPABE), Faculty of Sciences, Université Mohamed Premier, BV Mohammed VI BP 717, Oujda 60000, Morocco

[3] Laboratoire de Biologie des Ligneux et des Grandes Cultures, INRAE USC1328, Campus Eure et Loir, Orleans University, 28000 Chartres, France

[4] Laboratory of Chemistry of Natural Molecules, Gembloux Agro-Bio Tech, University of Liège, 5030 Gembloux, Belgium

* Correspondence: m.addi@ump.ac.ma (M.A.); a1.tahani@ump.ac.ma (A.T.); Tel.: +212-(0)641612183 (M.A.); +212-(0)667086196 (A.T.)

Abstract: The foremost objective of this work is to assess the microcapsules composition (polymer-based and polymer/clay-based) effect, on the release of rosemary essential oil into w/o medium and evaluate their antioxidant activity. Calcium alginate (CA) and calcium alginate/montmorillonite hybrid (CA-MTN) microcapsules were developed following an ionotropic crosslinking gelation and were used as host materials for the encapsulation of rosemary essential oil. The unloaded/loaded CA and hybrid CA-MTN microcapsules were characterized by Fourier transform infra-red (FT-ATR) spectroscopy, thermal analysis (TGA), scanning electron microscopy (SEM) and DPPH assay. The evaluation of the microcapsule's physicochemical properties has shown that the clay filling with montmorillonite improved the microcapsule's properties. The encapsulation efficiency improved significantly in hybrid CA-MTN microcapsules and exhibited higher values ranging from 81 for CA to 83% for hybrid CA-MTN and a loading capacity of 71 for CA and 73% for hybrid CA-MTN, owing to the large adsorption capacity of the sodic clay. Moreover, the hybrid CA-MTN microcapsules showed a time-extended release of rosemary essential oil compared to CA microcapsules. Finally, the DPPH assay displayed a higher reduction of free radicals in hybrid CA-MNT-REO (12.8%) than CA-REO (10%) loaded microcapsules. These results proved that the clay–alginate combination provides microcapsules with enhanced properties compared to the polymer-based microcapsules.

Keywords: encapsulation; hybrids; clay; sodium alginate; rosemary essential oil; controlled release

1. Introduction

Essential oils are liquids with concentrated molecules resulting from the metabolism of a plant. They are very popular in the cosmetics, perfumery, food preservation, and pharmaceutical sectors [1]. The oxidative stress that harms biological molecules can be diminished by essential oils, and many degenerative diseases, including diabetes, cardiovascular diseases, and neurological disorders, can also be successfully treated [2]. Despite the great biological potential that essential oils currently possess, they are under-utilized when their applications become difficult due to certain physicochemical characteristics provided by the compounds present in their compositions [3]. It is difficult to use essential oils in aqueous environments because they are generally highly volatile at room temperature, easily oxidized when exposed to oxygen and light, and above all have a high lipophilicity [4]. Aqueous solubilization and preservation of bioactive substances present in essential oils are crucial for their application, which requires the use of a system to promote their dispersion in aqueous media and regulate their release. These issues can be resolved through the incorporation of essential oils into various systems. Included in these systems are emulsions, beads, bioactive films, capsules, liposomes, nanocarriers, and inclusion complexes [5]. Encapsulation systems have attracted great interest in recent decades with a view to achieving an efficient drug delivery process [6]. Particular attention has been paid to finding an inert carrier where the drug is dispersed or incorporated. Recently, the preparation, characterization and application of controlled drug delivery systems prepared from biopolymer compounds has become popular owing to their distinctive properties such as a high encapsulation efficiency, biodegradability, nontoxicity, and controlled release characteristics [7]. The encapsulation of essential oils or their isolated components has been investigated by some researchers where they worked on the possibilities of using essential oils as a drug vehicle in delivery systems for lipophile drugs. Similarly, Rodriguez et al. [8] reviewed the possible strategies for essential oils encapsulation and showed the benefits of polymeric nanoencapsulation [9].

Out of the many biopolymers used for encapsulation purposes, alginate is widely used due to its biosafety, mild gelation properties, environmentally friendly [10], hydrophilicity, as well as its ability to be modified or combined with other biomaterials [11]. It is a hydrophilic polysaccharide, composed of two monomeric structures: β-D mannuronic acid (M) and α-L-guluronic acid (G) which are arranged in MM or GG blocks interspersed with MG blocks [12]. This structure has the unique property of forming water-insoluble calcium–alginate gel through ionotropic gelation with divalent cross-linking salt such as calcium chloride [13]. But the low efficiency of encapsulating water-soluble substances in alginate is one of the problems in developing a controlled delivery system [14]. In addition to the many advantages of polymeric encapsulation, it suffers from low strength. This disadvantage can be overcome by using natural clays as a filler; these systems are called hybrid capsules. Clay minerals have been proposed as fundamental constituents of various modified carriers and have different purposes. Among the promising clay materials, montmorillonite (MTN), which belongs to the smectites family, is of particular interest. The smectite clays are a layered aluminosilicate derived from the assembly of tetrahedral $[SiO_4]^{4-}$ and octahedral $[AlO_3(OH)_3]^{6-}$ sheets. Hybrid systems on the basis of an organic and inorganic materials amalgam have been developed to solve the problem of low chemical stability, low mechanical strength and the erratic release of the encapsulated substances. They are very attractive for diverse applications due to the features afforded by the combination of the organic and the inorganic materials. Among its popular applications is active packaging. In this technology, materials loaded with active agents as essential oils are incorporated to the packaging to inhibit food degradation by oxidation, contamination and many other factors [15,16]. Research work has dealt with nanohybrids valued as possible containers for the encapsulation of rosemary essential oil using pectins and halloysite nanotubes HNTs (a rare type of kaolinite) [17]. It was observed that the release of rosemary essential oil was much slower in the nanohybrid material than the release of the same essential oil blended into the pectin. In addition, the release from the composite filled with the clay shows multiple steps, a rapid diffusion in the first stage, followed by a decreased rate of release in the remaining stages. The antimicrobial activity showed a mold formation in the pectin composites after two weeks of storage, but was not detected in the hybrid composites even after three months. This shows that to elaborate efficient active packaging, it is necessary to entrap the active agent for as long as possible instead of enabling its fast release and that this characteristic can be provided by clay filling.

The preparation of a polymeric layered silicate composite offers the possibility of enhancing the properties of individual components. Different biocompatible and biodegradable polymers are suitable drug carriers that can release components at a constant rate [18]. To improve the active substance entrapment efficiency and thereby modulate the release of the active substance, it is desirable to incorporate materials such as clays. Thus, the alginate–clay composite formed would decrease the active substance release by increasing its loading capacity in the composite matrix. However, one of the main challenges facing the scientific community is the development of appropriate matrices for essential oils encapsulation and protection, as well as their controlled release under specific stimuli, which we call "smart carriers" [19,20].

The main objective of the present study is the elaboration of composites prepared in situ by ionotropic gelation of sodium alginate and the combination of the latter with natural sodium bentonite from the eastern region of Morocco for the encapsulation of rosemary essential oil (REO) and the pathway release as a function of microparticles formulation and REO concentration. The microcapsules were characterized by Fourier transform infrared spectroscopy (ATR-FTIR), thermogravimetric analysis (TGA), and scanning electron microscopy (SEM). In addition, the release profile of the composites in an aqueous medium was studied in vitro at an ambient temperature of 20 °C and the ability of the microcapsules to scavenge the DPPH free radicals was also examined.

2. Materials and Methods

2.1. Materials

Sodium alginate (SA) (15% loss during drying at 105 °C, 30% ignition residue and 0.004% heavy metal content) was purchased from Panreac Quimica (Barcelona, Spain). The polymer presented intrinsic viscosities of 1.03×10^3 mL/g and 5.39×10^3 mL/g in 0.1 M sodium chloride and distillate water, respectively, measured using an Ubbelohde capillary viscometer at 25 °C. The values of the typical molecular weight are 5.48×10^4 g/mol and 3.38×10^5 g/mol for 0.1 M sodium chloride and distillate water, respectively. In addition, the M/G ratio was determined using infrared spectra and was equal to 1.37. The raw bentonite was taken from the Azzouzet deposit (Nador, Morocco), previously characterized [21], purified and exchanged before the experiment to obtain a sodic clay. The main parameters of the raw and purified/exchanged clay are shown in Table 1.

Table 1. The main parameters determined for the raw and homosodic clay. CEC: is the cationic exchange capacity using the copper ethylene diamine complex [Cu (EDA)2]$^{2+}$ and BET: specific surface of clays.

Clay	CEC$_{EDA-Cu}$ (meq/100 g)	BET (m^2/g)	Pore Volume (cm^3/g)	Zeta Potential (mV)
Raw	68.74	-	0.009	−27.77
Homosodic	91.66	94.25	0.28	−24.72

Rosmarinus officinalis L. essential oil (REO) was obtained from a distillation unit in Jerrada (eastern region of Morocco) and was stored at −1 °C until examination. Tween 80 (density of 1.06 and viscosity of 300–500 mPa·s at 25 °C) was provided by Panreac Quimica (Barcelona, Spain) and calcium chloride (powder, 97% with a molar mass of 110.99 g/mole) was acquired from Riedel-de-Haën (Seelze, Germany).

2.2. Preparation of Rosemary Essential Oil Loaded Microcapsules

The preparation of the loaded microcapsules followed two steps as in Figure 1. Firstly, the preparation of emulsions containing alginate/montmorillonite/REO and alginate/REO was carried out, followed by the preparation of the microcapsules through the addition of those emulsions into a gelling bath.

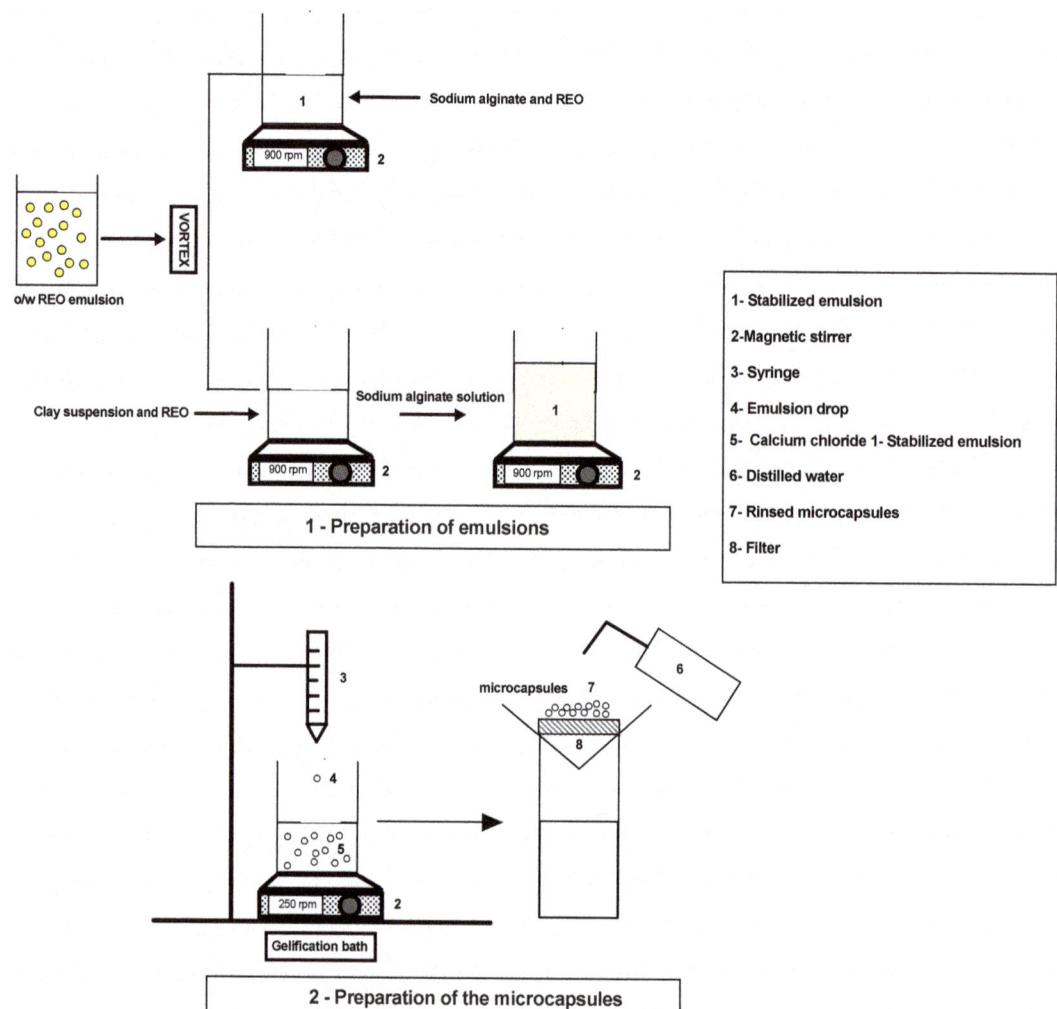

Figure 1. Illustration of the CA and hybrid CA-MTN microcapsules elaboration process.

For calcium alginate CA microcapsules, a sodium alginate solution (1% w/v) was dissolved in distillate water while being stirred magnetically (350 rpm) at room temperature (20 °C). An oil/water emulsion was formed by mixing REO with sodium alginate solution and left to stir overnight to reach a final concentration of 1, 2 and 3% of rosemary essential oil. For the CA-MTN hybrid microcapsules, a dispersion of sodium bentonite MTN (4%) in distillate water was left to stir overnight. Different concentrations of rosemary essential oil (1, 2 and 3%) were added to the sodium bentonite dispersion and left to stir overnight for the maximum adsorption of essential oil on the bentonite. The sodium MTN/REO emulsion was then blended with sodium alginate solution with a 1:2 ratio and left to stir for 5 h. The amount of alginate solution and bentonite dispersion used in hybrid CA-MTN microcapsules were calculated to obtain the same final essential oil concentration as in CA microcapsules. The CA and CA-MTN hybrid microcapsules were generated by adding the emulsions dropwise to calcium chloride solution (0.1 M) for 60 min under magnetic stirring.

The microcapsules were retrieved through filtration, repeatedly rinsed with distilled water and then put into storage at 4 °C.

2.3. Identification of Rosemary Essential Oil Chemotype

The gas chromatographic-MS analysis of *Rosmarinus officinalis* L. essential oil was performed via a Hewlett-Packard 6890 gas chromatograph interfaced with a Hewlett-Packard mass selective detector (Agilent Technologies, Santa Clara, CA, USA). An HP1 fused silica column (Phynel-methyl Siloxane 30 m × 0.25 mm i.d., film thickness 0.25 µm) was used. An interface temperature of 280 °C was used for the gas chromatography parameters, while for the mass spectrometry parameters an interface temperature of 250 °C. The MS source temperature was of 200 °C, the ionization energy and the ionization current were of 70 eV and 2A, respectively [22]. The carrier gas was helium and the flow rate along the column was 1.4 mL/min.

2.4. Characterization of Microcapsules

2.4.1. Attenuated Total Reflection Fourier Transform Infrared (ATR-FTIR)

The attenuated total reflectance-Fourier transform infrared (ATR-FTIR) analysis was carried out using a Shimadzu Jasco 4700-ATR spectrophotometer (Kyoto, Japan) to determine the functional groups in the rosemary essential oil, as well as unloaded and loaded CA and CA-MTN hybrid microcapsules in the wavelength region between 400 and 4000 cm^{-1}. The obtained spectra for each sample were affected by averaging 32 scans at a resolution of 4 cm^{-1}.

2.4.2. Thermogravimetric Analysis (TGA)

The thermogravimetric analysis was carried out via a SHIMADZU TA-60WS thermal analyzer (Kyoto, Japan) with an initial sample mass of 22 and 31.8 mg for CA and CA-MTN hybrid microcapsules, respectively, in an alumina sample holder at a heating rate of 20 °C/min in an N_2 atmosphere.

2.4.3. Particle Size and Morphology

The microcapsules sizes and shapes are related to their circularity, which indicates how closely the shape of the microparticle resembles a circle. Its value ranges from 0 to 1; the latter indicates the perfect sphere. The size was estimated using an Olympus SZ Stereomicroscope (Japan), equipped with a digital camera (D5000 Color Video Camera, Nikon (Tokyo, Japan), and the shape was characterized by aspect ratio and shape factor using an Image J Software program (NIH). The morphology analysis of the microcapsules was conducted by scanning electronic microscopy (SEM), realized in JNCASR (Bangalore, India), using a ZEISS Gemini model. Samples were dried using a critical point dryer and then fixed into a carbon tape for SEM imaging.

2.4.4. Loading Capacity and Encapsulation Efficiency

The amount of encapsulated rosemary essential oil was determined using a Rayleigh UV–VISIBLE 1800 spectrophotometer (Model Jasco 560, Pekin, China), following the method described by El Hosseini [23] with some modifications; 200 mg of the microcapsules were poured into a beaker containing distilled water and Tween 80 (1% w/v) then left to stir. The absorbance of the obtained solution was measured at 257 nm. The loading capacity and encapsulation efficiency were determined at the final concentration measured at the end of the release study and calculated using the rosemary essential oil calibration curve in the aqueous solution (1% Tween 80). The values were calculated following Equations (1) and (2) [24,25] for loading capacity and encapsulation efficiency, respectively.

$$\text{LC } (\%) = \frac{\text{Weight of essential oil in microcapsules}}{\text{Weight of microcapsules}} \times 100 \qquad (1)$$

$$\text{EE (\%)} = \frac{\text{Weight of essential oil in microcapsules}}{\text{Weight of essential oil added}} \times 100 \qquad (2)$$

2.5. Determination of Antioxidant Activity

The antioxidant activity of rosemary essential oil loaded and unloaded microcapsules was assessed using 1.1-diphenyl-2-picrylhydrazyl (DPPH), a free radical, following the procedure outlined by Ling et al. [26]. Briefly, a 0.1 mM solution of 1.1-diphenyl-2-picrylhydrazyl (DPPH) radical solution in 90% ethanol was prepared and 2.5 mL of this solution was mixed vigorously with different REO concentrations (25–7500 µg/mL in ethanol) and a mass of 100 mg of the loaded and unloaded microcapsules. After 30 min of incubation in the dark at room temperature, absorbance (A) was measured at 517 nm using a Rayleigh UV–VISIBLE 1800 spectrophotometer (Model Jasco 560, Pekin, China). The percentage of the radical scavenging capture (RSC) was calculated based on the following Equation (3), where A_{cont} and A_{sample} are the absorbance values of the control and the sample, respectively.

$$\text{RSC}(\%) = \frac{A_{cont} - A_{sample}}{A_{cont}} \times 100 \qquad (3)$$

2.6. REO Release Kinetics

The REO release study was carried out using a method described by El Hosseini et al. [23]. The REO-loaded microcapsules (250 mg) were placed in a flask containing 30 mL of the release medium (1% Tween 80 in distilled water). The suspensions were slowly stirred at ambient temperature (20 °C). At predetermined time intervals, samples from the release medium were taken. The REO concentration was determined by spectrophotometry at 257 nm using a Rayleigh UV–VISIBLE 1800 spectrophotometer (Model Jasco 560, Pekin, China). The concentration of rosemary essential oil in the release medium at different sampling times was determined using a calibration curve of free REO in 1% Tween 80. The cumulative percentage of the REO release (%CR) was calculated using Equation (4) [27]. To forecast and correlate the REO release behaviour from both microparticles, it is imperative to include the proper model. Therefore, the data from the release study were adjusted to an established empirical model by Korsmeyer and Peppas using Equation (5) [23].

$$\%CR = \sum_{t=0}^{t} \frac{M_t}{M_0} \times 100 \qquad (4)$$

$$m_t/m_\infty = k \times t^n \qquad (5)$$

where m_t/m_∞ is the percentage of the drug released at time t and for all time; k is the description of the macromolecular network system and n is the release exponent indicating the release mechanism. A linear form of the equation can be acquired through plotting ln (m_t/m_∞) against ln(t), whose linear coefficient is k and whose angular coefficient is n. The determined n value is used to discover which of the three mechanisms can describe the release behaviour: the n value is less than 0.43; the release follows the Fickian law and the n value is between 0.43 and 0.85; the non-Fickian release mechanism is established, and when the n value is greater than 0.85 it indicates the case II of transport release with a polymer chain relaxation [28]. Using OriginPro 2018 software, we calculated the squared correlation coefficient (R^2 was used to confirm the accuracy of the model).

3. Results

3.1. Identification of Rosemary Essential Oil Chemotype

Essential oil of *Rosmarinus officinalis* L. from the eastern region of Morocco was characterized by gas chromatography-mass spectroscopy (GC-MS) [22]. As can be seen in Table 2, 32 compounds were totally identified, representing 99.9% of the total oil content. The most abundant compounds of essential oil are mainly concentrated in both groups, monoterpenes hydrocarbons, and oxygenated monoterpenes. The main compounds are 1.8-cineole

(29.71%), α-thujene (14.17%), camphor (13.09%), β-pinene (9.94%), and camphene (6.46%). The high ratio of these compounds was previously revealed by some researchers [29,30] who worked on the same oil and according to the literature, the essential oil of *Rosmarinus officinalis* L. used in this work can be classified as a 1.8-Cineol chemotype. The volatility of these substances restricts their use even though they can slow down microbial growth or free radicals. Thus, encapsulating them in microcapsules is undoubtedly a better way to delay their volatilization, which can increase the range of their application.

Table 2. Chemical profile of *Rosmarinus officinalis* L. essential oil (R_T is the retention time).

Name of the Compound	R_T	Name of the Compound	R_T
α-Pinene	4.983	Camphor	8.308
Camphene	5.225	Borneol	8.625
β-Pinene	5.658	4-Terpineol	8.775
β-Myrcene	5.817	α-Terpieol	8.967
α-Terpinen	6.258	Isobornyl acetate	10.325
m-Cymene	6.383	Copaene	11.600
D-Limonene	6.458	Caryophyllene	12.217
1.8-Cineol	6.525	α-Humulene	12.650
γ-Terpinen	6.917	Isoledene	12.900
cis-β-Terpineol	7.067	γ-Murolene	13.383
Terpinolene	7.383	δ-Cadinene, (+)	13.475
Linalool	7.517	Caryophyllene oxide	14.283

3.2. Characterization of Microcapsules

3.2.1. Attenuated Total Reflection Fourier Transform Infrared (ATR-FTIR)

The attenuated total reflection Fourier transform infrared was performed on REO, REO-loaded microcapsules and non-loaded microcapsules to identify the characteristic bands of the chemical structures. The obtained spectra are shown in Figure 2. The REO spectrum presents characteristic bands from 1030 to 1135 cm^{-1} assigned to out-of-plane C-H wagging vibrations from terpenoids. The stretching vibrations of C–O present in the carbohydrates correspond to peaks from 1190 to 1035 cm^{-1}. The spectral region from 1700 to 1500 cm^{-1} is associated with C=C bending. The peaks attributed to the C–H bending (aliphatic domain) appear in the 2900 cm^{-1} frequency. Carboxylic acid (O–H) stretching has characteristic absorption between 3000–2500 cm^{-1}. The FTIR spectrum of the pure rosemary essential oil shows the expected characteristic C–H stretch (~2900 cm^{-1}), C=O stretch (~1700 cm^{-1}), broad O–H stretch (~3400 cm^{-1}), and C–O stretch (~1100 cm^{-1}) of terpenoid components.

The "fingerprint" characteristic bands of sodium alginate include the spectral region from 800 to 1000 cm^{-1} and corresponds to the C–C stretching of the alginate skeleton and deformation mode [31]. The bands that occur at 1417 and 1617 cm^{-1} correspond to its carboxylic groups, symmetric and asymmetric stretching, respectively [32], and have been observed in all the capsule spectra. The broad band around 3600 cm^{-1} in the ATR-FTIR spectrum of hybrid CA-MTN microcapsules is associated with the Al–OH and Mg–OH vibrations. The intensive peak at a wave number of 1036 cm^{-1} represents the Si–O stretching vibrations, which are mainly associated with montmorillonite clay. The peak at 1640 cm^{-1} is attributed to the bending vibration of water that is physically adsorbed [33]. The broad band at 3400 cm^{-1} corresponds to the O–H stretching of interlayer adsorbed water. Regarding the spectra of the REO-loaded microcapsules, some peaks characteristic of rosemary essential oil aromatic domain (2600–3200 cm^{-1}) appeared on the spectra of loaded microcapsules and do not appear on the non-loaded microcapsules. Some peaks typical of calcium alginate (1417–1617 cm^{-1}) were intensified compared to the non-loaded microcapsules suggesting interactions between the polymer, clay and rosemary essential oil. In the ATR-FTIR spectra, the peak of sodium alginate carboxyl anions was shifted from 1627 cm^{-1} for CA-MTN microcapsules to 1636 cm^{-1} for CA-MTN-REO and from 1630 cm^{-1} to 1651 cm^{-1} for CA and CA-REO, respectively. This can be explained by the REO incorporation. Our presumption is that REO molecules were adsorbed on the clay surface and its interlayer space which induced interaction and the creation of new bonding. In the work of Volic et al. [31], the combination of soy protein, sodium alginate and thyme oil created an aggregation of the soy protein involving a hydrogen bonds disruption.

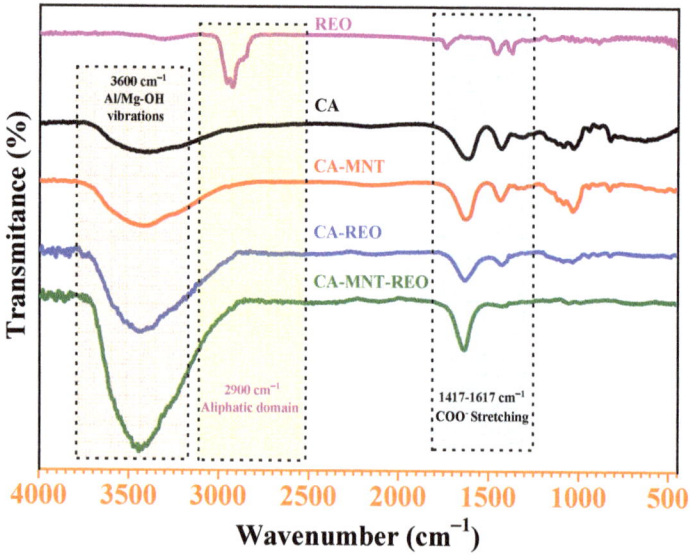

Figure 2. ATR-FTIR spectral of rosemary essential oil (REO), calcium alginate (CA), calcium alginate-montmorillonite (CA-MTN), calcium alginate-rosemary essential oil (CA-REO), and calcium alginate-montmorillonite-rosemary essential oil (CA-MTN-REO).

3.2.2. Thermogravimetric Analysis (TGA)

Thermal analysis of the unloaded and loaded microcapsules showed the weight loss of microcapsules related to moisture. According to Wang et al. [34], alginate polymer has two sorts of water; unbounded water whose thermal event occurs below 100 °C and bounded water whose thermal event occurs above 100 °C. Thermogravimetric studies have shown that the stability of several antioxidants such as essential oils decreases with the increase in temperature [35]. The thermograms (Figure 3) present the weight loss pattern of CA, CA-MTN microcapsules, CA-REO and CA-MTN-REO microcapsules. REO is sensitive to heat and therefore the thermal property was carried out up to 150 °C.

All of the microcapsules show thermal stability up to 42 °C. A weight loss then takes place and slows at a constant rate to 83 °C. This weight decrease of 30% is merely due to the REO present on the capsule's surface and a part of the external moisture losses [36]. Between 83 and 117 °C, a fast weight loss appears in both loaded microcapsules compared to the non-loaded microcapsules. Until 117 °C, the evaporation of the outer mist and the diffusion of the REO within the core capsule occur faster in CA microcapsules than in CA-MTN hybrid microcapsules. At that stage, the loss is important and exceeds 48%. The study continues until total decomposition. The results have shown that the presence of montmorillonite clay improved the thermal stability of the hybrid microcapsules compared with CA microcapsules. This improvement in the thermal stability is related to the reduced motions of alginate networks following the clay filling which slows the diffusion of the inner particles through the capsule membrane [37]. Therefore, REO could be better preserved in the CA-MTN hybrid microcapsules for encapsulation purposes.

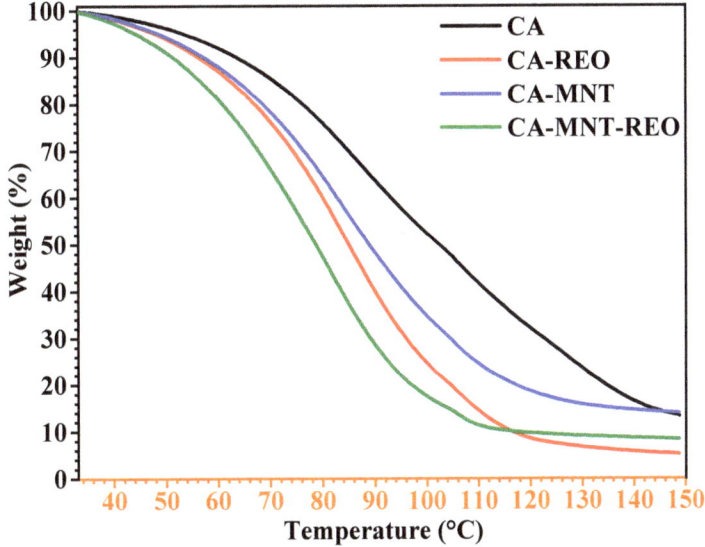

Figure 3. TGA curves of calcium alginate (CA), calcium alginate-montmorillonite (CA-MTN), calcium alginate-rosemary essential oil (CA-REO), and calcium-alginate-montmorillonite-rosemary essential oil (CA-MTN-REO) microcapsules.

3.2.3. Particle Size and Morphology

The size and morphological characterization of the microcapsules are important because they can have an impact on the physicochemical properties such as water sorption and release of the encapsulated agent, as well as the aesthetic quality which could be a desirable feature for pharmaceutical and food products [38]. Because the microcapsules were generated by ionotropic gelation of the emulsion through a needle of 1 mm diameter into a calcium gelling bath under magnetic stirring they differ in size and shape. The average size of hydrated microcapsules is shown in Figure 4 and Table 3 and were of 1.71 ± 0.1 mm for CA and 1.64 ± 0.5 mm for CA-MTN. The size of the microcapsules indicates that they are micrometric. The shape and morphology of the microcapsules were determined from the aspect ratio and circularity factor. The circularity values equal or closely equal to zero indicate a perfect sphere. The values are of 0.92 and 0.94 for CA and CA-MTN microcapsules, respectively, and show the morphological characteristics of a sphere.

Table 3. Size and morphology parameters of CA and hybrid CA-MTN microcapsules.

Microcapsule	CA Microcapsules	CA-MTN Hybrid Microcapsules
Diameter (mean ± SD) (mm)	1.73 ± 0.01	1.64 ± 0.05
Circularity (mean ± SD)	0.92 ± 0.01	0.94 ± 0.01
Aspect Ratio (mean ± SD)	1.05 ± 0.01	1.17 ± 0.06

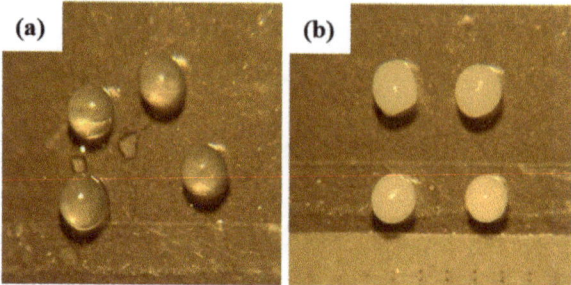

Figure 4. Photograph of CA (**a**) and hybrid CA-MTN (**b**) microcapsules. Magnification ×10.

The SEM micrographs (Figure 5) show that the CA microcapsule (Figure 5a) has an irregular form and a wrinkled surface. High magnificence SEM micrographs; 1220× and 1370× (Figure 5c,e), we can observe that an alignment of lines attributed to the polymer chains and its external surface is covered with a network of fissures and cracks. However, the hybrid CA-MTN (Figure 5b) microcapsules show an improved spherical form, a more compact surface and less irregularity. On the micrographs Figure 5d,f, the polymeric chains in the hybrid CA-MTN microcapsules are associated with the montmorillonite aggregates to form a crumpled network. The hybridation improved the aesthetic aspect of the microcapsules and the polymer matrix strength by filling its interstitial voids and therefore increasing the "time hosting" of the encapsulated agent in the hybrid CA-MTN microcapsules.

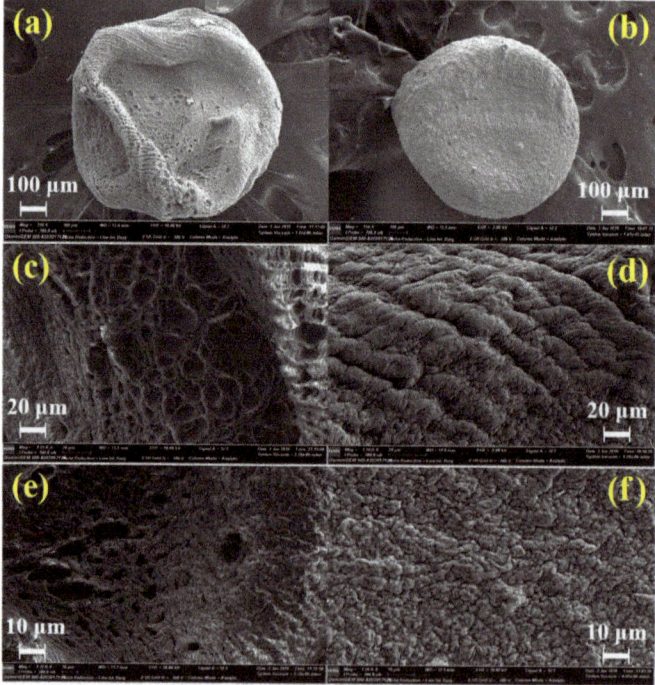

Figure 5. SEM micrographs of CA ((**a**): 229 X, (**c**): 1.22 K X, (**e**): 1.37 K X) and CA-MTN ((**b**): 150 X, (**d**): 1.59 K X, (**f**): 2.24 K X) microcapsules.

3.2.4. Loading Capacity and Encapsulation Efficiency

The results in Table 4 showed that the loading capacity and encapsulation efficiency are influenced by the essential oil content. The loading capacity increased when we increased the rosemary essential oil concentration (3%), 71% and 73% for CA and hybrid CA-MTN microcapsules, respectively, whereas the encapsulation efficiency decreases with an increase in the essential oil amount. The loading capacity depends on the weight of the REO in the microcapsules and the weight of the microcapsules. Our assumption is that the clay filling densifies the microcapsule's matrix network, creating more space for the oil to fit and consequently improving the loading capacity by 2% for the 3% REO concentration. However, the encapsulation efficiency decreases with the increase in essential oil concentration for CA microcapsules. This can be explained by the fact that the polymer network is insufficient to host a higher amount of REO, suggesting a saturation of the material capacity. While for the hybrid CA-MTN microcapsules, thanks to the combination alginate-montmorillonite, they showed a better entrapment of 2% REO concentration equal to 94% compared to CA microcapsules, which is equal to 81%. These findings are supported by El Hosseini et al. [23] who worked on the encapsulation of sunflower in sodium alginate microparticles. The results indicated the highest encapsulation efficiency for the lowest sunflower concentration and the highest loading capacity for the highest sunflower concentration. Piornos et al. [39] reported that the encapsulation efficiency is related to the system of encapsulation strength, which can explain the improvement from adding the montmorillonite clay in this work.

Table 4. Loading capacity (LC) and encapsulation efficiency (EE) for CA and hybrid CA-MTN microcapsules.

Microcapsule	CA Microcapsules			CA-MTN Hybrid Microcapsules		
REO Concentration	1%	2%	3%	1%	2%	3%
LC (%)	26.62 ± 0.19	49.93 ± 0.03	71.36 ± 0.1	28.58 ± 0.21	54.83 ± 0.12	73.6 ± 0.14
EE (%)	92.73 ± 0.14	81.946 ± 0.04	81.9 ± 0.04	83.58 ± 0.23	94.59 ± 0.12	83.59 ± 0.13

3.3. Determination of the Antioxydant Activity

The microcapsules' ability as antioxidants was tested. In the present study, the elaborated unloaded microcapsules, CA and CA-MNT, as well as the loaded microcapsule CA-MNT–3% REO were tested for their capability to reduce the free radical 1.1-diphenyl-2-picrylhydrazyl (DPPH); the REO was also evaluated for its antioxidant activity. The results of the scavenging activity performed are highlighted in Figure 6. As can be seen from these results, the unloaded microcapsules showed the lowest scavenging activity, presenting a value of 8.6% and 4.8% for CA and hybrid CA MNT, respectively. These results can be explained by the fact that calcium alginate has the ability to react with the free radical compared to bentonite clay, which increases the free radical inhibition. Furthermore, there was a slight increase in the free radical scavenging activity of CA-REO compared to the control CA (from 8.6 to 10%); on the other hand, the CA-MNT-REO exhibited the highest scavenging activity, reaching 12.8%, which was ten times higher than that of CA-MNT (4.8%).

Based on the literature data, the same results were found by Singh et al. [40] reporting that the chitosan-gelatin-REO microcapsules exhibited a higher antioxidant property than the unloaded microcapsules. Similarly, Teixeira-Costa et al. [41] revealed that the essential oil encapsulated in the system chitosan/alginate polyelectrolyte complexes inhibits greatly compared to the control polyelectrolyte complexes. These results are mainly due to the presence of oxygenate monoterpenes and hydrocarbon monoterpenes which have significant redox properties and play crucial roles in scavenging free radicals and in the breakdown of peroxide. Our results demonstrated that the combination alginate/montmorillonite microcapsules can contribute to the protection of the essential oil against oxidation, leading to the formation of an oxygen barrier and increasing its stability.

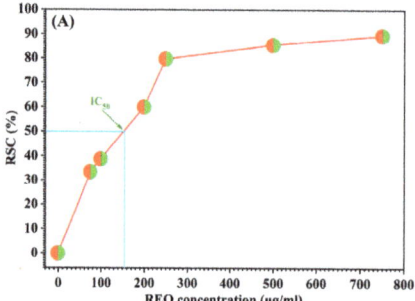
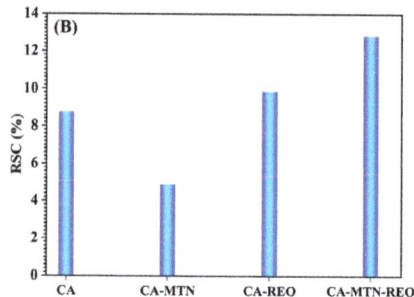

Figure 6. DPPH-scavenging activities of rosemary essential oil (**A**) and of unloaded and REO loaded CA and hybrid CA-MTN microcapsules (**B**).

3.4. REO Release Kinetics

The rosemary essential oil release from the microcapsules depends on the medium of release; the latest penetrates into the microcapsules to dissolve the entrapped essential oil which diffuses through the microcapsules' membrane. The release studies were performed to show the difference in the release pathway as a function of the microparticle formulation and the REO concentration. The experiment was carried out in 1% Tween 80-water to avoid water saturation due to the low solubility of rosemary essential oil in water. As shown in Figure 7, two phases were discerned: an initial phase of fast release correlated to the REO adsorbed on or near the surface of the microcapsules. Contrariwise, the second phase corresponds to the slow release of the REO and stays nearly constant over time. At that stage, the release can be due to the diffusion of the REO dispersed into the microcapsules [42]. The profiles suggest that the release rate of microcapsules with higher REO concentrations is slower than that of microcapsules with lower REO concentrations. The amount of REO released from CA microcapsules (Figure 7A) was approximately equal to the amount released from hybrid CA-MTN microcapsules (Figure 7B), but the time of release differed from 28 to 85 h for CA and hybrid CA-MTN, respectively.

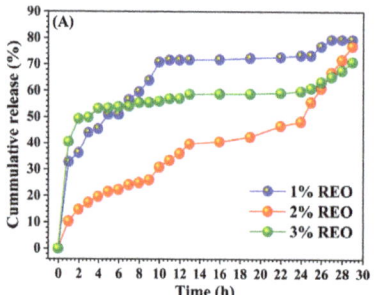
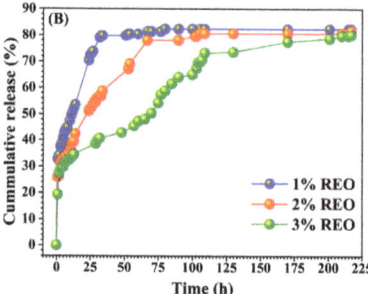

Figure 7. Kinetical release profiles of rosemary essential oil from CA-REO (**A**) and hybrid CA-MTN-REO (**B**) microcapsules in w/o medium.

To explain the mechanism of the REO release from the microcapsules, the release data CR (%) were fitted to the Korsmeyer and Peppas semi-empirical model. The obtained kinetic parameters are listed in Table 5. Release-kinetics correlation coefficients (R^2) of both microparticles were close to 0.9. The diffusional exponent (n) ranged between 0.28 and 0.29 for hybrid CA-MTN and CA microcapsules, respectively. These data indicate that the rosemary essential oil release pathway follows a Fickian diffusion, in which the release is governed by a diffusional behaviour. Therefore, the diffusional process of essential oil in microcapsules can be influenced by the presence of clay particles which prevents the movement of the polymer chains and limits the rate of the internal diffusion of the essential oil in the hybrid microcapsules compared to CA microcapsules.

Table 5. Release kinetic parameters (n and k) for the CA and hybrid CA-MTN microcapsules; n is the release exponent related to the drug release mechanisms, k is the rate constant and R^2 is the coefficient of correlation.

Kinetic Parameters	CA-REO Microcapsules	CA-MTN-REO Hybrid Microcapsules
n	0.29	0.28
k	0.18	0.12
R^2	0.93	0.90

4. Conclusions

Based on the previous results, it can be said that the process of encapsulation is a well-established and efficient method to entrap and maintain molecules. It proposes a plethora of benefits and can be achieved by a myriad of techniques whose main objective is to sustain the release of encapsulated components such as essential oils. REO-loaded microcapsules were prepared by o/w emulsions drop wised into a calcium gelling bath. The TGA thermograms show that hybrid microcapsules CA-MTN have higher thermal stability than CA microcapsules due to the presence of the clay. The rate loss of rosemary essential oil is greater in CA than in hybrid CA-MTN microcapsules which proves the high entrapment capacity of the hybrid system. The microcapsule size and morphology analysis have shown that the obtained particles are micrometric and spherical. Concerning the loading capacity and the encapsulation efficiency, the studies indicate that with a higher value of REO concentration, the loading capacity increased while the encapsulation efficiency decreased for both systems. The REO release is three times slower in hybrid CA-MTN microcapsules than in CA microcapsules which suggests that the hybrid CA-MTN have a higher retention capacity due to the wide specific area of the clay. The kinetic release is of a diffusional type and is correlated with the Korsmeyer–Peppas kinetic model. This work's main result is to establish a clear influence of the clay filling with sodium montmorillonite from eastern Morocco to obtain the hybrid formulation CA-MTN which shows better thermal stability, higher loading capacity/encapsulation efficiency and slower release of rosemary essential oil using the simple and inexpensive method of ionotropic gelation without including any harmful substance in the whole process. These developed microcapsule systems loaded with rosemary essential oil can be used in active food packaging to protect the food from external factors, and in cosmetics and beauty care products as well as in non-eco-toxic pesticides.

Author Contributions: Conceptualization, D.B. and A.T.; methodology, A.T.; software, C.H.; validation, A.T. and M.-L.F.; formal analysis, K.E.; investigation, D.B.; resources, A.T.; data curation, D.B.; writing—original draft preparation, D.B.; writing—review and editing, A.T., C.H. and M.A.; visualization, M.A.; supervision, A.T.; funding acquisition, A.T. All authors have read and agreed to the published version of the manuscript.

Funding: The authors sincerely thank CNRST, MESRSFC-Morocco and Mohamed Premier University for their financial support of this research through the PPR15-17 and the ANPMA/CNRST/UMP-247/20 projects. Part of this study was funded by Conseil Départemental d'Eure-et-Loir et Région Centre Val de Loire.

Institutional Review Board Statement: Not applicable.

Informed Consent Statement: Not applicable.

Data Availability Statement: Not applicable.

Acknowledgments: We are thankful to the analytical platform in the chemistry department in the Faculty of Sciences of Oujda for the analysis services. The authors are also grateful to Eswaramoorthy Muthusamy, head of the Nanocomposites-Catalysis laboratory in JNCASR in India, for the opportunity to work with his team.

Conflicts of Interest: The authors declare no conflict of interest.

References

1. Shetta, A.; Kegere, J.; Mamdouh, W. Comparative study of encapsulated peppermint and green tea essential oils in chitosan nanoparticles: Encapsulation, thermal stability, in-vitro release, antioxidant and antibacterial activities. *Int. J. Biol. Macromol.* **2019**, *126*, 731–742. [CrossRef]
2. Silva Damasceno, E.T.; Almeida, R.R.; de Carvalho, S.Y.B.; de Carvalho, G.S.G.; Mano, V.; Pereira, A.C.; de Lima Guimarães, L.G. Lippia origanoides Kunth. essential oil loaded in nanogel based on the chitosan and p-coumaric acid: Encapsulation efficiency and antioxidant activity. *Ind. Crops Prod.* **2018**, *125*, 85–94. [CrossRef]
3. Delshadi, R.; Bahrami, A.; Tafti, A.G.; Barba, F.J.; Williams, L.L. Micro and nano-encapsulation of vegetable and essential oils to develop functional food products with improved nutritional profiles. *Trends Food Sci. Technol.* **2020**, *104*, 72–83. [CrossRef]
4. Mamusa, M.; Resta, C.; Sofroniou, C.; Baglioni, P. Encapsulation of volatile compounds in liquid media: Fragrances, flavors, and essential oils in commercial formulations. *Adv. Colloid Interface Sci.* **2021**, *298*, 102544. [CrossRef]
5. Kfoury, M.; Auezova, L.; Greige-Gerges, H.; Fourmentin, S. Encapsulation in cyclodextrins to widen the applications of essential oils. *Environ. Chem. Lett.* **2018**, *17*, 129–143. [CrossRef]
6. Calderón-Santoyo, M.; Iñiguez-Moreno, M.; Barros-Castillo, J.C.; Miss-Zacarías, D.M.; Díaz, J.A.; Ragazzo-Sánchez, J.A. Microencapsulation of citral with Arabic gum and sodium alginate for the control of Fusarium pseudocircinatum in bananas. *Iran. Polym. J.* **2022**, *31*, 665–676. [CrossRef]
7. Alardhi, S.M.; Albayati, T.M.; Alrubaye, J.M. Adsorption of the methyl green dye pollutant from aqueous solution using mesoporous materials MCM-41 in a fixed-bed column. *Heliyon* **2020**, *6*, e03253. [CrossRef]
8. Rodríguez, J.; Martín, M.J.; Ruiz, M.A.; Clares, B. Current encapsulation strategies for bioactive oils: From alimentary to pharmaceutical perspectives. *Food Res. Int.* **2016**, *83*, 41–59. [CrossRef]
9. Martinez Rivas, C.J.; Tarhini, M.; Badri, W.; Miladi, K.; Greige-Gerges, H.; Nazari, Q.A.; Galindo Rodriguez, S.A.; Roman, R.A.; Fessi, H.; Elaissari, A. Nanoprecipitation process: From encapsulation to drug delivery. *Int. J. Pharm.* **2017**, *532*, 66–81. [CrossRef] [PubMed]
10. Busatto, C.A.; Taverna, M.E.; Lescano, M.R.; Zalazar, C.; Estenoz, D.A. Preparation and Characterization of Lignin Microparticles-in-Alginate Beads for Atrazine Controlled Release. *J. Polym. Environ.* **2019**, *27*, 2831–2841. [CrossRef]
11. Hiremani, V.D.; Gasti, T.; Masti, S.P.; Malabadi, R.B.; Chougale, R.B. Polysaccharide-based blend films as a promising material for food packaging applications: Physicochemical properties. *Iran. Polym. J.* **2022**, *31*, 503–518. [CrossRef]
12. Tondervik, A.; Klinkenberg, G.; Aachmann, F.L.; Svanem, B.I.; Ertesvag, H.; Ellingsen, T.E.; Valla, S.; Skjak-Braek, G.; Sletta, H. Mannuronan C-5 epimerases suited for tailoring of specific alginate structures obtained by high-throughput screening of an epimerase mutant library. *Biomacromolecules* **2013**, *14*, 2657–2666. [CrossRef] [PubMed]
13. Manso, S.; Cacho-Nerin, F.; Becerril, R.; Nerín, C. Combined analytical and microbiological tools to study the effect on Aspergillus flavus of cinnamon essential oil contained in food packaging. *Food Control* **2013**, *30*, 370–378. [CrossRef]
14. Tran, P.H.; Tran, T.T.; Park, J.B.; Lee, B.J. Controlled release systems containing solid dispersions: Strategies and mechanisms. *Pharm. Res.* **2011**, *28*, 2353–2378. [CrossRef]
15. Ju, J.; Chen, X.; Xie, Y.; Yu, H.; Guo, Y.; Cheng, Y.; Qian, H.; Yao, W. Application of essential oil as a sustained release preparation in food packaging. *Trends Food Sci. Technol.* **2019**, *92*, 22–32. [CrossRef]
16. Luis, A.; Ramos, A.; Domingues, F. Pullulan Films Containing Rockrose Essential Oil for Potential Food Packaging Applications. *Antibiotics* **2020**, *9*, 681. [CrossRef]
17. Gorrasi, G. Dispersion of halloysite loaded with natural antimicrobials into pectins: Characterization and controlled release analysis. *Carbohydr. Polym.* **2015**, *127*, 47–53. [CrossRef] [PubMed]
18. Pathania, D.; Gupta, D.; Kothiyal, N.C.; Sharma, G.; Eldesoky, G.E.; Naushad, M. Preparation of a novel chitosan-g-poly(acrylamide)/Zn nanocomposite hydrogel and its applications for controlled drug delivery of ofloxacin. *Int. J. Biol. Macromol.* **2016**, *84*, 340–348. [CrossRef] [PubMed]
19. Montoya, C.; Roldan, L.; Yu, M.; Valliani, S.; Ta, C.; Yang, M.; Orrego, S. Smart dental materials for antimicrobial applications. *Bioact. Mater.* **2023**, *24*, 1–19. [CrossRef] [PubMed]
20. Nastyshyn, S.; Stetsyshyn, Y.; Raczkowska, J.; Nastishin, Y.; Melnyk, Y.; Panchenko, Y.; Budkowski, A. Temperature-Responsive Polymer Brush Coatings for Advanced Biomedical Applications. *Polymers* **2022**, *14*, 4245. [CrossRef]
21. El Miz, M.; Akichoh, H.; Berraaouan, D.; Salhi, S.; Tahani, A. Chemical and Physical Characterization of Moroccan Bentonite Taken from Nador (North of Morocco). *Am. J. Chem.* **2017**, *7*, 105–112. [CrossRef]
22. Ribeiro, R.; Fernandes, L.; Costa, R.; Cavaleiro, C.; Salgueiro, L.; Henriques, M.; Rodrigues, M.E. Comparing the effect of Thymus spp. essential oils on Candida auris. *Ind. Crops Prod.* **2022**, *178*, 114667. [CrossRef]

23. Hosseini, S.M.; Hosseini, H.; Mohammadifar, M.A.; Mortazavian, A.M.; Mohammadi, A.; Khosravi-Darani, K.; Shojaee-Aliabadi, S.; Dehghan, S.; Khaksar, R. Incorporation of essential oil in alginate microparticles by multiple emulsion/ionic gelation process. *Int. J. Biol. Macromol.* **2013**, *62*, 582–588. [CrossRef] [PubMed]
24. Rawat, M.; Singh, D.; Saraf, S.; Saraf, S. Nanocarriers: Promising vehicle for bioactive drugs. *Biol. Pharm. Bull.* **2006**, *29*, 1790–1798. [CrossRef] [PubMed]
25. Couvreur, P.; Vauthier, C. Nanotechnology: Intelligent design to treat complex disease. *Pharm. Res.* **2006**, *23*, 1417–1450. [CrossRef] [PubMed]
26. Ling, Q.; Zhang, B.; Wang, Y.; Xiao, Z.; Hou, J.; Xiao, C.; Liu, Y.; Jin, Z. Chemical Composition and Antioxidant Activity of the Essential Oils of Citral-Rich Chemotype Cinnamomum camphora and Cinnamomum bodinieri. *Molecules* **2022**, *27*, 7356. [CrossRef]
27. Reboucas, L.M.; Sousa, A.C.C.; Sampaio, C.G.; Silva, L.M.R.; Costa, P.M.S.; Pessoa, C.; Brasil, N.; Ricardo, N. Microcapsules based on alginate and guar gum for co-delivery of hydrophobic antitumor bioactives. *Carbohydr. Polym.* **2023**, *301*, 120310. [CrossRef]
28. Essifi, K.; Brahmi, M.; Berraaouan, D.; Ed-Daoui, A.; El Bachiri, A.; Fauconnier, M.-L.; Tahani, A.; Fernandez-Sanchez, J.F. Influence of Sodium Alginate Concentration on Microcapsules Properties Foreseeing the Protection and Controlled Release of Bioactive Substances. *J. Chem.* **2021**, *2021*, 5531479. [CrossRef]
29. Turasan, H.; Sahin, S.; Sumnu, G. Encapsulation of rosemary essential oil. *LWT—Food Sci. Technol.* **2015**, *64*, 112–119. [CrossRef]
30. Yeddes, W.; Djebali, K.; Aidi Wannes, W.; Horchani-Naifer, K.; Hammami, M.; Younes, I.; Saidani Tounsi, M. Gelatin-chitosan-pectin films incorporated with rosemary essential oil: Optimized formulation using mixture design and response surface methodology. *Int. J. Biol. Macromol.* **2020**, *154*, 92–103. [CrossRef]
31. Volic, M.; Pajic-Lijakovic, I.; Djordjevic, V.; Knezevic-Jugovic, Z.; Pecinar, I.; Stevanovic-Dajic, Z.; Veljovic, D.; Hadnadjev, M.; Bugarski, B. Alginate/soy protein system for essential oil encapsulation with intestinal delivery. *Carbohydr. Polym.* **2018**, *200*, 15–24. [CrossRef]
32. Sankalia, M.G.; Mashru, R.C.; Sankalia, J.M.; Sutariya, V.B. Papain entrapment in alginate beads for stability improvement and site-specific delivery: Physicochemical characterization and factorial optimization using neural network modeling. *AAPS PharmSciTech* **2005**, *6*, E209–E222. [CrossRef] [PubMed]
33. He, Y.; Wu, Z.; Tu, L.; Han, Y.; Zhang, G.; Li, C. Encapsulation and characterization of slow-release microbial fertilizer from the composites of bentonite and alginate. *Appl. Clay Sci.* **2015**, *109–110*, 68–75. [CrossRef]
34. Wang, L.; Shelton, R.M.; Cooper, P.R.; Lawson, M.; Triffitt, J.T.; Barralet, J.E. Evaluation of sodium alginate for bone marrow cell tissue engineering. *Biomaterials* **2003**, *24*, 3475–3481. [CrossRef]
35. Santos, N.A.; Cordeiro, A.M.T.M.; Damasceno, S.S.; Aguiar, R.T.; Rosenhaim, R.; Carvalho Filho, J.R.; Santos, I.M.G.; Maia, A.S.; Souza, A.G. Commercial antioxidants and thermal stability evaluations. *Fuel* **2012**, *97*, 638–643. [CrossRef]
36. Zhou, G.; Xing, Z.; Tian, Y.; Jiang, B.; Ren, B.; Dong, X.; Yi, L. An environmental-friendly oil-based dust suppression microcapsules: Structure with chitosan derivative as capsule wall. *Process Saf. Environ. Prot.* **2022**, *165*, 453–462. [CrossRef]
37. Chang, J.-H.; An, Y.U.; Cho, D.; Giannelis, E.P. Poly(lactic acid) nanocomposites: Comparison of their properties with montmorillonite and synthetic mica (II). *Polymer* **2003**, *44*, 3715–3720. [CrossRef]
38. Alfatama, M.; Lim, L.Y.; Wong, T.W. Alginate-C18 Conjugate Nanoparticles Loaded in Tripolyphosphate-Cross-Linked Chitosan-Oleic Acid Conjugate-Coated Calcium Alginate Beads as Oral Insulin Carrier. *Mol. Pharm.* **2018**, *15*, 3369–3382. [CrossRef]
39. Piornos, J.A.; Burgos-Díaz, C.; Morales, E.; Rubilar, M.; Acevedo, F. Highly efficient encapsulation of linseed oil into alginate/lupin protein beads: Optimization of the emulsion formulation. *Food Hydrocoll.* **2017**, *63*, 139–148. [CrossRef]
40. Singh, N.; Sheikh, J. Novel Chitosan-Gelatin microcapsules containing rosemary essential oil for the preparation of bioactive and protective linen. *Ind. Crops Prod.* **2022**, *178*, 114549. [CrossRef]
41. Teixeira-Costa, B.E.; Silva Pereira, B.C.; Lopes, G.K.; Tristão Andrade, C. Encapsulation and antioxidant activity of assai pulp oil (*Euterpe oleracea*) in chitosan/alginate polyelectrolyte complexes. *Food Hydrocoll.* **2020**, *109*, 106097. [CrossRef]
42. Faisant, N.; Siepmann, J.; Richard, J.; Benoit, J.P. Mathematical modeling of drug release from bioerodible microparticles: Effect of gamma-irradiation. *Eur. J. Pharm. Biopharm.* **2003**, *56*, 271–279. [CrossRef] [PubMed]

Disclaimer/Publisher's Note: The statements, opinions and data contained in all publications are solely those of the individual author(s) and contributor(s) and not of MDPI and/or the editor(s). MDPI and/or the editor(s) disclaim responsibility for any injury to people or property resulting from any ideas, methods, instructions or products referred to in the content.

Article

Grafted Microparticles Based on Glycidyl Methacrylate, Hydroxyethyl Methacrylate and Sodium Hyaluronate: Synthesis, Characterization, Adsorption and Release Studies of Metronidazole

Aurica Ionela Gugoasa [1], Stefania Racovita [2], Silvia Vasiliu [2] and Marcel Popa [1,3,*]

[1] Department of Natural and Synthetic Polymers, Faculty of Chemical Engineering and Environmental Protection, "Gheorghe Asahi" Technical University of Iasi, Prof. Dr. Docent Dimitrie Mangeron Street No. 73, 700050 Iasi, Romania
[2] "Petru Poni" Institute of Macromolecular Chemistry, Grigore Ghica Voda Alley No. 41A, 700487 Iasi, Romania
[3] Academy of Romanian Scientists, Splaiul Independentei Street No. 54, 050085 Bucharest, Romania
* Correspondence: marpopa2001@yahoo.fr

Citation: Gugoasa, A.I.; Racovita, S.; Vasiliu, S.; Popa, M. Grafted Microparticles Based on Glycidyl Methacrylate, Hydroxyethyl Methacrylate and Sodium Hyaluronate: Synthesis, Characterization, Adsorption and Release Studies of Metronidazole. *Polymers* 2022, *14*, 4151. https://doi.org/10.3390/polym14194151

Academic Editors: Lorenzo Antonio Picos Corrales, Angel Licea-Claverie and Grégorio Crini

Received: 27 August 2022
Accepted: 25 September 2022
Published: 3 October 2022

Publisher's Note: MDPI stays neutral with regard to jurisdictional claims in published maps and institutional affiliations.

Copyright: © 2022 by the authors. Licensee MDPI, Basel, Switzerland. This article is an open access article distributed under the terms and conditions of the Creative Commons Attribution (CC BY) license (https://creativecommons.org/licenses/by/4.0/).

Abstract: Three types of precursor microparticles based on glycidyl methacrylate, hydroxyethyl methacrylate and one of the following three crosslinking agents (mono-, di- or triethylene glycol dimethacrylate) were prepared using the suspension polymerization technique. The precursor microparticles were subsequently used to obtain three types of hybrid microparticles. Their synthesis took place by grafting sodium hyaluronate, in a basic medium, to the epoxy groups located on the surface of the precursor microparticles. Both types of the microparticles were characterized by: FTIR spectroscopy, epoxy groups content, thermogravimetric analysis, dimensional analysis, grafting degree of sodium hyaluronate, SEM and AFM analyses, and specific parameters of porous structures (specific surface area, pore volume, porosity). The results showed that the hybrid microparticles present higher specific surface areas, higher swelling capacities as well as higher adsorption capacities of antimicrobial drugs (metronidazole). To examine the interactions between metronidazole and the precursor/hybrid microparticles the adsorption equilibrium, kinetic and thermodynamic studies were carried out. Thus, it was determined the performance of the polymer systems in order to select a polymer–drug system with a high efficiency. The release kinetics reflect that the release mechanism of metronidazole in the case of hybrid microparticles is a complex mechanism characteristic of anomalous or non-Fickian diffusion.

Keywords: grafted microparticles; sodium hyaluronate; crosslinked microparticles; metronidazole adsorption; adsorption isotherm; adsorption kinetics; metronidazole release

1. Introduction

The dynamics of medical research around the world are being directed towards developing new materials to solve and remedy various health problems. In recent years, dentistry, a part of the medical field, has been seeking to solve a series of problems related to the prevention of dental caries, periodontal disease and the treatment of bone tissue loss [1]. It is known that periodontitis (periodontal disease) is a chronic inflammatory condition of the gingiva and the tissues supporting teeth on their arches (gingiva, periodontal ligament, alveolar bone) [2,3]. The World Health Organization considers periodontal disease as one of the most common diseases of the oral cavity, statistically affecting three quarters of the world's population.

The treatment of periodontal disease is complex and generally aims to slow down the evolution of the disease. In the early stages of the disease, local treatment is instituted [3], which is generally antimicrobial and consists of: scaling, which helps remove tartar and bacteria from the tooth surface and under the gingiva; and the use of antibiotics.

The development of polymer science and polymer characterization methods has led to the use of natural and synthetic polymers to produce polymeric materials in the form of microparticles to improve the prevention, diagnosis and treatment of injured tissues.

Among the natural polymers, chitosan, pectin, alginate, starch and dextran are used to prepare microparticles loaded with different drugs to treat periodontal diseases [4–6]. Another important polysaccharide, hyaluronic acid, which is used in the cosmetic industry, prepares the scaffold for tissue engineering for drug delivery, disease diagnoses and biomedical imaging; it represents an ideal candidate to obtain polymeric materials for the dental field [7–10]. The use of polymeric microparticles in dentistry has been increasing due to the excellent properties of both polymeric materials (good surface, biological and mechanical properties, low manufacturing cost, simple synthesis methods) as well as of microparticles (high surface/volume ratio, good accessibility to the site of action due to their size, stability in biological fluids, drug transport capacity, decreased frequency and intensity of adverse effects) [11,12]. Polymer-based microparticles can be achieved through several processes, the most important of which is polymerization. Among the polymerization methods used to obtain spherical microparticles (bulk polymerization, precipitation polymerization, dispersion polymerization, emulsion polymerization, etc.), aqueous suspension polymerization can be used because of the technical and economic advantages it confers. Since in suspension polymerization a small number of reagents and simple reaction equipment are used, the purification of the resulting products takes place by simple methods and the cost of obtaining the products is low; thus, it can be concluded that this method is suitable for obtaining polymeric materials with special properties that can be successfully applied both in the field of environmental protection and in the medical or pharmaceutical fields [13–15].

The main aim of this work is to develop polymeric materials in the form of porous microparticles with special architectures that are obtained by combining simple, economical and environmentally friendly methods that combine the properties of synthetic polymers (chemical, mechanical and thermal stability) with those of natural polymers (biocompatibility, biodegradability, non-toxicity) and have potential applications in the dental field. The choice of sodium hyaluronate, a derivative of hyaluronic acid, as a natural component in the production of the hybrid microparticles was based on the following considerations: hyaluronic acid is an essential component of the periodontal ligament matrix; it plays important roles in cell migration, adhesion and differentiation by binding proteins and cell receptors; and it has been studied as a metabolite or diagnostic marker of inflammation in gingival crevicular fluid [16,17]. Thus, by grafting hyaluronic acid onto the surface of precursor microparticles, it was intended to obtain a system for the delivery of chemotherapeutic agents (antimicrobial and anti-inflammatory drugs) to treat dental diseases. For obtaining polymer–drug systems, the biological active principle chosen was metronidazole, which is a bactericide that can be administered orally together with other antibiotics with an increased efficacy in treating periodontal disease.

2. Materials and Methods

2.1. Materials

All reagents used in the preparation and characterization of the precursor/hybrid microparticles were purchased from Sigma-Aldrich (Schnelldorf, Germany). Glycidyl methacrylate (GMA) and hydroxyethyl methacrylate (HEMA) were distilled before use under reduced pressure to remove the inhibitor. Dimethacrylic monomers [ethylene glycol dimethacrylate (EGDMA), diethylene glycol dimethacrylate (DEGDMA) and triethylene glycol dimethacrylate (TEGDMA)], the initiator [benzoyl peroxide (BOP)], porogenic agent [butyl acetate], poly(vinyl alchohol) (PVA, M_w = 67,000 g/mol, degree of hydrolysis, 88%), gelatine, NaCl, sodium hyaluronate (HA), HBr, NaOH, glacial acetic acid, Crystal violet, n-heptane, methanol, ethanol and metronidazole (M_w = 171.064 g/mol) are of analytical grade and were used as received.

2.2. Methods

2.2.1. Synthesis of Precursor Microparticles

The precursor microparticles based on GMA, HEMA and EGDMA, DEGDMA or TEGDMA denoted AE, AD and AT were synthesized by the suspension polymerization technique in a 250 cm^3 cylindrical reactor fitted with mechanical stirrer, thermometer and reflux condenser. The reaction mixture is formed by two phases:

- Aqueous phase containing a polymeric stabilizer (2 wt% mixture of PVA and gelatine) and NaCl (3 wt%);
- Organic phase is formed by GMA (70% mol), HEMA (20% mol), crosslinking agents (10% mol of EGDMA, DEGDMA or TEGDMA), BOP and butyl acetate at a dilution of D = 0.6.

The copolymerization reaction was conducted under a nitrogen atmosphere for 8 h at 78 °C and 1 h at 90 °C with a stirring rate of 400 rpm. After the copolymerization reactions were completed, the precursor microparticles were separated by decantation and washed with hot water. Then, to remove traces of residual monomers and porogenic agent, the precursor microparticles were extracted with methanol or ethanol in a Soxhlet apparatus.

2.2.2. Synthesis of Hybrid Microparticles

The hybrid microparticles denoted AEHA, ADHA and ATHA were prepared by grafting the sodium hyaluronate to the epoxy groups situated on the surface of the precursor microparticles. A known quantity of precursor microparticles was immersed in a solution of various concentrations of sodium hyaluronate (0.2–1%) obtained by dissolving the polysaccharide in water with the addition of NaOH (pH = 7–10). The reaction took place at temperatures between 30 and 60 °C over a period of time ranging between 2 and 10 h. After the aforementioned time, the hybrid microparticles were filtered off, washed with water to remove unreacted polysaccharide and NaOH and then dried under a vacuum at 40 °C for 24 h.

2.2.3. Infrared Spectroscopy

The precursor/hybrid microparticles were characterized by FTIR spectroscopy using a Bruker Vertex FT-IR Spectrometer at a resolution of 2 cm^{-1} in the range of 4000–400 cm^1 by KBr pellet technique. In order to obtain FT-IR spectra, 0.03 g the precursor/hybrid microparticles or HA were mixed and ground with potassium bromide.

2.2.4. Epoxy Group Content

The epoxy group content was determined by ASTM D1652 (standard test method for epoxy content of epoxy resins) [18]. This method consisted of direct titration with a standard solution of HBr in glacial acetic acid. Experimental values of epoxide equivalent weight (EEW) were determined using the following equation:

$$EEW = \frac{100 \cdot w}{N \cdot V} \; (mol \cdot g^{-1}) \qquad (1)$$

where w—grams of precursor/hybrid microparticles; N—normality of the HBr in acetic acid (mol·L^{-1}); and V—volume of HBr solution used for titration (mL).

2.2.5. Thermogravimetric Analysis (TGA)

The thermal behavior of the precursor/hybrid microparticles (4 mg of sample) was performed at a heating rate of 10 °C·min^{-1} in nitrogen atmosphere, using a Mettler Toledo TGA 851 Derivatograph.

2.2.6. Scanning Electron Microscopy (SEM)

The surface morphology of the precursor/hybrid microparticles was analyzed with a Quanta 200 environmental scanning electron microscope at 25 kV.

2.2.7. Atomic Force Microscopy (AFM)

AFM images for the precursor/hybrid microparticles were performed using a Scanning Probe Microscope Solver Pro-M platform (NT-MDT, Moscow, Russia) with a rectangular silicon cantilever NSG 10 and 203 kHz oscillation frequency, in air at ambient temperature (23 °C). The latest version of NT-MDT NOVA software was used for the analysis and calculation of the microparticle surface characteristic parameters (arithmetic mean deviation of the surface (S_a); root-mean-square deviation of the surface (S_q); surface skewness (S_{sk}); and surface kurtosis (S_{ku})) and AFM imaging. The equations of the three-dimensional roughness parameters used in the current study are presented below [19]:

$$S_a = \frac{1}{MN} \sum_{j=1}^{N} \sum_{i=1}^{M} |z(x_i, y_j) - \bar{z}| \tag{2}$$

$$S_q = \left[\frac{1}{MN} \sum_{j=1}^{N} \sum_{i=1}^{M} |z(x_i, y_j) - \bar{z}|^2 \right]^{1/2} \tag{3}$$

$$S_{sk} = \frac{1}{MNS_q^3} \sum_{j=1}^{N} \sum_{i=1}^{M} |z(x_i, y_j) - \bar{z}|^3 \tag{4}$$

$$S_{ku} = \frac{1}{MNS_q^4} \sum_{j=1}^{N} \sum_{i=1}^{M} |z(x_i, y_i) - \bar{z}|^4 \tag{5}$$

where N—number of points along of scan line; M—number of lines; S_q—the root mean square roughness; and z—the height of each point of coordinates x_i and x_j.

Additionally, shape and elongation factors are two important parameters in pore structure analysis and can be calculated with the following equations:

$$f_{shape} = \frac{4 \cdot \pi \cdot A}{1.064 \cdot P^2} \tag{6}$$

$$f_{elongation} = \frac{D_{min}}{D_{max}} \tag{7}$$

where A—pore area; P—pore perimeter; D_{min}—minimum Feret diameter; and D_{max}—maximum Feret diameter.

2.2.8. Dimensional Analysis of Precursor/Hybrid Microparticles

The number average diameter (D) was obtained using a WingSALD 7001 laser diffraction particle size analyzer (UK). The measurements were performed by suspending the precursor microparticles or hybrid microparticles in methanol (non-solvent). The experimental data were recorded and processed using WingSALD software.

2.2.9. Specific Parameters for the Characterization of the Morphology of Porous Structure

The morphology of the porous structure of the precursor/hybrid microparticles can be characterized using the following parameters: porosity (P, %), pore volume (PV, mL·g^{-1}) and specific surface area (S_{sp}, m^2·g^{-1}). The pore volume and the porosity of the precursor/hybrid microparticles were calculated as follows:

$$PV = \frac{1}{\rho_{ap}} - \frac{1}{\rho_{sp}} \tag{8}$$

$$\%P = 100 \cdot \left(1 - \frac{\rho_{ap}}{\rho_{sp}}\right) \tag{9}$$

where ρ_{ap}—apparent density (g·cm^{-3}); and ρ_{sp}—skeletal density (g·cm^{-3}).

The apparent and skeletal densities of precursor/hybrid microparticles were measured by pycnometric methods with mercury and n-heptane, respectively [20] and were calculated with the following equations:

$$\rho_{ap} = \frac{m_1}{V_P - (m_3 - m_2)/\rho_{Hg}} \quad (10)$$

$$\rho_{sp} = \frac{m_1}{V_f - (m_s - m_4)/h} \quad (11)$$

where m_1—mass of the sample (precursor/hybrid microparticles) (g); V_P—volume of the pycnometer (cm^3); V_f—volume of the volumetric flask (cm^3); m_2—mass of the pycnometer with the sample (g); m_3—mass of the pycnometer with mercury and the sample (g); ρ_{Hg}—density of mercury (g·cm^{-3}); m_4—mass of the volumetric flask with the sample (g); m_s—mass of the volumetric flask with the sample and n-heptane (g); and h—density of n-heptane (g·cm^{-3}).

The specific surface area was determined by dynamic vapor sorption using the Brunauer, Emmet and Teller (BET) method [21] and the sorption–desorption curves recorded for the precursor/hybrid microparticles. Sorption–desorption isotherms were registered using the fully automated gravimetric analyzer IGAsorp produced by Hiden Analytical, Warrington (UK).

2.2.10. Swelling Studies

The swelling capacities of the precursor/hybrid microparticles were determined in aqueous solution at pH 1.2 and 5.5, respectively, using the gravimetric method. A known amount of dried precursor/hybrid microparticles (0.2 g) was immersed in 10 mL aqueous solution at 25 °C. At a certain period of time ranging between 10 and 1440 min, the precursor/hybrid microparticles were removed, centrifuged at 500 rpm for 10 min and weighed. The swelling capacity of the precursor/hybrid microparticles was calculated using the following equation:

$$S_w (\%) = \frac{w_s - w_d}{w_d} \cdot 100 \quad (12)$$

where w_s—amount of swollen precursor/hybrid microparticles (g); and w_d—amount of dry precursor/hybrid microparticles (g).

2.2.11. Bach Adsorption Studies

The adsorption of metronidazole on precursor/hybrid microparticles was investigated in a batch system. Metronidazole adsorption was realized as follows: 0.1 g of precursor/hybrid microparticles of known moisture were introduced in 50 mL conical flasks filled with 10 mL metronidazole solution with various initial concentrations (0.25–1 mg·mL^{-1}). The conical flasks were placed in a thermostatic shaker bath (Memmert M00/M01, Germany) and shaken at 180 rpm and 25, 30 and 40 °C for different periods of time ranging from 10 to 1440 min. After the specified period of time, the precursor/hybrid microparticles were removed quantitatively from the metronidazole solution by centrifugation at 1000 rpm for 10 min. The concentration of metronidazole in the supernatant solution before and after adsorption was determined using a UV–VIS spectrophotometer (UV–VIS SPEKOL 1300, Analytik Jena, Jena, Germany) at a wavelength of 277 nm based on the calibration curve obtained with various drug solutions of known conditions. The amounts of metronidazole at equilibrium, q_e (mg·g^{-1}), and at any time, q_t (mg·g^{-1}), were calculated from the following equations:

$$q_e = \frac{(C_0 - C_e) \cdot V}{w} \quad (13)$$

$$q_t = \frac{(C_0 - C_t) \cdot V}{w} \quad (14)$$

where C_0—initial concentration of metronidazole solution (mg·g^{-1}); C_e—concentration of metronidazole at equilibrium (mg·g^{-1}); C_t—concentration of metronidazole at any time (mg·g^{-1}); V—volume of drug solution (L); and w—amount of precursor/hybrid microparticles (g).

2.2.12. Drug Release Studies

In vitro drug release studies were carried out as follows: 100 mg of the drug-microparticle systems were introduced in 10 mL of buffer solution of pH = 1.2 (stimulated gastric solution) at 37 °C, over a period of 8 h, under gentle shaking (50 rpm) using a thermostatic shaker bath (Memmert M00/M01). Very small volumes of the release medium (1 µL) were collected with microsyringes at different intervals of time. The amount of metronidazole was determined spectrophotometrically (Nanodrop ND100, Wilmington, DE, USA) at a wavelength of 277 nm using a calibration curve.

3. Results and Discussion

3.1. Synthesis of Precursor/Hybrid Microparticles

The synthesis of the hybrid microparticles took place in two steps.

In the first step, precursor microparticles based on glycidyl methacrylate, hydroxyethyl methacrylate and dimethacrylic monomers (EGDMA, DEGDMA and TEGDMA) were synthesized using the suspension polymerization technique. The reaction to obtain the precursor microparticles (AE, AD and AT) is shown in Figure 1.

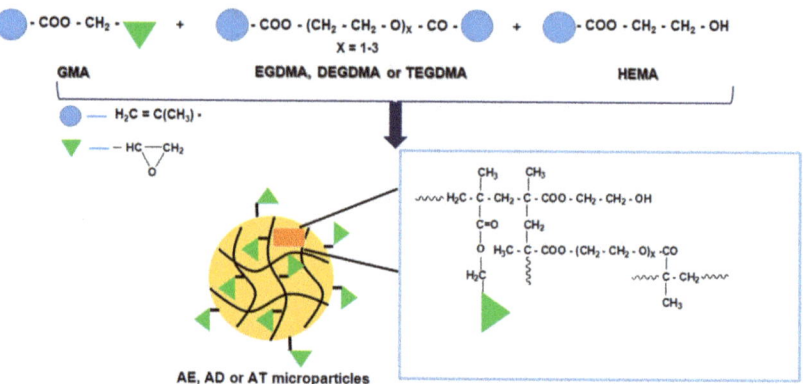

Figure 1. Schematic representation of the reaction to obtain precursor microparticles (AE, AD or AT).

Table 1 shows the experimental conditions required for the synthesis of precursor microparticles.

Table 1. The experimental conditions for the synthesis of precursor microparticles.

Sample Code	GMA (% mol)	HEMA (% mol)	EGDMA (% mol)	DEGDMA (% mol)	TEGDMA (% mol)	Porogenic Agent	Dilution	Reaction Yield (%)
AE	70	20	10	-	-	butyl acetate	0.6	93
AD	70	20	-	10	-	butyl acetate	0.6	90
AT	70	20	-	-	10	butyl acetate	0.6	96

From Table 1 it can be seen that suspension polymerization produces high-yield microparticles, regardless of the type of crosslinker used. Similar results were observed for grafting chitosan onto three-dimensional networks based on glycidyl methacrylate and dimethacrylic esters when the grafting reaction occurred directly in the suspension polymerization process [13].

In the second step, hybrid microparticles were synthesized by grafting sodium hyaluronate to the epoxy groups located on the surface of the precursor microparticles; the reaction was carried out in a basic medium under gentle stirring and in a nitrogen atmosphere (Figure 2).

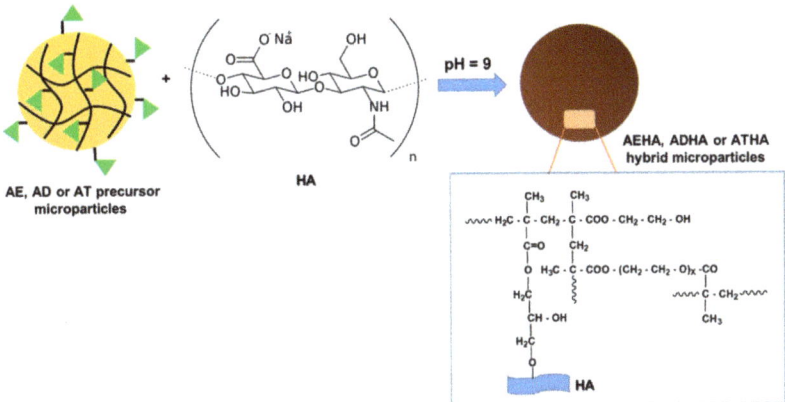

Figure 2. Schematic representation of the reaction to obtain hybrid microparticles (AEHA, ADHA or ATHA).

The amount of grafted HA was determined gravimetrically and calculated using the following relationship:

$$Q\,(\%) = \frac{W_1 - W_0}{W_2} \cdot 100 \qquad (15)$$

where W_0—the initial amount of microparticles (g); W_1—the amount of grafted microparticles after purification (g); and W_2—the initial amount of HA (g).

Optimization of the Grafting Reaction

The optimal conditions for obtaining hybrid microparticles with the highest degree of HA grafting were determined by changing only one reaction parameter (HA concentration, temperature, pH, reaction time) while keeping the other parameters constant. Thus, the influences of the mentioned reaction parameters on the grafting degree of HA are shown in Figures 3 and 4.

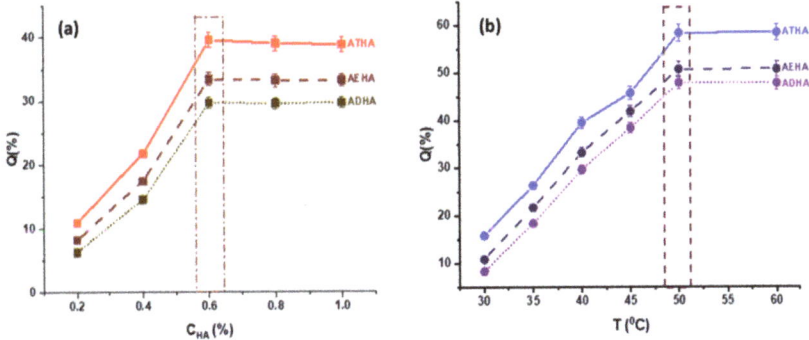

Figure 3. (a) The influence of HA concentration on amount of grafted polymer (T = 35 °C, t = 6 h, pH = 9); and (b) the influence of temperature on amount of grafted polymer (C_P = 0.6%, t = 6 h, pH = 9).

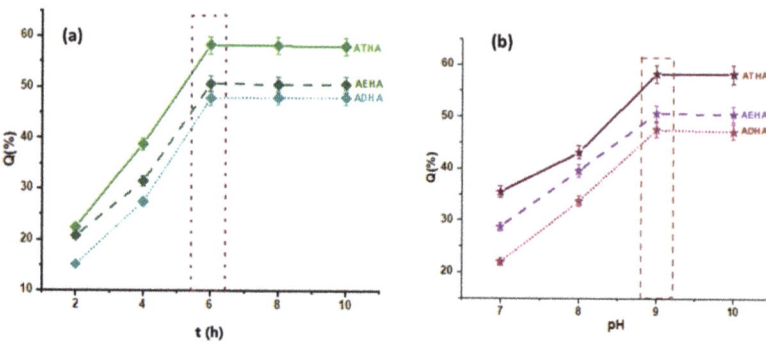

Figure 4. (a) The influence of the reaction time on the amount of grafted polymer ($C_P = 0.6\%$, $T = 50$ °C, pH = 9); and (b) the influence of the pH of the HA solution on amount of grafted polymer ($C_P = 0.6\%$, $T = 50$ °C, $t = 6$ h).

From the graphical representations presented above, it can be seen that:
- The amount of grafted HA increases with the concentration of the polymer solution up to a value of 0.6%, after which equilibrium is reached. This behavior is due to the fact that as the concentration of the HA solution increases, the number of hydroxyl groups that will react in a basic medium with the epoxy groups increases, leading to the formation of a covalent ether -CH_2-O-HA bond. Additionally, at low concentrations, the viscosity of HA solutions is lower, ensuring a more uniform stirring and better accessibility for the epoxy groups on the surface of the precursor microparticles;
- Increasing the temperature of the reaction medium has the effect of decreasing the viscosity of the reaction medium and increasing the mobility of the polymer chains, leading to a better interaction between the -OH groups belonging to HA and the epoxy groups on the surface of the precursor microparticles, and thus to a higher amount of grafted HA. Temperature is also known to increase the reaction rate and to favor higher yields for most chemical reactions;
- Increasing the reaction time to 6 h resulted in an increase in the amount of grafted HA;
- Another important parameter of the grafting process is the pH of the reaction. In a basic medium, the epoxy ring opening reaction by the -OH group proceeds by an SN2 mechanism and the -OH group is formed at the most substituted atom in the ring. In an acidic environment, the reaction proceeds through the SN1 mechanism, leading to the formation of -OH at the methylene group of the ring, and the rest of the HA molecule, which is huge in volume, encounters significant steric hindrances, making it difficult to bind to the secondary carbon atom of the ring. Thus, the epoxide cycle opening reaction resulting in the grafting of HA to the polymer particles will be increasingly favored by the increasing pH, which intensifies the nucleophilic attack (SN2) of -OH from the polysaccharide to the epoxy ring.

The chemical structure of the crosslinker used to obtain the precursor microparticles also influences the grafting yield. As can be seen, the amount of grafted HA increases in the following order: ADHA < AEHA < ATHA, so the highest amount of grafted HA is recorded in the ATHA hybrid microparticles that were obtained in the presence of TEGDMA as a crosslinker. Thus, increasing the chain length between the two methacrylic groups leads to the formation of microparticles with larger mesh sizes, thus allowing a better interaction between the two reaction partners (precursor microparticles and sodium hyaluronate). A special case is the ADHA microparticles that were obtained in the presence of DEGDMA as a crosslinking agent. In this case, the lower amount of grafted HA is probably due to the more compact structure of the microparticles, which is the result of the complexity of the crosslinking polymerization reaction, a reaction that is often accompanied by internal cyclisation processes.

In conclusion, the most favorable conditions for the synthesis of hybrid AEHA, ADHA and ATHA microparticles are as follows: the concentration of the HA solution = 0.6%, T = 50 °C, t = 6 h, pH = 9, and the ratio of the amount of microparticles: HA = 1:0.6 (g·g^{-1}).

3.2. Structural Characterization

3.2.1. FTIR Spectroscopy

FTIR spectroscopy was used to highlight HA grafting on precursor microparticles. The infrared spectra of precursor/hybrid microparticles are presented in Figure 5 as well as in Figures S1 and S2.

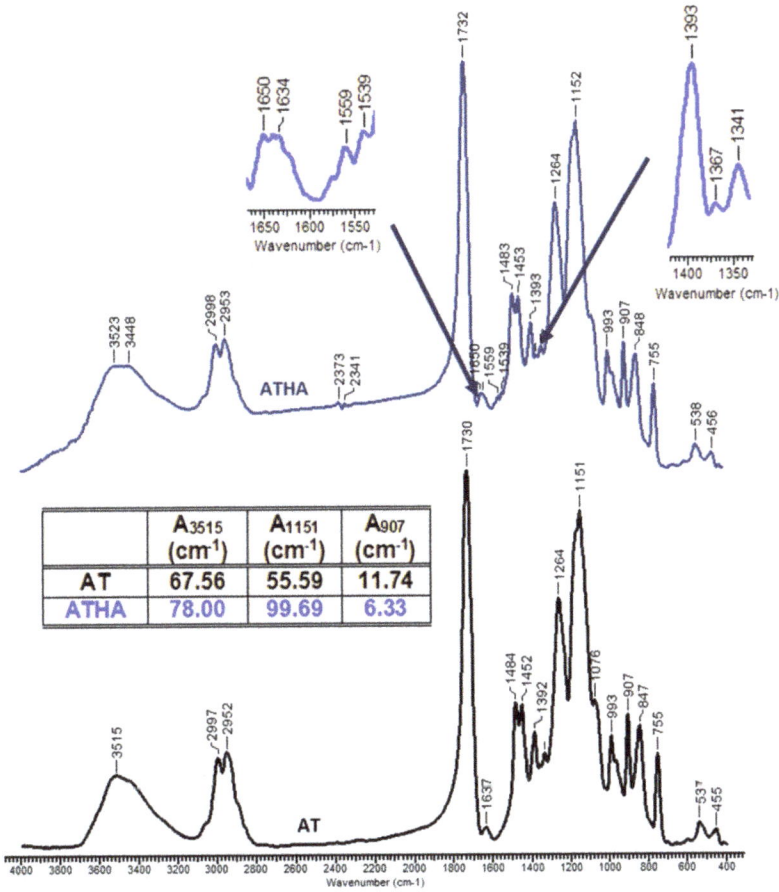

Figure 5. The infrared spectra of AT and ATHA microparticles.

The characteristic absorption bands of AE, AD and AT microparticles are located at: 3452, 3482 and 3515 cm^{-1}, characteristic of valence vibrations ν_{O-H} belonging to the HEMA monomer; 2991–2997 cm^{-1} and 2953 cm^{-1} are specific to the symmetric and asymmetric stretching vibrations of the -CH$_3$, >CH$_2$ and >CH- groups; 1633, 1634 and 1637 cm^{-1} are attributed to the >C=C< bond; 1730 cm^{-1} is characteristic of the >C=O bond of the ester group, which is present in all the chemical structures of the three types of precursor microparticles; 1481 and 1484 cm^{-1} are specific to the bending vibration of the methylene group (δ_{CH2}); and 907 cm^{-1} is attributed to the stretching vibrations of the epoxy groups.

The appearance of new absorption bands at 1559, 1539, 1367 and 1341 cm^{-1} indicates the presence in the structures of the hybrid microparticles (AEHA, ADHA and ATHA) of the carboxylate group characteristic of the polysaccharide.

In addition, by comparing the corresponding areas of AE, AD and AT microparticles at wavenumbers 3450, 1152 and 907 cm^{-1} with the specific areas of similar absorption bands for AEHA, ADHA and ATHA microparticles, the following can be observed:

- The values of the specific absorption band areas at the wavenumber 3450 cm^{-1} are higher for the hybrid microparticles (A_{AEHA} = 46.95 cm^{-1}, A_{ADHA} = 28.95 cm^{-1}, A_{ATHA} = 78.00 cm^{-1}) compared to the values of similar absorption band areas corresponding to the precursor microparticles (A_{AE} = 20.97 cm^{-1}, A_{AD} = 13.73 cm^{-1} and A_{AT} = 67.56 cm^{-1}). The higher values of the absorption band of the hybrid microparticles are due to the presence of sodium hyaluronate, which has several hydroxyl groups in its structure;
- The increase in the values of the adsorption band area at 1151 cm^{-1} in the case of the hybrid microparticles is due to the formation of new ether bonds by grafting HA to the epoxy groups from the GMA structure;
- In the case of the hybrid microparticles, a decrease in the values of the specific areas of the absorption bands is observed from the wavenumber 907 cm^{-1}, due to the grafting reaction of HA by the opening of the epoxy ring in a basic medium.

Based on the data obtained from the FTIR spectra, it can be concluded that the grafting reaction of HA on the surface of the precursor microparticles took place successfully.

3.2.2. Dimensional Analysis of Precursor/Hybrid Microparticles

In the case of suspension polymerization, the size and size distribution of the microparticles are influenced by various parameters: the shape of the reaction vessel, the type of stirrer, the stirring speed, the temperature, the chemical structure of the crosslinker or the thermodynamic quality of the porogenic agent used.

The particle size distributions as well as the diameter values of the precursor/hybrid microparticles analyzed using laser diffractometry are shown in Figure S3 and Table 2, respectively.

Table 2. Diameters of precursor/hybrid microparticles.

	AE	AEHA	AD	ADHA	AT	ATHA
D_m (μm)	124	135	165	173	184	194

As can be seen from Table 2, the precursor microparticles are micrometric in size, their diameter being influenced by the chemical structure of the crosslinker, i.e., they increase with increasing chain length between the methacrylic groups in the crosslinking agent. Additionally, the hybrid microparticles have larger diameters than the precursor microparticles, leading to the idea that HA has reacted with the epoxy groups to form a layer on the surface covering these microparticles, generating a core-shell structure.

3.2.3. Thermogravimetric Analysis

Thermogravimetric studies were carried out to obtain additional information about the precursor/hybrid microparticles.

Table 3 shows the thermogravimetric characteristics of the precursor/hybrid microparticles, namely: the degradation steps, temperature range for each degradation step, residual mass, activation energy (E_a) and reaction order (n).

Table 3. Thermogravimetric characteristics of precursor/hybrid microparticles.

Sample Code	Decomposition Temperature			Residual Mass (%)	E_a (kJ·mol^{-1})	n	R^2
	T_i (°C)	T_m (°C)	T_f (°C)				
AE	180	240	260	5.181	121	1.7	0.992
	270	332	375		179	1.8	0.992
	380	411	430		327	1.3	0.998
AD	188	209	222	1.57	149	1.7	0.996
	249	321	349		171	1.8	0.994
	349	417	439		223	1.8	0.994
AT	142	150	154	22.88	114	1.9	0.996
	212	242	308		173	1.8	0.992
	381	410	442		192	1.7	0.993
AEHA	180	231	240	14.68	133	1.4	0.991
	280	341	360		194	1.7	0.992
	380	416	460		425	2.6	0.994
ADHA	185	199	252	1.38	159	1.7	0.993
	252	291	342		174	1.9	0.994
	342	411	432		256	1.7	0.994
ATHA	72	94	115	36.30	87	1.7	0.997
	213	253	273		239	1.8	0.996
	292	299	321		323	1.8	0.997
	361	407	434		476	1.8	0.996
HA	65	102	135	34.82	62	1.4	0.993
	225	263	309		127	1.7	0.997
	309	411	510		310	1.9	0.997

The thermal behavior of precursor microparticles and sodium hyaluronate is characterized by three stages of thermal decomposition. The first degradation step between 65 and 135 °C (HA), 180–260 °C (AE), 188–222 °C (AD) and 142–154 °C (AT) is characterized by weight losses of 6.20% (HA), 15% (AE), 9.88% (AD) and 6.32% (AT), which are associated in the case of precursor microparticles with the loss of solvents retained in the crosslinked mesh of their structure. The second stage of degradation occurs in the temperature ranges of 225–263 °C (HA), 270–375 °C (AE), 249–349 °C (AD) and 212–308 °C (AT) and is characterized by the highest amount of weight loss: 38.16% (HA), 73.59% (AE), 76.21% (AD) and 38.37% (AT). In this stage, the breakage of the labile bonds occurs first, followed by the destruction of the crosslinked network by the cleavage of the macromolecular chains. The third stage of degradation is in the temperature ranges. 309–510 °C (HA), 380–411 °C (AE), 349–439 °C (AD) and 381–442 °C (AT) with mass losses of 18.21% (HA), 6.46% (AE), 12.34% (AD) and 31.76% (AT).

In the case of the hybrid microparticles, the presence of sodium hyaluronate leads to a slight increase in thermal stability compared to that of the precursor microparticles. The thermal degradation of the microparticles occurs in three steps for AEHA and ADHA microparticles and in four steps for ATHA microparticles. Additionally, as in the case of the precursor microparticles, the second degradation step is characterized by the highest amount of weight loss: 61.27% (AEHA), 67.9% (ADHA) and 28.55% (ATHA). The fourth degradation step specific only to the ATHA microparticles located in the temperature range 361–434 °C is characterized by a weight loss of 19.80%.

The kinetic parameters (activation energy and reaction order) for each thermal decomposition step were determined using the Urbanovici–Segal integral method [22]. If we consider for comparison the second degradation step, which is the step characterized by the highest weight loss, it can be observed that the activation energies for the precursor microparticles have close values (179 kJ·mol^{-1} (AE), 171 kJ·mol^{-1} (AD) and 173 kJ·mol^{-1}

(AT)), so the chemical structure of the crosslinker does not influence the way the microparticle degradation takes place. In the case of the hybrid microparticles, however, the activation energies are different (194 kJ·mol^{-1} (AEHA), 174 kJ·mol^{-1} (ADHA), 239 kJ·mol^{-1} (ATHA)) leading to the idea that HA grafting to epoxy groups produces polymeric materials with different chemical structures and thermal stabilities depending on the degree of grafting of the polysaccharide.

3.2.4. Determination of Epoxy Groups

Since hybrid microparticles are obtained by grafting HA to the epoxy groups found in the precursor microparticle structure, it is important to determine their content in the microparticle structure before and after the grafting reaction. The HBr-glacial acetic acid titrimetric method was chosen for the determination of the epoxy groups. The reaction between halogenated acids and the epoxide group results in the opening of the three-atom ring and the formation of a hydroxyl functional group. Table 4 shows the results of the titrimetric method for the determination of the epoxy groups.

Table 4. The values of epoxy groups obtained theoretically and experimentally by titration.

Sample Code	Epoxy Groups			
	Theoretical		Experimental	
	mmol·g^{-1}	%	mmol·g^{-1}	%
AE	4.80	20.71	2.82	12.17
AD	4.67	20.10	2.55	10.98
AT	4.54	19.52	3.22	13.86
AEHA			1.01	4.32
ADHA			1.56	6.71
ATHA			0.30	1.30

From Table 4, it can be seen that the theoretical values obtained for the epoxy groups are higher than those obtained experimentally by titration. By titration, only the epoxide groups can be determined, which are accessible to the HBr reaction, especially those on the surface and those on the layers which are very close to the surface, which partly explains these differences. Another factor to be taken into account is the difference in the chemical composition of the copolymers compared to the starting monomers. For AEHA, ADHA and ATHA microparticles, the number of epoxy groups is reduced due to HA grafting by the ring-opening reaction of the epoxy groups found on the microparticle surface or in the surface layers. However, a small percentage of these epoxy groups remain unreacted, probably due to their reduced accessibility of HBr. From the data in Table 4, it can be seen that grafting was best performed on AT microparticles, the results of which are also in agreement with the amount of grafted HA calculated by the gravimetric method.

3.3. Morphological Characterization
3.3.1. Scanning Electron Microscopy

The size, shape and surface morphology of the synthesized polymeric materials were analyzed using scanning electron microscopy. Figures 6 and 7 show micrographs of the precursor/hybrid microparticles, and for easy comparison, SEM images were taken at the same magnification (×5000 for surface structures (small images) and ×500 for microparticle overview (large images)).

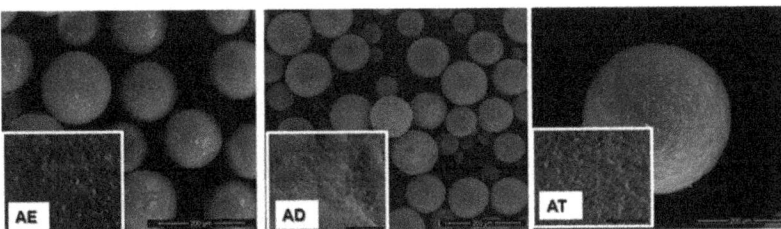

Figure 6. SEM micrographs of precursor microparticles.

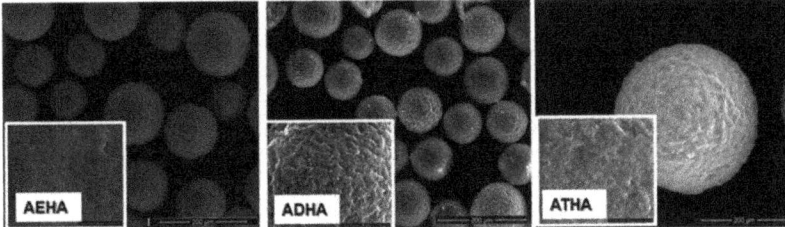

Figure 7. SEM micrographs of hybrid microparticles.

The SEM micrographs show that by the chosen synthesis method, spherical particles of micrometric dimensions are obtained and that the surface morphology is influenced by the chemical structure of the crosslinking agent used. Thus, as the alkyl chain between the two methacrylic groups increases, precursor microparticles with a more pronounced porous structure and a rougher surface are obtained. The interaction of the precursor microparticles with sodium hyaluronate results in hybrid microparticles that retain their spherical shape, but the surface morphology changes due to the deposition of a polymer layer on the surface of the precursor microparticles, confirming once again that the grafting reaction of the polysaccharide to the epoxy groups has taken place.

3.3.2. Atomic Force Microscopy

Atomic force microscopy was also used to investigate the surface morphology of the precursor/hybrid microparticles, obtaining information on surface roughness, pore size and geometry. AFM images for AE and AEHA microparticles are shown in Figure 8 as an example.

The AFM images correlate well with those from the scanning electron microscopy, revealing differences in surface morphology between the precursor and hybrid microparticles. Table 5 shows the values of the parameters characteristic of microparticle surfaces.

From the data in Table 5, it can be seen that grafting HA onto precursor microparticles has the effect of decreasing the surface roughness, but also the size of existing pores on the surface. The negative values of S_{sk}, a statistical parameter that gives us information about the degree of asymmetry of the distribution of heights on the surface [23], indicate that the two types of microparticles analyzed show porous structures. Additionally, S_{ku} values lower than three confirm that the microparticles present irregular surfaces with various roughness. All these observations reinforce and confirm the conclusions drawn from the SEM analysis. Shape and elongation factors are two important parameters, with which pore structures can be analyzed [24]. The values of these parameters shown in Table 5 indicate that the precursor/hybrid microparticles have elliptical-shaped pores with an irregular outline.

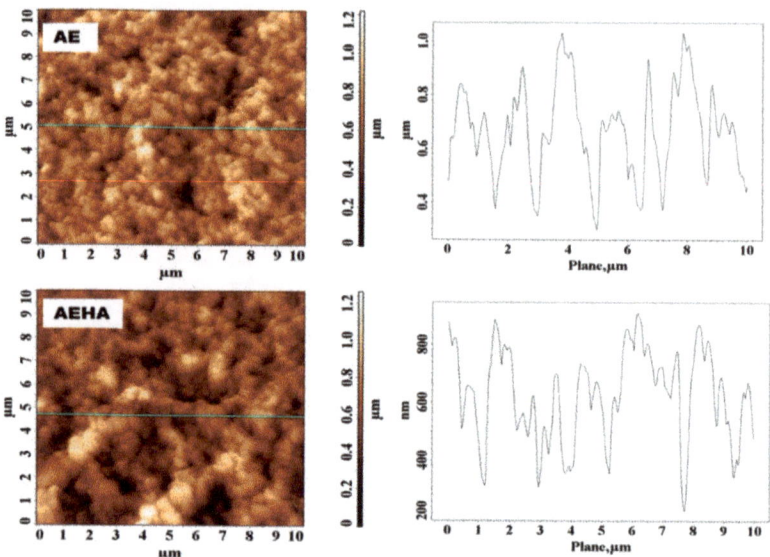

Figure 8. AFM images (2D) and cross-section profile of AE and AEHA microparticles.

Table 5. Surface profile parameters for precursor/hybrid microparticles.

Sample Code	S_a (nm)	S_q (nm)	S_{sk}	S_{ku}	d_{med} (nm)	f_{shape}	$f_{elongation}$
AE	124.2	131.3	−0.686	0.310	314	0.363	0.485
AD	143.3	156.1	−0.423	0.486	369	0.391	0.426
AT	156.8	172.5	−0.143	0.532	265	0.426	0.375
AEHA	95.4	110.2	−0.514	0.195	238	0.174	0.228
ADHA	112.9	129.9	−0.365	0.371	280	0.248	0.269
ATHA	126.6	143.7	−0.089	0.498	176	0.323	0.280

3.3.3. Specific Parameters for Characterizing the Morphology of Porous Structures

The introduction of a pore-forming substance, known as a porogenic agent or diluent, into the organic phase of the reaction system specific to suspension polymerization leads to the formation of permanently heterogeneous structures, i.e., structures containing pores after drying. A highly effective porogenic agent should not react during polymerization but should remain within the microparticle structure until the end of the reaction. When it is removed by extraction, the sites occupied by the porogen become the pores of the crosslinked networks.

Table 6 shows the values of the specific parameters determined to characterize the morphology of the precursor/hybrid microparticles.

Table 6. Porosity parameters of precursor/hybrid microparticles.

Sample Code	VP (mL·g^{-1})	P (%)	S_{sp} (m^2·g^{-1})
AE	0.7073	45	78
AD	0.5076	37	54
AT	0.9363	52	92
AEHA	0.6004	41	120
ADHA	0.4461	34	86
ATHA	0.7668	47	160

From the data presented in Table 6, it can be seen that the pore volume and porosity of the hybrid microparticles increase with the increasing alkyl chain, except for the AD microparticles, whose values decrease. When DEGDMA is used as a crosslinking agent, probably during the crosslinking radical polymerization process, a decrease in the apparent reactivity of the double pendant groups occurs due to steric factors, resulting in the appearance of internal cyclizations and thus the formation of microparticles with more compact structures characterized by lower values of both porosity and pore volume. In the case of the AEHA, ADHA and ATHA microparticles, decreases in pore volume and porosity values compared to those of precursor microparticles are observed, which is due to HA grafting to the epoxy groups on the microparticle surface. Through the grafting reaction, HA coats part of the pores or penetrates into the pores, decreasing their size. This is confirmed by the information obtained by the AFM method as well as by the SEM micrographs. The decrease in pore size was also observed by the AFM method as follows:

- from 314 to 238 nm for the AE–AEHA microparticle system;
- from 369 to 280 nm for the AD–ADHA microparticle system;
- and from 265 to 176 nm for the AT–ATHA microparticle system.

It is also observed that the specific surface area values are higher for hybrid microparticles. This can be explained by the fact that S_{sp} was determined by the dynamic vapor sorption method, and hybrid microparticles due to their hydrophilic structure have a higher capacity to absorb water than precursor microparticles. Additionally, the higher values of the specific surface are due to the fact that the hybrid microparticles have smaller pore sizes than the corresponding precursor microparticles.

The morphology of the pore structures is influenced by the structure and concentration of the monomers, the nature of the porogenic agent and in particular the amount of porogenic agent used. Thus, Figure 9 shows graphical representations of the specific surface area and pore volume values, depending on the amount of porogenic agent used to obtain the AT and ATHA microparticles.

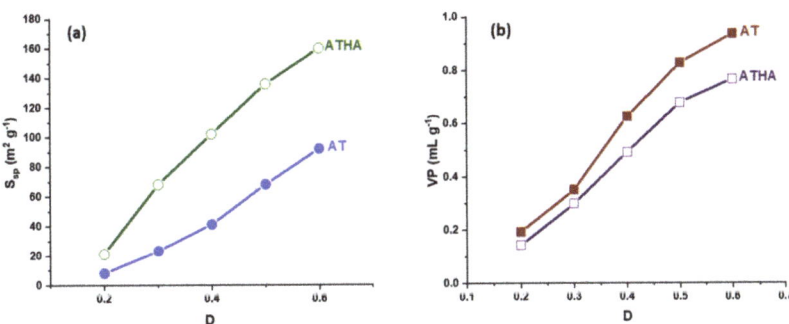

Figure 9. The influence of the amount of porogenic agent on the specific surface area (**a**) and the pore volume (**b**) for AT and ATHA microparticles.

From Figure 9, it can be seen that microparticles with higher porosity structures and more specific surface values are obtained when the amount of porogen agent is increased. For this reason, for the preparation of the precursor microparticles, the dilution was chosen to be 0.6.

3.4. Swelling Capacity of Precursor/Hybrid Microparticles in Aqueous Media

Graphical representations of the degree of swelling of precursor/hybrid microparticles versus time in aqueous media with different pH values are shown in Figure 10.

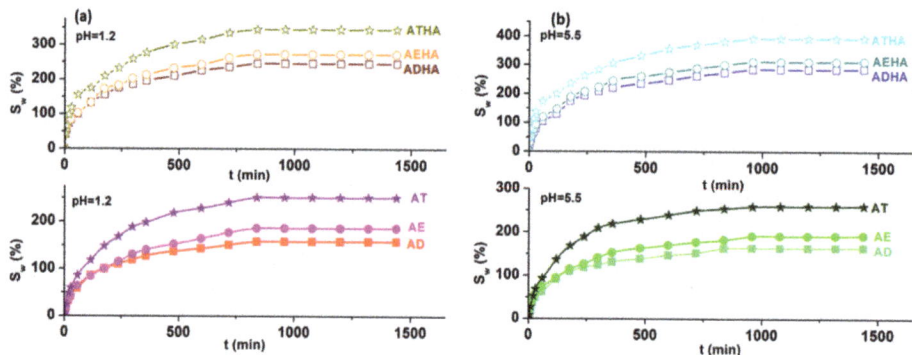

Figure 10. Time dependence of the degree of swelling in pH = 1.2 (**a**) and pH = 5.5 (**b**) for precursor/hybrid microparticles.

Figure 10 shows that the swelling process is carried out in three stages. In the first stage, the rapid absorption of aqueous solutions of different pH values into the structure of the precursor/hybrid microparticles takes place. The second stage is characterized by a slower absorption, and in the third stage the equilibrium of the swelling process is reached. For AE, AD and AT microparticles, an equilibrium is reached after 960 min and for hybrid microparticles, an equilibrium is reached after 840 min.

The swelling degree of precursor and hybrid microparticles determined with Equation (12) depends on the pore structure and the presence of HA on the microparticle surface. Thus, the degree of swelling for precursor microparticles is not influenced by the pH value of the swelling medium and increases in the order $S_{W,AD} < S_{W,AE} < S_{W,AT}$, similar to the increase in the specific surface values (Table 6). Thus, AT microparticles characterized by a high specific surface area (160 m$^2 \cdot$g^{-1}) show the highest degree of swelling. In the case of the hybrid microparticles, it is observed that the degree of swelling is higher than that corresponding to the precursor microparticles. This is explained by the presence of HA on the surface of the hybrid microparticles, which is a hydrophilic polymer having several -OH groups in its structure. The swelling degrees of the hybrid microparticles in pH = 1.2 are lower than in pH = 5.5. This behavior can be explained by the fact that the COO- groups of the HA molecule in the acidic medium are transformed into COOH groups, thus reducing the electrostatic repulsion between them, and consequently the polymer matrix swells less.

Two mathematical models were used to describe the swelling mechanism in media with different pH values, namely:

1. The second-order kinetic model using the equations [25]:

$$\frac{t}{SR} = \frac{1}{K_S \cdot S_{eq}^2} + \frac{t}{S_{eq}} \tag{16}$$

$$SR\left(g \cdot g^{-1}\right) = \frac{W_t - W_0}{W_0} \tag{17}$$

$$S_{eq}\left(g \cdot g^{-1}\right) = \frac{W_{eq} - W_0}{W_0} \tag{18}$$

where W_0, W_t and W_{eq} are the amount of precursor/hybrid microparticles at time $t = 0$, $t = t$ and at equilibrium, respectively [26]. The straight-line plots of t/SR versus t gave the slope of S_{eq} and intercept K_S.

2. Korsmeyer–Peppas model. The linear form of the Korsmeyer–Peppas equation [27] is given as:

$$lnF = lnK + n \cdot lnt \tag{19}$$

where $F = M_t/M_\infty$, M_t—the amount of water uptake at time t; M_∞—the amount of water uptake at time approaching infinity; K—the swelling rate constant; and n—diffusion exponent characteristic for the transport mechanism. The values of K and n were determined from the linear plots of lnF versus lnt.

The values of the kinetic parameters obtained by applying the two models are shown in Table 7.

Table 7. Kinetic parameters of the swelling process in various aqueous solutions.

Sample Code	Second-Order Model				Korsmeyer–Peppas Model		
	S_{exp} (g·g^{-1})	K_S (g·g^{-1})	S_{eq} (g·g^{-1})	R^2	K	n	R^2
pH = 1.2							
AE	1.86	0.0033	2.07	0.997	0.282	0.172	0.999
AD	1.58	0.0026	1.73	0.998	0.212	0.164	0.998
AT	2.51	0.0045	2.80	0.997	0.315	0.225	0.998
AEHA	2.73	0.0035	2.93	0.997	0.302	0.145	0.998
ADHA	2.46	0.0030	2.61	0.997	0.216	0.141	0.999
ATHA	3.45	0.0050	3.69	0.997	0.325	0.239	0.999
pH = 5.5							
AE	1.92	0.0028	2.09	0.997	0.234	0.200	0.997
AD	1.64	0.0025	1.77	0.997	0.205	0.195	0.999
AT	2.60	0.0030	2.84	0.998	0.266	0.237	0.998
AEHA	3.12	0.0039	3.35	0.997	0.248	0.188	0.997
ADHA	3.08	0.0030	2.86	0.997	0.209	0.174	0.997
ATHA	3.93	0.0054	4.19	0.998	0.294	0.242	0.998

From the data in Table 7, it can be seen that there is a good correlation between the experimental (S_{exp}) and calculated (S_{eq}) values and the correlation coefficient R^2 values are greater than 0.997. These results suggest that the swelling mechanism of precursor/hybrid microparticles in aqueous media with different pH values follows the second-order kinetic model. The values of n were in the range 0.141–0.242, indicating that the most likely swelling mechanism is Fickian.

3.5. Metronidazole Adsorption and Release Studies

In order to achieve an optimal system with a controlled drug release, the influence of the following parameters must be taken into account: pH; contact time; temperature; and initial drug concentration.

Since metronidazole dissolves in acidic pH, the adsorption process of metronidazole on the precursor/hybrid microparticles was performed from an aqueous solution with pH = 1.2.

Additionally, the effect of contact time is very important for assessing the affinity of precursor/hybrid microparticles for the model drug. Figure 11 shows the influence of contact time on the adsorption capacity of metronidazole on precursor/hybrid microparticles for a metronidazole concentration of 0.5×10^{-3} g·mL^{-1} at a temperature of 25 °C.

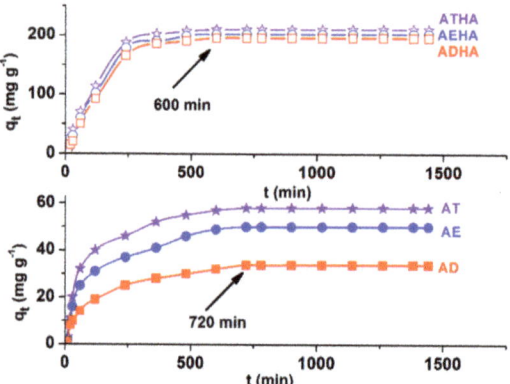

Figure 11. Influence of contact time on the adsorption capacity of metronidazole on precursor/hybrid microparticles.

For precursor microparticles, the contact time for reaching the equilibrium is 720 min, while for hybrid microparticles, the equilibrium is reached at 600 min. Above these contact time values, the amount of drug adsorbed on the precursor/hybrid microparticles remains constant. The shorter time to reach equilibrium indicates a better affinity of the hybrid microparticles for metronidazole compared to that of the precursor microparticles.

Temperature is another important parameter to be taken into account when adsorbing drugs on different polymeric supports, with metronidazole adsorption studies being carried out at 25, 30 and 35 °C (Figure 12a).

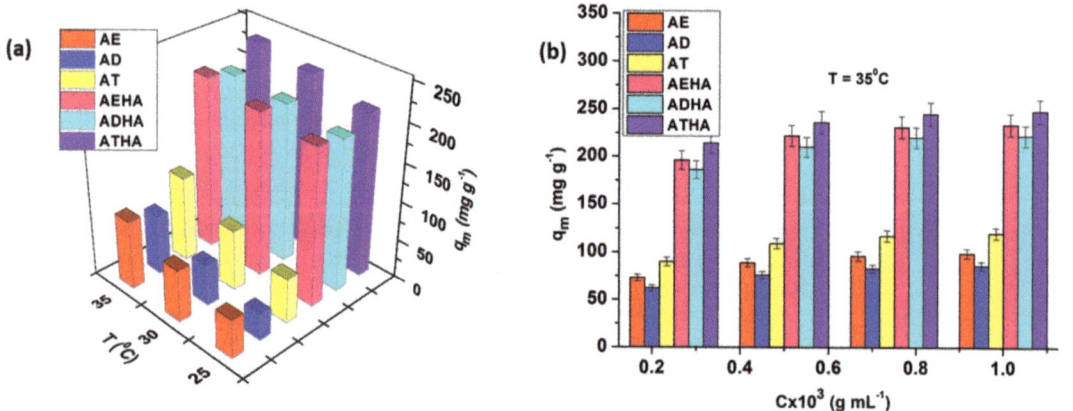

Figure 12. Influence of temperature (a) and the initial concentration of metronidazole (b) on the adsorption capacity of the drug onto precursor/hybrid microparticles.

Analyzing the graphical representation in Figure 12a, it can be seen that the adsorption of the drug is favored by increasing temperature, an effect that is absolutely expected since the process is of a diffusional nature, causing an increase in the degree of swelling and thus in the diffusion rate of metronidazole into the pores of the precursor/hybrid microparticles.

Increasing the concentration of the drug has the effect of increasing the rate of adsorption. It was also observed that drug adsorption was achieved in higher amount in the case of hybrid microparticles compared to that of precursor microparticles (Figure 12b). This phenomenon is explained by the presence of sodium hyaluronate, which has the role of enhancing the hydrophilicity of the microparticles, leading to a higher degree of swelling and consequently to the adsorption of a higher amount of the drug. In the case of metronidazole adsorption, in an acidic pH mainly physical interactions take place, such as hydrogen bonding between the OH group of metronidazole and the -COOH and -OH groups of the hybrid microparticle structure. The greatest amount of immobilized drug was obtained in the case of the ATHA hybrid microparticles.

In-depth studies of the adsorption process were carried out considering two physicochemical aspects, namely: the adsorption equilibrium by means of adsorption isotherms, which quantify the interaction between the drug and the support; and adsorption kinetics, which can explain the mechanism of drug adsorption on the solid supports.

3.5.1. Adsorption Equilibrium Studies

For efficient polymer–drug systems, it is important to know how the adsorbate (drug solution) and adsorbent (precursor/hybrid microparticles) interact. For this purpose, the description of the adsorption equilibrium of metronidazole on precursor/hybrid microparticles was performed using the mathematical models of Langmuir, Freundlich, Dubinin–Radushkevich (two-parameter models), Sips and Khan (three-parameter models) isotherms.

The nonlinear forms of the isotherms used can be written as follows:

1. Two parameter isotherm models:
   - Langmuir isotherm [28]:
   $$q_e = \frac{q_m \cdot K_L \cdot C_e}{1 + K_L \cdot C_e} \quad (20)$$

   - Freundlich isotherm [29]:
   $$q_e = K_F \cdot C_e^{\frac{1}{n_f}} \quad (21)$$

   - Dubinin–Radushkevich isotherm [30]:
   $$q_e = q_m \exp\left(-K_D \cdot \epsilon_D^2\right) \quad (22)$$

   where q_e is the metronidazole amount adsorbed at equilibrium (mg·g^{-1}); q_m is the maximum adsorption capacity (mg·g^{-1}); K_L is the Langmuir constant that reflects the affinity between the adsorbate and the adsorbent (L·g^{-1}); K_F is the adsorption capacity for a unit's equilibrium concentration (L·g^{-1}); $1/n_F$ is a constant that suggests the favorability and capacity of the adsorbent–adsorbate system; ϵ is the Polanyi potential; and K_D is the constant which is related to the calculated average sorption energy E (kJ·mol^{-1}). The constant K_D can give the valuable information regarding the mean energy of adsorption by the equation:
   $$E = (-2K_D)^{1/2} \quad (23)$$

2. Three-parameter isotherm models:
   - Sips isotherm [31]:
   $$q_e = \frac{q_m \cdot K_S \cdot C_e^{n_S}}{1 + K_S \cdot C_e^{n_S}} \quad (24)$$

   - Khan isotherm [32]:
   $$q_e = q_m \frac{b_K \cdot C_e}{(1 + b_K \cdot C_e)^{n_K}} \quad (25)$$

   where K_S is the Sips constant (L·mg^{-1}); n_S is the Sips model exponent; b_K is the Khan model constant; and n_K is the Khan model exponent.

Sips and Khan isotherms represent the combined features of the Langmuir and Freundlich isotherm equations. Thus, in the case of a low adsorbent concentration, the Sips isotherm is reduced to the Freundlich isotherm, while at high adsorbate concentrations, it shows the characteristics of the Langmuir isotherm [33] Additionally, if $n_K = 1$, Equation (24) can be simplified to the Langmuir isotherm equation, whereas if $n_K \cdot C_e \gg 1$, Equation (25) can be approximated by the Freundlich type isotherm equation [32].

The isotherm model plots of metronidazole adsorption onto precursor/hybrid microparticles are illustrated in Figure 13, while the parameters and the statistical error functions values (R^2 and χ^2) are presented in Tables 8 and 9.

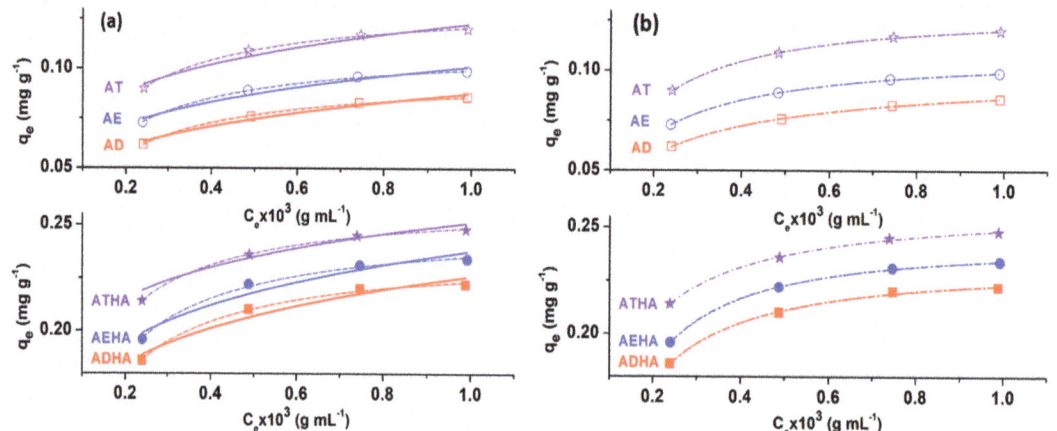

Figure 13. The results of nonlinear fits of Langmuir (dash line); Freundlich (solid line); Dubinin–Radushkevich (short dot line) (**a**); Sips (dash dot line); and Khan (short dash line). (**b**) Isotherm for metronidazole adsorption on precursor/hybrid microparticles at T = 35 °C.

Table 8. Two- and three-parameter isotherm values for adsorption of metronidazole onto AE, AD and AT microparticles.

Isotherm Model	Parameter	AE			AD			AT		
		25 °C	30 °C	35 °C	25 °C	30 °C	35 °C	25 °C	30 °C	35 °C
Langmuir	q_m (mg·g^{-1})	72.93	94.66	112.25	55.56	66.23	92.22	84.89	103.02	128.43
	K_L (L·g^{-1})	5.24	5.98	6.79	4.07	4.64	5.17	6.31	7.63	9.24
	R^2	0.997	0.996	0.998	0.999	0.997	0.996	0.995	0.998	0.999
	χ^2	0.143	0.566	0.163	0.025	0.218	0.356	0.891	0.360	0.161
Freundlich	K_F (L·g^{-1})	60.70	80.25	101.08	43.16	62.32	87.65	70.28	95.68	122.57
	$1/n_F$	0.32	0.29	0.21	0.38	0.28	0.23	0.32	0.31	0.20
	R^2	0.949	0.947	0.952	0.972	0.969	0.968	0.941	0.956	0.941
	χ^2	52.148	83.909	64.88	14.212	25.898	36.104	79.554	101.484	107.284
Dubinin-Radushkevich	q_m (mg·g^{-1})	65.39	85.82	106.01	47.16	66.23	92.28	75.73	103.02	128.43
	E (kJ·mol^{-1})	3.80	4.06	4.86	3.47	3.91	4.67	3.86	4.12	4.97
	R^2	0.999	0.999	0.999	0.996	0.997	0.997	0.998	0.998	0.999
	χ^2	0.293	0.285	0.252	0.257	0.186	0.162	0.335	0.064	0.061
Sips	q_m (mg·g^{-1})	67.14	87.98	108.98	55.18	72.91	100.08	75.82	108.28	129.30
	K_S (L·mg^{-1})	5.31	6.03	7.17	4.25	4.88	5.25	7.30	8.94	10.30
	n_S	1.15	1.13	1.13	1.01	0.95	0.98	1.15	1.16	1.14
	R^2	0.999	0.998	0.998	0.999	0.999	0.997	0.999	0.997	0.998
	χ^2	0.393	0.209	0.094	0.214	0.275	0.239	0.250	0.619	0.154
Khan	q_m (mg·g^{-1})	85.51	100.91	120.95	57.11	71.42	95.51	94.88	113.90	144.25
	b_K	2.11	3.17	4.38	1.05	2.46	3.41	2.66	3.03	5.88
	n_K	1.09	1.01	1.04	1.01	0.99	0.99	1.10	1.10	1.08
	R^2	0.998	0.999	0.998	0.998	0.999	0.997	0.999	0.998	0.997
	χ^2	0.143	0.097	0.168	0.177	0.043	0.125	0.037	0.142	0.166

Table 9. Two- and three-parameter isotherm values for adsorption of metronidazole onto AEHA, ADHA and ATHA microparticles.

Isotherm Model	Parameter	AEHA 25 °C	AEHA 30 °C	AEHA 35 °C	ADHA 25 °C	ADHA 30 °C	ADHA 35 °C	ATHA 25 °C	ATHA 30 °C	ATHA 35 °C
Langmuir	q_m (mg·g^{-1})	227.75	237.60	250.99	221.75	230.22	238.39	233.79	251.62	262.05
	K_L (L·g^{-1})	21.20	24.71	29.17	12.98	14.68	16.74	23.08	28.32	34.58
	R^2	0.995	0.996	0.998	0.995	0.996	0.998	0.997	0.998	0.998
	χ^2	0.721	0.579	0.369	0.608	0.256	0.149	0.142	0.134	0.156
Freundlich	K_F (L·g^{-1})	216.32	225.40	237.72	209.33	218.16	225.80	223.80	240.63	251.10
	$1/n_F$	0.13	0.12	0.12	0.13	0.13	0.12	0.11	0.10	0.10
	R^2	0.934	0.927	0.923	0.923	0.929	0.930	0.950	0.920	0.947
	χ^2	153.564	181.197	228.701	180.312	166.542	190.317	92.568	171.073	123.686
Dubinin-Radushkevich	q_m (mg·g^{-1})	282.36	300.36	312.01	256.49	279.21	295.56	296.41	311.17	320.38
	E (kJ·mol^{-1})	6.24	6.31	6.38	6.00	6.27	6.33	6.71	6.77	6.97
	R^2	0.999	0.998	0.997	0.998	0.997	0.998	0.999	0.997	0.998
	χ^2	0.087	0.603	0.207	0.066	0.289	0.125	0.212	0.209	0.057
Sips	q_m (mg·g^{-1})	222.61	230.17	242.42	213.49	223.37	232.16	233.96	244.28	260.19
	K_S (L·mg^{-1})	22.67	26.47	30.10	12.09	14.23	17.39	23.75	29.90	35.34
	n_S	1.14	1.22	1.15	1.17	1.19	1.14	0.99	1.06	1.07
	R^2	0.997	0.999	0.998	0.998	0.999	0.997	0.997	0.997	0.997
	χ^2	0.348	0.130	0.267	0.072	0.182	0.391	0.254	0.186	0.172
Khan	q_m (mg·g^{-1})	233.19	240.25	254.73	227.07	232.27	241.19	241.44	257.32	269.98
	b_K	2.53	3.86	4.92	1.76	2.90	3.84	3.79	4.52	6.95
	n_K	1.03	1.04	1.05	1.06	1.04	1.04	1.00	1.04	1.01
	R^2	0.997	0.999	0.998	0.998	0.999	0.997	0.998	0.997	0.997
	χ^2	0.323	0.136	0.191	0.137	0.074	0.185	0.126	0.153	0.167

From the analysis of the data presented in Tables 8 and 9, it can be seen that:

- The values of the maximum adsorption capacity, q_m, calculated based on the Langmuir, Dubinin–Radushkevich, Sips and Khan models, are close to the experimental values;
- With increasing temperatures, the saturation capacity increases, indicating a better accessibility to the adsorption centers on the surface of the precursor/hybrid microparticles;
- The Langmuir constant values increase with increasing temperatures, thus showing a higher metronidazole adsorption efficiency at higher temperatures;
- The highest value of K_L was obtained for the ATHA microparticles;
- The values of the constant $1/n_F$ are in the range between 0 and 1, which would indicate that the Freundlich isotherm is favorable for metronidazole adsorption on precursor/hybrid microparticles;
- The values of E are in the range of 3.47–6.97 kJ/mol, indicating that the adsorption process of metronidazole on both the precursor and hybrid microparticles is of a physical nature;
- The values of the exponents n_S and n_K are very close to unity, which provides a further argument that the adsorption process of metronidazole on precursor/hybrid microparticles is better suited to the Langmuir model than to the Freundlich model;
- The value of the Khan constant, b_K, increases with increasing temperatures and has the highest values when using the ATHA hybrid microparticles;
- The values close to unity for the correlation coefficient R^2 that are associated with low values of the χ^2 test indicate that the Langmuir, Dubinin–Radushkevich, Sips and Khan isotherms apply quite well to the experimental data obtained for metronidazole adsorption on precursor/hybrid microparticles;
- Lower values of R^2 and higher values of χ^2 obtained from the application of the Freundlich isotherm indicate that this isotherm does not describe the experimental data well.

3.5.2. Kinetic Studies

In order to investigate the mechanism of metronidazole adsorption on precursor/hybrid microparticles, the experimental data were interpreted using four mathematical models, namely: the Lagergren model (pseudo-first order kinetic model), the Ho model (pseudo-second order kinetic model), the Elovich model and the Weber–Morris intraparticle diffusion model. The nonlinear forms of the Lagergren (Equation (26)) [34], Ho (Equation (27)) [35] and Elovich models (Equation (28)) [36], as well as the linear form of the Weber–Morris model (Equation (29)) [37] are written below:

$$q_t = q_e\left(1 - e^{-k_1 t}\right) \quad (26)$$

$$q_t = \frac{k_2 \cdot q_e^2 \cdot t}{1 + k_2 \cdot q_e \cdot t} \quad (27)$$

$$q_t = \frac{1}{\beta}\ln(1 + \alpha \cdot \beta \cdot t) \quad (28)$$

$$q_t = k_{id} \cdot t^{0.5} + C_i \quad (29)$$

where k_1 is the rate constant of the pseudo-first order model (min^{-1}); k_2 is the rate constant of the pseudo-second order model (g·mg^{-1}·min^{-1}); α is the initial adsorption rate (mg·g^{-1}·min^{-1}); β is the desorption constant (g·mg^{-1}); k_{id} is the intraparticle diffusion rate constant (g·mg^{-1}·min$^{-0.5}$); and C_i is the constant that gives information about the thickness of the boundary layer.

Figure 14a,b presents the nonlinear plots of the Lagergren, Ho and Elovich models as well as the straight-line plots of the Weber–Morris model in case of metronidazole adsorption ($C_{metronidazole}$ = 1 × 10^{-3} g·mL^{-1}) on the precursor/hybrid microparticles at 35 °C.

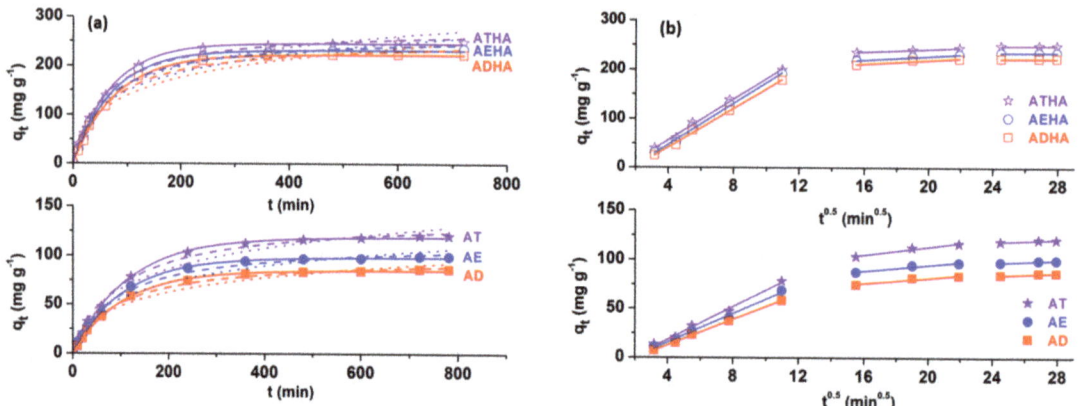

Figure 14. Lagergren (solid line), Ho (dash line), Elovich (dot line) models (**a**) and Weber–Morris intraparticle diffusion model (**b**) for metronidazole adsorption on precursor/hybrid microparticles ($C_{metronidazole}$ = 1 × 10^{-3} g·mL^{-1}, T = 35 °C).

The kinetic parameters obtained from the Lagergren, Ho, Elovich and Weber–Morris models are presented in Tables 10 and 11.

Table 10. Kinetic parameters for adsorption of metronidazole onto AE, AD and AT microparticles.

	AE			AD			AT		
	25 °C	30 °C	35 °C	25 °C	30 °C	35 °C	25 °C	30 °C	35 °C
$q_{e,exp}$ (mg·g^{-1})	59.00	78.00	99.00	42.00	61.00	86.00	68.00	93.00	120.00
				Lagergren model					
$q_{e,calc}$ (mg·g^{-1})	57.97	76.99	97.92	41.78	60.44	84.60	67.28	91.68	118.53
$k_1 \times 10^2$ (min^{-1})	0.953	0.989	1.00	0.80	0.89	0.95	1.01	1.13	1.23
R^2	0.998	0.998	0.998	0.997	0.998	0.998	0.998	0.999	0.999
χ^2	1.228	1.317	1.257	1.248	1.209	1.228	0.656	0.311	0.431
				Ho model					
$q_{e,calc}$ (mg·g^{-1})	66.53	89.09	113.94	48.65	72.01	98.64	76.47	106.10	139.07
$k_2 \times 10^5$ (g·mg^{-1}·min^{-1})	10.88	12.01	22.68	9.54	10.84	19.51	7.16	13.01	18.92
R^2	0.991	0.995	0.994	0.989	0.996	0.996	0.987	0.995	0.995
χ^2	4.629	4.128	4.554	9.230	4.201	6.215	9.396	6.081	9.206
				Elovich model					
α (mg·g^{-1}·min^{-1})	1.37	1.62	1.93	0.80	0.89	1.62	1.83	1.93	2.11
β (g·mg^{-1})	0.06	0.04	0.03	0.08	0.05	0.04	0.06	0.04	0.03
R^2	0.964	0.975	0.974	0.963	0.981	0.977	0.953	0.975	0.979
χ^2	59.661	24.428	40.029	50.971	11.864	27.106	34.217	33.678	46.254
				Weber–Morris intraparticle diffusion model					
k_{id} (mg·g^{-1}·min$^{-0.5}$)	1.10	1.18	1.27	0.44	0.47	0.54	1.42	1.49	1.65
C_{i2}	25.10	36.30	44.37	22.62	33.21	40.90	29.03	39.94	48.96
R^2	0.994	0.995	0.996	0.993	0.994	0.994	0.995	0.995	0.996

Table 11. Kinetic parameters for adsorption of metronidazole onto AEHA, ADHA and ATHA microparticles.

	AEHA			ADHA			ATHA		
	25 °C	30 °C	35 °C	25 °C	30 °C	35 °C	25 °C	30 °C	35 °C
$q_{e,exp}$ (mg·g^{-1})	213.00	222.00	234.00	206.00	215.00	221.00	221.00	237.00	248.00
				Lagergren model					
$q_{e,calc}$ (mg·g^{-1})	212.67	221.74	230.96	207.12	213.36	220.58	218.68	234.06	244.72
$k_1 \times 10^2$ (min^{-1})	1.16	1.28	1.34	0.93	0.96	1.10	1.35	1.38	1.43
R^2	0.999	0.998	0.998	0.998	0.998	0.998	0.997	0.998	0.998
χ^2	1.057	1.287	1.268	1.146	1.131	1.164	1.391	1.379	1.354
				Ho model					
$q_{e,calc}$ (mg·g^{-1})	253.13	256.88	262.92	246.05	251.24	253.98	249.82	266.95	276.84
$k_2 \times 10^5$ (g·mg^{-1}·min^{-1})	4.11	5.12	6.16	4.09	4.88	5.82	5.89	6.21	6.89
R^2	0.990	0.989	0.990	0.991	0.988	0.988	0.987	0.993	0.993
χ^2	76.336	89.173	79.617	62.628	92.856	88.002	93.383	63.624	59.873
				Elovich model					
α (mg·g^{-1}·min^{-1})	3.77	5.22	7.27	3.55	4.61	6.14	6.44	7.12	8.58
β (g·mg^{-1})	0.02	0.02	0.02	0.01	0.01	0.01	0.02	0.02	0.02
R^2	0.967	0.960	0.958	0.971	0.959	0.957	0.954	0.962	0.964
χ^2	247.758	325.663	347.937	212.145	312.975	338.627	352.700	319.802	331.668
				Weber–Morris intraparticle diffusion model					
k_{id} (mg·g^{-1}·min$^{-0.5}$)	1.36	1.41	1.56	0.70	0.72	0.86	1.63	1.73	1.81
C_{i2}	34.60	49.61	52.27	31.29	46.44	51.45	40.37	52.30	57.03
R^2	0.995	0.995	0.995	0.994	0.994	0.995	0.995	0.995	0.996

From the data presented in Tables 10 and 11, it can be seen that the calculated adsorption capacity values based on the first-order kinetic model are very close to the experimental values for metronidazole adsorption on the precursor/hybrid microparticles. The values of the rate constant k_1 increase with increasing temperatures, indicating a higher adsorption rate of the drug at higher temperatures. It is also observed that values of R^2 are very close to unity and are associated with low values of χ^2, showing that the first-order kinetic model describes the experimental data quite well. These results suggest that the adsorption of metronidazole on precursor/hybrid microparticles is of a physical nature. By applying the second-order kinetic model, it can be seen that the values of $q_{e,calc}$ are not as close to the values of $q_{e,exp}$ as obtained in the case of applying the first-order kinetic model. The relatively high values of R^2 associated with the high values of χ^2 indicate that the second-order kinetic model does not describe the experimental data well in the case of metronidazole adsorption on the precursor/hybrid microparticles. Additionally, the value of the rate constant k_2 increases with increasing temperatures, again asserting that the rate of drug adsorption is higher at higher temperatures. The lower values of R^2 were correlated with higher values of χ^2 obtained for metronidazole adsorption on precursor/hybrid microparticles; hence, the application of the Elovich model provides a further argument that the metronidazole adsorption is not chemical in nature.

Additionally, from Tables 10 and 11, it can be seen that the C_{i2} values increase with increasing temperatures, thus indicating increasing boundary layer thickness associated with decreasing external mass transfer and increasing internal mass transfer. The highest values of C_{i2} were obtained for hybrid microparticles, confirming that they are good adsorbents. The results obtained by applying the Weber–Morris model lead us to the conclusion that intraparticle diffusion is not the only process influencing the adsorption rate.

3.5.3. Thermodynamic Studies

The adsorbent performance of the precursor/hybrid microparticles was also demonstrated by thermodynamic studies. For this purpose, the following thermodynamic parameters were calculated: Gibbs free energy changes (ΔG), enthalpy change (ΔH) and entropy change (ΔS).

The values of ΔH and ΔS were estimated using Van't Hoff equation [38]:

$$\ln K = \frac{\Delta S}{R} - \frac{\Delta H}{RT} \tag{30}$$

where K—the Langmuir adsorption equilibrium constant obtained at different temperature values [39]; R—the ideal gas constant (8.314 J·mol^{-1}·K^{-1}); and T—temperature in Kelvin.

The ΔG value was calculated using the thermodynamic equation:

$$\Delta G = \Delta H - T \cdot \Delta S \tag{31}$$

According to the data in the literature, the values of ΔH and ΔG can provide information about the type of adsorption process [40]. The linear plot of lnK versus $1/T$ gives us the thermodynamic parameters of the adsorption process of metronidazole on precursor/hybrid microparticles, and their values are shown in Table 12.

Table 12. Thermodynamic parameters.

Sample Code	ΔH (kJ·mol^{-1})	ΔS (J·mol^{-1})	ΔG (kJ·mol^{-1})			R^2
			25 °C	30 °C	35 °C	
AE	19.81	80.28	−4.10	−4.50	−4.90	0.999
AD	18.36	73.31	−3.48	−3.85	−4.21	0.997
AT	29.17	113.21	−4.56	−5.12	−5.69	0.999
AEHA	24.34	107.07	−7.56	−8.09	−8.63	0.997
ADHA	19.40	86.42	−6.34	−6.77	−7.21	0.998
ATHA	30.82	129.53	−7.77	−8.42	−9.07	0.999

From the data presented in Table 12, it can be seen that:
- ΔH values < 40 kJ·mol^{-1} indicate that the interactions between precursor/hybrid microparticles and metronidazole are physical in nature.
- The positive enthalpy value, ΔH, demonstrates that the adsorption process studied is endothermic.
- The negative values of ΔG indicate that the adsorption processes of metronidazole on the precursor/hybrid microparticles are spontaneous and as the temperatures increase, the negative value of the parameter increases in absolute value, which demonstrates that the adsorption of the drug is favorable at higher temperatures.
- The positive values of entropy, ΔS, suggest the affinity of precursor/hybrid microparticles for metronidazole, and this affinity increases with increasing temperatures.

3.5.4. Release Studies

The goal of the research was to obtain a microparticulate system capable of the controlled/sustained release of the adsorbed drug. For this reason, after loading the precursor/hybrid microparticles with metronidazole, kinetic release studies were performed. Release studies have been conducted for the precursor–drug microparticle and respectively for the hybrid–drug microparticle systems containing the highest amount of the included drug. The release profiles are represented in Figure 15.

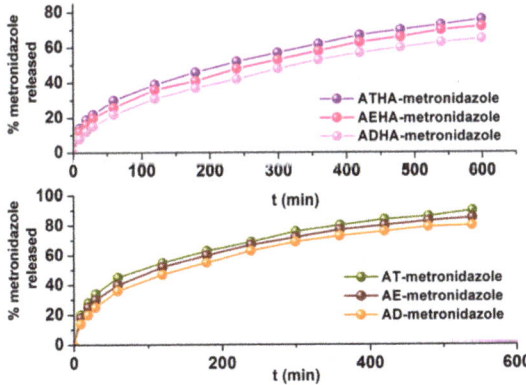

Figure 15. Metronidazole release profiles from precursor/hybrid microparticles (pH = 1.2).

From the graphical representations, it can be seen that the release process of metronidazole from the precursor microparticles occurs at a higher rate than that for the hybrid microparticles.

The interpretation of the metronidazole release kinetics from precursor/hybrid microparticle–drug systems was performed using three mathematical models:
- Higuchi model [41]:

$$Q_t = k_H \cdot t^{0.5} \tag{32}$$

- Korsmeyer–Peppas model [27]:

$$\frac{M_t}{M_\infty} = k_r \cdot t^n \tag{33}$$

- Baker–Lansdale model [42]:

$$\frac{3}{2}\left[1 - \left(1 - \frac{M_t}{M_\infty}\right)^{\frac{2}{3}}\right] - \frac{M_t}{M_\infty} = k_{BL} \cdot t \tag{34}$$

where Q_t—the amount of drug released at time t; k_H—the Higuchi dissolution constant; M_t/M_∞—the fraction of drug released at time t; k_r—the release rate constant that is characteristic for drug–polymeric interactions; n—the diffusion exponent that is characteristic for the release mechanism; and k_{BL}—the release constant.

The values of the release parameters are presented in Table 13.

Table 13. Kinetic release parameters of metronidazole pH = 1.2.

Sample Code	Higuchi Model		Korsmeyer–Peppas Model			Baker–Lansdale Model	
	k_H (min$^{-0.5}$)	R^2	k_r (min^{-n})	n	R^2	K_{BL}	R^2
AE	4.101	0.991	0.082	0.40	0.998	0.047	0.990
AD	3.879	0.990	0.075	0.35	0.997	0.040	0.991
AT	4.258	0.994	0.087	0.42	0.998	0.051	0.989
AEHA	3.052	0.989	0.062	0.61	0.997	0.041	0.989
ADHA	2.736	0.990	0.037	0.57	0.998	0.034	0.988
ATHA	3.208	0.989	0.075	0.63	0.999	0.046	0.989

The rate constants obtained by applying the three kinetic models indicate that the release rate of metronidazole from the precursor microparticles is higher than that for the hybrid microparticles. The different amounts of drug released can be explained by the physical interactions of metronidazole with the functional groups belonging to the chemical structure of the hybrid microparticles. In the case of precursor microparticles, the drug is retained in a larger quantity on the surface and for this reason can be released at a higher rate.

The value of the diffusion exponent n calculated on the basis of the Korsmeyer–Peppas model further argues that:

- In the case of precursor microparticles, the value of n < 0.43 indicates that the release mechanism of metronidazole is a Fick-type diffusion mechanism;
- In the case of hybrid microparticles, the value of n is in the range of 0.57–0.63, indicating that the release mechanism of metronidazole is a complex mechanism, controlled by both diffusion and swelling processes characteristic of an anomalous or non-Fickian diffusion;
- The values of the n parameter are less than 0.85, leading to the conclusion that the microparticles swelled but did not undergo any disintegration or erosion.

Similar results have been found in the literature for other microparticulate systems. For example, in case of microparticles based on gelatin and poly(ethylene glycol) coated with ethyl cellulose, the metronidazole release rates and transport parameters have suggested the non-Fickian mechanism [43]. Additionally, the release kinetics of the metronidazole from the hydrogel containing crosslinked chitosan microparticles best fit the Higuchi model [44].

4. Conclusions

By aqueous suspension polymerisation, three series of porous microparticles were obtained based on GMA, HEMA and one of the following crosslinking agents: EGDMA, DGDMA and TEGDMA. By the grafting reaction of sodium hyaluronate to the existent epoxy groups on the surface of AE, AD and AT microparticles, hybrid porous microparticles were obtained.

Precursor/hybrid microparticles were structurally characterized by appropriate techniques: FTIR spectroscopy, epoxy groups content, thermogravimetric analysis, dimensional analysis, and grafting degree of HA. From a morphological point of view the precursor/hybrid microparticles were characterized by: scanning electron microscopy, atomic force microscopy, and specific parameters for the characterization of the morphology of porous structures.

The information from the data acquired using the above-mentioned techniques showed that spherical microparticles of micrometer size with different surface morphologies depending on the synthesis conditions were obtained by suspension polymerization, and the grafting reaction of HA on the surface of the precursor microparticles in a basic medium was a success.

The swelling ability of precursor/hybrid microparticles in aqueous media with different pH values was studied, and the mechanism by which the swelling of precursor/hybrid microparticles in aqueous solutions with different pH values occurred is Fick-type and follows the second-order kinetic model.

The adsorptive performance of the precursor/hybrid microparticles has been shown by kinetic, thermodynamic and equilibrium studies. The experimental data obtained in the case of the metronidazole adsorption on precursor/hybrid microparticles were described using the nonlinear forms of Langmuir, Freundlich, Dubinin–Radushkevich, Sips and Khan isotherms. Adsorption isotherms demonstrate that the adsorption of metronidazole on precursor/hybrid microparticles occurs according to a monolayer adsorption.

To explain the mechanism of metronidazole adsorption on precursor/hybrid microparticles, the experimental data were modelled using four kinetic models, namely: first-order kinetic model, second-order kinetic model, Elovich model and Weber–Morris intraparticle diffusion model. The first-order kinetic model describes the experimental data quite well for metronidazole adsorption on both precursor and hybrid microparticles.

The release kinetics reflect that the release mechanism of metronidazole is a Fick-type diffusion mechanism in the case of precursor microparticles, while in the case of hybrid microparticles, it is a complex mechanism characteristic of anomalous or non-Fickian diffusion.

Supplementary Materials: The following supporting information can be downloaded at: https://www.mdpi.com/article/10.3390/polym14194151/s1, Figure S1: The infrared spectra of AE and AEHA microparticles; Figure S2: The infrared spectra of AD and ADHA microparticles; Figure S3: Particle size distributions of precursor/hybrid microparticles.

Author Contributions: Conceptualization, A.I.G. and S.R.; methodology, S.R., A.I.G. and S.V.; software, A.I.G.; validation, S.R., S.V. and M.P.; formal analysis, A.I.G.; investigation, A.I.G. and S.R.; data curation, A.I.G., S.R. and S.V.; writing—original draft preparation, A.I.G.; writing—review and editing, S.V. and M.P.; visualization, S.V.; supervision, M.P. All authors have read and agreed to the published version of the manuscript.

Funding: This research received no external funding.

Institutional Review Board Statement: Not applicable.

Data Availability Statement: The data presented in this study are available on request from the corresponding author.

Conflicts of Interest: The authors declare no conflict of interest.

References

1. Ficai, D.; Sandulescu, M.; Ficai, A.; Andronescu, E.; Yetmez, M.; Agrali, O.B.; Elemek, E.; Gunduz, O.; Sahin, Y.M.; Oktar, F.N. Drug Delivery Systems for Dental Applications. *Curr. Org. Chem.* **2017**, *21*, 64–73. [CrossRef]
2. Kina, J.R.; Suzuki, T.Y.U.; Kina, E.F.U.; Kina, J.; Kina, M. Non-Inflammatory Destructive Periodontal Disease. *Open Dent. J.* **2016**, *10*, 50–57.
3. Sopi, M.; Kocani, F.; Bardhoshi, M.; Meqa, K. Clinical and Biochemical Evaluation of the Effect of Diode Laser Treatment Compared to the Non-surgical and Surgical Treatment of Periodontal Diseases. *Open Dent. J.* **2020**, *14*, 281–288. [CrossRef]
4. Braga, G.K.; Oliveira, W.P. Manufacturing Drug Loaded Chitosan Microspheres by Spray Drying: Development, Characterization and Potential Use in Dentistry. *Dry. Technol.* **2007**, *25*, 303–310. [CrossRef]
5. Dias, R.J.; Havaldar, V.D.; Ghorpade, V.S.; Mali, K.K.; Gailwad, V.K.; Kumbhar, D.M. Development and Evaluation of In-Situ Gel Containing Ornidazole LoadedMicrospheres for Treatment of Periodontitis. *J. Appl. Pharm. Sci.* **2016**, *6*, 200–209. [CrossRef]
6. Scholz, M.; Reske, T.; Bohmer, F.; Hornung, A.; Grabow, N.; Lang, H. In vitro chlorhexidine release from alginate based microbeads for periodontal therapy. *PLoS ONE* **2017**, *12*, e0185562.
7. Ma, Y.; Liu, J.; Cui, X.; Hou, J.; Yu, F.; Wang, J.; Wang, X.; Chen, C.; Tong, L. Hyaluronic Acid Modified Nanostructured Lipid Carrier for Targeting Delivery of Kaempferol to NSCLC: Preparation, Optimization, Characterization and Performance Evaluation In Vitro. *Molecules* **2022**, *27*, 4553. [CrossRef]
8. Ghost, S.; Dutta, S.; Sarkar, A.; Kundu, M.; Sil, P.C. Targeted delivery of curcumin in breast cancer cells via hyaluronic acid modified mesoporous silica nanoparticle to enhance anticancer efficiency. *Colloids Surf. B Biointerfaces* **2021**, *197*, 111404.
9. Al-Khateeb, R.; Olszewska-Czyz, I. Biological molecules in dental applications: Hyaluronic acid as a companion biomaterial for diverse dental applications. *Heliyon* **2020**, *6*, e03722. [CrossRef]
10. Sanchez, D.C.; Ocampo, B.R.Y.; Chirino, C.A.E. Use of hyaluronic acid as an alternative for reconstruction of interdental papilla. *Rev. Odontol. Mex.* **2017**, *21*, 205–213.
11. Liang, J.; Peng, X.; Zhou, X.; Zou, J.; Chen, L. Emerging Applications of Drug Delivery Systems in Oral Infectious Diseases Prevention and Treatment. *Molecules* **2020**, *25*, 516. [CrossRef]
12. Joshi, D.; Garg, T.; Goyal, A.K.; Rath, G. Advanced drug delivery approaches against periodontitis. *Drug Deliv.* **2016**, *23*, 363–377. [CrossRef]
13. Vasiliu, S.; Lungan, M.A.; Gugoasa, I.; Drobota, M.; Popa, M.; Mihai, M.; Racovita, S. Design of Porous Microparticles Based on Chitosan and Methacrylic Monomers. *ChemistrySelect* **2019**, *4*, 9331–9338. [CrossRef]
14. Cigu, T.A.; Vasiliu, S.; Racovita, S.; Lionte, C.; Sunel, V.; Popa, M.; Cheptea, C. Adsorption and release studies of new cephalosporin from chitosan-g-poly(glycidyl methacrylate) microparticles. *Eur. Polym. J.* **2016**, *82*, 132–152. [CrossRef]
15. Vasiliu, S.; Lungan, M.A.; Racovita, S.; Popa, M. Pourous microparticles based on methacrylic copolymers and gellan as drug delivery systems. *Polym. Int.* **2020**, *69*, 1066–1080. [CrossRef]
16. Casale, M.; Moffa, A.; Vella, P.; Sabatino, L.; Capuano, F.; Salvinelli, B.; Lopez, M.A.; Carinci, F.; Salvinelli, F. Hyaluronic acid: Pespectives in dentistry. A systematic review. *Int. J. Immunopathol. Pharmacol.* **2016**, *29*, 577–582. [CrossRef]
17. Dahiya, P.; Kamal, P. Hyaluronic acid: A boon in periodontal therapy. *N. Am. J. Med. Sci.* **2013**, *5*, 309–315. [CrossRef]
18. Barde, M.; Adhikari, S.; Via, B.K.; Auad, M.L. Synthesis and characterization of epoxy resin from fast pyrolysis bio-oil. *Green Mater.* **2018**, *6*, 76–84. [CrossRef]
19. Lungan, M.A.; Popa, M.; Racovita, S.; Hitruc, G.; Doroftei, F.; Desbrieres, J.; Vasiliu, S. Surface characterization and drug release from porous microparticles based on methacrylic monomers and xanthan. *Carbohydr. Polym.* **2015**, *125*, 323–333. [CrossRef]
20. Vlad, C.D.; Vasiliu, S. Crosslinking polymerization of polyfunctional monomers by free radical mechanism. *Polimery* **2010**, *55*, 181–185. [CrossRef]
21. Ng, E.P.; Mintova, S. Nanoporous materials with enhanced hydrophilicity and high water sorption capacity. *Microporous Mesoporous Mater.* **2008**, *114*, 1–26. [CrossRef]
22. Urbanovici, E.; Segal, E. New formal relationship to describe the kinetics of crystallization. *Thermochim. Acta* **1990**, *171*, 87–94. [CrossRef]
23. Ba, E.C.T.; Dumont, M.R.; Martins, P.S.; Drumond, R.M.; Martins da Cruz, M.P.; Vieira, V.F. Investigation of the effects of skewness Rsk and Kurtosis Rku on tribological behavior in a pin-on-disc test of surfaces machined by conventional milling and turning processes. *Mater. Res.* **2021**, *24*, e20200435. [CrossRef]
24. Marcu Puscas, T.; Signorini, M.; Molinari, A.; Straffelini, G. Image analysis investigation of the effect of the process variables on the porosity of sintered chromium steels. *Mater. Charact.* **2003**, *50*, 1–10. [CrossRef]
25. Zauro, S.A.; Vishalakshi, B. Amphoteric gellan gum-based terpolymer-montmorillonite composite: Synthesis, swelling and dye adsorption studies. *Int. J. Ind. Chem.* **2017**, *8*, 345–362. [CrossRef]
26. Atta, S.; Khaliq, S.; Islam, A.; Javeria, I.; Jamil, T.; Athar, M.M.; Shafiq, M.I.; Gaffar, A. Injectable biopolymer based hydrogels fordrug delivery applications. *Int. J. Biol. Macromol.* **2015**, *80*, 240–245. [CrossRef] [PubMed]
27. Korsmeyer, R.W.; Gurny, R.; Doelker, E.; Buri, P.; Peppas, N.A. Mechanisms of solute release from porous hydrophilic polymers. *Int. J. Pharm.* **1983**, *15*, 25–35. [CrossRef]
28. Langmuir, I. The adsorption of gases on plane surfaces of glass, mica and platinum. *J. Am. Soc.* **1918**, *40*, 1361–1368. [CrossRef]
29. Freundlich, H.M.F. Over the adsorption in solution. *J. Phys. Chem.* **1906**, *57*, 385–470.

30. Dubinin, M.M.; Zaverina, E.D.; Radushkevich, L.V. Sorption and structure of active carbons. I. Adsorption of organic vapors. *Zhurnal Fizicheskoi Khimii* **1947**, *21*, 1351–1362.
31. Sips, R. On structure of a catalyst surface. *J. Chem. Phy.* **1948**, *16*, 490–495. [CrossRef]
32. Khan, A.R.; Ataullah, R.; Al-Haddad, A. Equilibrium adsorption studies of some aromatic pollutants from dilute aqueous solutions on active carbon at different temperature. *J. Colloid Interface Sci.* **1997**, *194*, 154–165. [CrossRef]
33. Ho, Y.S.; Porter, J.F.; McKay, G. Equilibrium Isotherm Studies for the Sorption of Divalent Metal Ions onto Peat: Copper, Nickel and Lead Single Component Systems. *Water Air Soil Pollut.* **2002**, *141*, 1–33. [CrossRef]
34. Lagergren, S. About the theory of so-called adsorption of soluble substance. *Kungliga Svenska Vetenskapsakademiens Handlingar* **1898**, *24*, 1–39.
35. Ho, Y.S.; McKay, G. The kinetics of sorption of basic dyes from aqueous solutions by *Sphagnum moss peat*. *Can.J. Chem. Eng.* **1998**, *76*, 822–827. [CrossRef]
36. Hameed, B.H.; Tan, I.A.W.; Ahmad, A.L. Adsorption isotherm, kinetic modeling and mechanism of 2,4,6-trichlorophenol on count husk- based activated carbon. *Chem. Eng. J.* **2009**, *144*, 235–244. [CrossRef]
37. Sarici-Ozdemir, C.; Onal, Y. Equilibrium, kinetic and thermodynamic adsorption of the environmental pollutant tannic acid onto activated carbon. *Desalination* **2010**, *251*, 146–152. [CrossRef]
38. Rodriquez, A.; Garcia, J.; Ovejero, G.; Mestanza, M. Adsorption of anionic and cationic dyes on activated carbon from aqueous solutions: Equilibrium and kinetics. *J. Hazard. Mater.* **2009**, *172*, 1311–1320. [CrossRef]
39. Hefne, J.A.; Mekhemer, W.K.; Alandis, N.M.; Aldayel, O.A.; Alajyan, T. Kinetic and thermodynamic study of the adsorption of Pb (II) from aqueous solution to the naturel and treated bentonite. *Int. J. Phys. Sci.* **2008**, *3*, 281–288.
40. Yu, Y.; Zhuang, Y.Y.; Wang, Z.H.; Qui, M.Q. Adsorption of water-soluble dyes onto modified resin. *Chemosphere* **2004**, *54*, 425–430. [CrossRef]
41. Higuchi, T. Diffusional models useful in biopharmaceutics drug release rate processes. *J. Pharm. Sci.* **1967**, *56*, 315–324. [CrossRef]
42. Costa, P.; Sousa Lobo, J.M. Modeling and comparison of dissolution profiles. *Eur. J. Pharm. Sci.* **2001**, *13*, 123–133. [CrossRef]
43. Phadke, K.V.; Manjeshwar, L.S.; Aminabhavi, T.M. Microspheres of Gelatin and Poly(ethylene glycol) Coated with Ethyl Cellulose for Controlled Release of Metronidazole. *Ind. Eng. Chem. Res.* **2014**, *53*, 6575–6584. [CrossRef]
44. Zupancic, S.; Potrc, T.; Baumgartner, S.; Kocbek, P.; Kristl, J. Formulation and evaluation of chitosan/polyethylene oxide nanofibers loaded with metronidazole for local infections. *Eur. J. Pharm. Sci.* **2016**, *95*, 152–160. [CrossRef]

Article

Immobilization Systems of Antimicrobial Peptide Ib−M1 in Polymeric Nanoparticles Based on Alginate and Chitosan

Carlos Enrique Osorio-Alvarado [1,*], Jose Luis Ropero-Vega [1], Ana Elvira Farfán-García [2] and Johanna Marcela Flórez-Castillo [1,*]

1. Universidad de Santander, Facultad de Ciencias Naturales, Ciencias Básicas y Aplicadas para la Sostenibilidad-CIBAS, Calle 70 No. 55-210, Bucaramanga 680003, Colombia; jose.ropero@udes.edu.co
2. Universidad de Santander, Facultad de Ciencias Médicas y de la Salud, Instituto de Investigación Masira, Calle 70 No. 55-210, Bucaramanga 680003, Colombia; afarfan@udes.edu.co
* Correspondence: c.osorio1202@gmail.com (C.E.O.-A.); johanna.florez@udes.edu.co (J.M.F.-C.); Tel.: +57-7606-516500 (ext. 1665) (J.M.F.-C.)

Abstract: The development of new strategies to reduce the use of traditional antibiotics has been a topic of global interest due to the resistance generated by multiresistant microorganisms, including *Escherichia coli*, as etiological agents of various diseases. Antimicrobial peptides are presented as an alternative for the treatment of infectious diseases caused by this type of microorganism. The Ib−M1 peptide meets the requirements to be used as an antimicrobial compound. However, it is necessary to use strategies that generate protection and resist the conditions encountered in a biological system. Therefore, in this study, we synthesized alginate and chitosan nanoparticles (Alg−Chi NPs) using the ionic gelation technique, which allows for the crosslinking of polymeric chains arranged in nanostructures by intermolecular interactions that can be either covalent or non-covalent. Such interactions can be achieved through the use of crosslinking agents that facilitate this binding. This technique allows for immobilization of the Ib−M1 peptide to form an Ib−M1/Alg−Chi bioconjugate. SEM, DLS, and FT-IR were used to determine the structural features of the nanoparticles. We evaluated the biological activity against *E. coli* ATCC 25922 and Vero mammalian cells, as well as the stability at various temperatures, pH, and proteases, of Ib−M1 and Ib−M1/Alg-Chi. The results showed agglomerates of nanoparticles with average sizes of 150 nm; an MIC of 12.5 μM, which was maintained in the bioconjugate; and cytotoxicity values close to 40%. Stability was maintained against pH and temperature; in proteases, it was only evidenced against pepsin in Ib−M1/Alg-Chi. The results are promising with respect to the use of Ib−M1 and Ib−M1/Alg−Chi as possible antimicrobial agents.

Keywords: nanoparticles; alginate; chitosan; Ib-M peptides; *E. coli*; peptide stability

1. Introduction

Owing to their biocompatibility, biodegradability, bioavailability, and low-toxicity properties, biopolymers, such as alginate and chitosan, are widely used in the biomedical and pharmaceutical industries [1]. They have a wide range of medical applications in tissue engineering, implants, and drug delivery [2–5]. Additionally, they have reactive functional groups that allow for the conjugation of peptides and proteins. [5–10].

Different strategies have been used to explore the preparation of polymeric nanoparticles, including coprecipitation, chemical crosslinking, thermodecomposition, coacervation, emulsification, and ionic gelation [11–14]. The latter is widely used for the preparation of nanoparticles of alginate and chitosan with the aim of bioconjugating them with antimicrobial peptides [7,15].

Antimicrobial peptides (AMPs) are small molecules composed of a length of 12 to 50 amino acids and are usually positively charged and amphiphilic [16]. They are among the body's first line of defense against the inactivation of pathogens, such as Gram-negative and Gram-positive bacteria, fungi, viruses, and parasites. The positive charge of AMPs allows for initial binding to the membrane via electrostatic interaction [17].

Despite their high antimicrobial activity, most AMPs are not widely used in clinical settings due to limitations such as toxicity and stability. Thus, AMPs present considerable challenges when considering the type of administration. For example, in oral administration, the pH and proteases in the gastrointestinal tract can inhibit the action of AMPs by hydrolysis or denaturation [18]. The conjugation of AMPs with nanoparticles (NPs) has been proposed to increase the local concentration of the peptide and improve its antimicrobial activity [19–22]. Alginate and chitosan nanoparticles (Alg–Chi NPs) have been used to achieve such strategies, as they preserve their structure and can therefore enhance bioactivity [14].

Chitosan is a linear polysaccharide composed of D-glucosamine and N-acetyl-D-glucosamine. It is a deacetylated form of chitin, the structural component of the exoskeletons of crustaceans [23]. The amino (-NH2) and hydroxyl (-OH) groups of the polymer guarantee high reactivity and charge, allowing for its union with different biomolecules [23]. In addition, alginate is a non-toxic polysaccharide composed of one to four linked β-D-mannuronate and α-l-guluronate blocks and can form chitosan-crosslinked gels. Owing to its properties, it is qualified as one of the best drug delivery systems and bioapplications, particularly in the immobilization of peptides, affording bioconjugates with equal or improved biological activity [24–26].

The Ib–M1 peptide is part of the group of Ib-M peptides synthesized by Flórez-Castillo et al. [27], which exhibit very good antimicrobial activity at low inhibitory concentrations against clinical and reference strains of *Escherichia coli* [27–29]. Ib–M1 peptides have a net charge of +6 and an isoelectric point of 12.5. Ib-M peptides are promising for clinical applications [28–30].

In this study, we immobilized the Ib–M1 peptide on polymeric nanoparticles of alginate and chitosan to maintain the antibacterial activity of this peptide and increase its stability with respect to proteases, changes in pH, and temperature. *E. coli* ATCC 25922 was used as a reference microorganism to evaluate antibacterial activity. In this experimental study, we report, for the first time, the effect of the immobilization of the Ib–M1 peptide on Alg–Chi nanoparticles to establish the scope of its clinical use.

2. Materials and Methods

2.1. Materials

Peptide Ib–M1 (sequence EWGRRMMGRGPGRRMMRWWR-NH2 [27]) was purchased through the commercial company Biomatik USA (Wilmington, DE, USA). Chitosan (Sigma-Aldrich, St. Louis, MO, USA, ≥75%), alginate (Sigma-Aldrich, St. Louis, MO, USA), sodium tripolyphosphate (TPP, Sigma-Aldrich, St. Louis, MO, USA), acetic acid (CH_3COOH, Merck KGaA, Darmstadt, Germany), 2-(1H-Benzotriazole-1-yl)-1,1,3,3-tetraethylammonium tetrafluoroborate (TBTU, Sigma-Aldrich, St. Louis, MO, USA), N, N-diisopropylethylamine (DIPEA, Sigma-Aldrich, St. Louis, MO, USA); *Escherichia coli* strain ATCC 25922, Vero ATCC CCL-81 cell line, Müller–Hinton broth (MH—Scharlau), Luria–Bertani broth (LB Broth, Oxoid, Basingstoke, England), trypsin, phosphate-buffered saline (PBS), dimethyl sulfoxide (DMSO, Sigma-Aldrich, St. Louis, MO, USA), bromide-3 (4,5-dimethylthiazole- 2-yl) 2,5-diphenyltetrazolium (MTT), streptomycin, Dulbecco's Modified Eagle Medium (DMEM), fetal bovine serum.

2.2. Preparation of Alginate-Chitosan Nanoparticles (Alg–Chi NPs)

Alg–Chi NPs were prepared by the ionic gelation method using pentasodium tripolyphosphate (TPP) as a crosslinking agent [31,32]. A 0.1% w/v chitosan solution (1% acetic acid and pH 3.5) and aqueous solutions of 1% w/v alginate and 1% w/v TPP were prepared independently. Subsequently, 1 mL of the TPP solution was added dropwise and under constant stirring to 9 mL of the chitosan solution. The obtained suspension was stirred for 2 h; then, 10 mL of type I water was added. Immediately afterward, 20 mL of the alginate solution was added dropwise and under constant stirring. The obtained suspension was left to stand for 18 h. Finally, the suspension was agitated for 30 min at 1500 rpm and centrifuged for 10 min at 5000 rpm to remove larger agglomerates. The supernatant was stored for characterization tests.

2.3. Structural Characterization of Alg–Chi Nanoparticles

Scanning electron microscopy (SEM) was used to explore the morphology of the Alg–Chi NPs with a Quanta field emission gun (Model 650) operated at 20.0 kV. Images were obtained in secondary electron mode.

To determine the size of the Alg–Chi obtained NPs, dynamic light scattering (DLS) measurements were used with a Zetasizer ZS90 (Malvern) equipped with a helium-neon gas laser with a wavelength of 632 nm.

Fourier transforms infrared (FTIR) measurements were used to verify the presence of the polymers in the synthesized Alg–Chi NPs, which identified the presence of the functional groups of each compound. A Bruker Tensor II spectrometer equipped with a platinum ATR cell and a cooled deuterated triglycine sulfate (DTGS) detector were used.

2.4. Preparation of Ib–M1/Alg–Chi Bioconjugate

The Ib–M1 peptide in the Alg–Chi NPs was immobilized according to the methodology of Ropero-Vega [33]. Briefly, the carboxyl group in the Glu-1 residue of the peptide was activated. To this end, 44.2 mg of TBTU and 24.2 µL of DIPEA were added to a 2000 µM solution of Ib–M1 in Tris HCl buffer (10 mM pH 7.4), allowing the reaction to take place for 20 min under constant stirring. Subsequently, this reaction mixture was added to 0.4 mg/mL of nanoparticles and left under constant stirring for 2 h, allowing for the formation reaction of the peptide bond between the activated carboxyl group and the amino groups of the nanoparticles. Finally, the Ib–M1/Alg–Chi bioconjugate was separated by centrifugation at 15,000 rpm for 45 min at 18 °C.

To determine the amount of immobilized peptide, the concentration of peptide in the supernatant was determined using the Bradford method [34] with a Quick Start™ Bradford protein assay kit from BioRad at 595 nm. Before the preparation of the immobilized peptide, the free peptide solution was measured under the same conditions. With these data and using the following formula, it was possible to determine the amount of peptide immobilized in the nanoparticles forming the Ib-M/Alg–Chi bioconjugate:

$$\%immobilization = \frac{Free\ peptide\ abs - supernatant\ abs}{Free\ peptide\ abs} * 100$$

where Free peptide abs is the absorbance of the peptide activated with TBTU and DIPEA before being combined with the nanoparticles for immobilization, and supernatant abs is the absorbance of the supernatant from the first centrifugation [33].

2.5. Antimicrobial Activity against E. coli ATCC 25922

Escherichia coli strain ATCC 25922 was cyropreserved at −80 °C in Luria–Bertani broth (LBB) with 15% glycerol. For the reactivation of the microorganism, 50 µL of cryopreserved material was added to 5 mL of LB and incubated at 35 ± 2 °C for 18 to 24 h before each assay.

The minimum inhibitory concentration (MIC) was determined using the microdilution method as described in the Clinical and Laboratory Standards Institute protocol M07A9 [35]. Briefly, twofold serial dilutions of the Ib–M1 peptide (100 and 0.78 µM concentration) were incubated in 96-well, round-bottom plates at 200 µL/well final volume for 24 h at 37 °C with shaking at 150 rpm with 5×10^5 colony forming units per mL (CFU/mL) of bacterial inoculum. The bacterial suspension was adjusted from a concentration of 1×10^6 CFU/mL. Absorbance at 595 nm was determined every hour for 8 h, with a final measurement at 24 h with a microplate reader kit (Bio-Rad, iMark). Mueller–Hinton (MH) broth and *E. coli* in MH broth were taken as negative and growth positive controls, respectively. The MIC was defined as the lowest peptide concentration that inhibited bacterial growth after 24 h. Data represent at least two independent experiments.

2.6. Cytotoxicity of Ib−M1 Peptides and Ib−M1/Alg−Chi Bioconjugate

The indirect cytotoxicity of the membranes against Vero cell cultures was analyzed using the MTT (3-(4,5-Dimethyl-2-thiazolyl)-2,5-diphenyl-2H-tetrazolium bromide) reduction method. VERO cells at a concentration of 3×10^4 cells/mL were cultured in 96-well, flat-bottom plates with DMEM medium supplemented with 10% inactivated fetal bovine serum and incubated at 37 °C in a 5% CO_2 atmosphere until reaching a confluence greater than 90%. Cells were then exposed to serial 1:2 dilutions of peptide Ib−M1 at concentrations in the range of 200 µM to 0.78 µM, with concentrations of the MIC and 1/2 MIC in the bioconjugates and the Alg−Chi NPs at 0.4 mg/mL for 24 h. Subsequently, 20 µL of MTT in PBS pH 7.4 was added to each well at a concentration of 5 mg/mL and incubated for 4 more hours under the same conditions described above, after which time the culture medium was removed from the wells, and 100 µL/well of DMSO was added to solubilize the formazan crystals, which were measured by the absorbances obtained in the spectrophotometric readings at 570 nm to calculate the percentage of cytotoxicity of the compounds. Cells in culture medium receiving no treatment were employed as a negative control.

2.7. Stability of Peptide Ib−M1 and Ib−M1/Alg−Chi Bioconjugate

The stability of the free peptide and the Ib−M1/Alg−Chi bioconjugate was determined under the conditions listed in Table 1. Subsequently, the antimicrobial activity against *E. coli* was determined, as mentioned in the previous section.

Table 1. Conditions used to evaluate the stability of peptides Ib−M1 and Ib−M1/Alg-Chi.

Condition	Features
pH	2 and 11
Temperature	4 and 100 °C
Proteases	Trypsin and Pepsin

2.7.1. pH Stability

The peptide and the bioconjugate were dissolved in glycine buffer solution at pH 2 and pH 11 for stability tests. These solutions were left for 30 min at 37 °C under stirring at 120 rpm; then, the pH was adjusted to 7, and the antimicrobial activity was determined as indicated in Section 2.5 [36].

2.7.2. Temperature Stability

For the stability tests at various temperatures, the Ib−M1 peptide and the Ib−M1/Alg-Chi bioconjugate were left for 90 min at the indicated temperatures. These tests were carried out on agitation plates. Then, antimicrobial activity was determined, as previously mentioned in Section 2.5 [37].

2.7.3. Trypsin and Pepsin Stability

Ib−M1 peptide and Ib−M1/Alg−Chi bioconjugate were exposed to pepsin and trypsin in a ratio (enzyme: peptide) of 27:1 and 1:20, respectively, in a beaker. This solution was stirred at 120 rpm at 37 °C for 90 min, taking a sample every 30 min, starting with an initial sample at 0 min. For an activity with pepsin, it was necessary to adjust the pH of the solution to 2 in order to simulate gastric conditions and allow for the action of the enzyme [38]. Assays with bioconjugates were performed in the same way as for a free peptide, adjusting concentrations to maintain the enzyme: peptide ratio. Then, antimicrobial activity was determined, as previously mentioned in Section 2.5.

2.8. Statistical Analysis

Except where indicated otherwise, all results are representative of two experiments, and each experiment was replicated three times. Arithmetic means values ± standard deviations are reported for each case. All analyses and graphics were generated using Origin (Pro) version 2019 (OriginLab Corporation, Northampton, MA, USA).

3. Results

3.1. Preparation and Characterization of Alg–Chi NPs

Figure 1 shows preparation scheme of Alg–Chi NPs. TPP was used as a crosslinking agent due to its high negative charge, allowing it to interact with the amino groups of chitosan. Similarly, the negative charge of the alginate interacts with the positive surface charges of the chitosan-TPP aggregates, allowing for the formation of the Alg–Chi NPs.

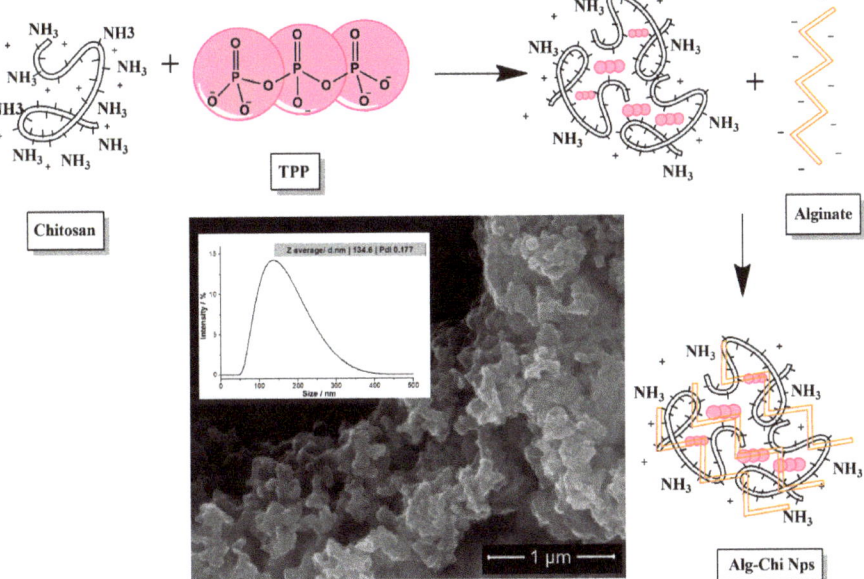

Figure 1. Scheme: synthesis of Alg-Chit NPs by ionic gelation. Inset: scanning electron micrograph of Alg–Chi NPs at 40.000X and size distribution of Alg–Chi NPs. Alg–Chit NPs: alginate-chitosan nanoparticles.

The obtained NPs present with high agglomeration, as evidenced by the SEM micrograph presented in the inset of Figure 1. This aggregation is responsible for the size of the nanoparticles, which was determined to be 134.6 nm by DLS. Despite this aggregation, the dispersion index of the nanoparticle suspension corresponded to 0.177, which indicates that the obtained size distribution is homogeneous.

The Figure 2 shows the FTIR spectrum of Alg–Chi NPs; the stretching presented at 3267 cm^{-1} corresponds to the O-H and N-H groups of the polymers that are present in their structures. The band at 2916 cm^{-1} is attribuided to C-H strectching vibrations. The band at 1589 cm^{-1} corresponds to the carbonyl group of Alg [39], whereas the band at 1430 cm^{-1} is assigned to the amino group of chitosan. The intense band that appears at 1040 cm^{-1} is associated with molecules present in the polymers. Bands associated with TPP can also be seen in the spectrum; for example, the band at 1290 cm^{-1} is associated with p = C strectching, that at 1084 cm^{-1} is associated with symmetric and antisymmetric stretching vibrations in the PO$_2$ group, that at 1020 cm^{-1} is associated with symmetric and antisymmetric stretching vibrations in the PO_3 group, and that at 820 cm^{-1} corresponds to antisymmetric stretching of the P-O-P bridge [40].

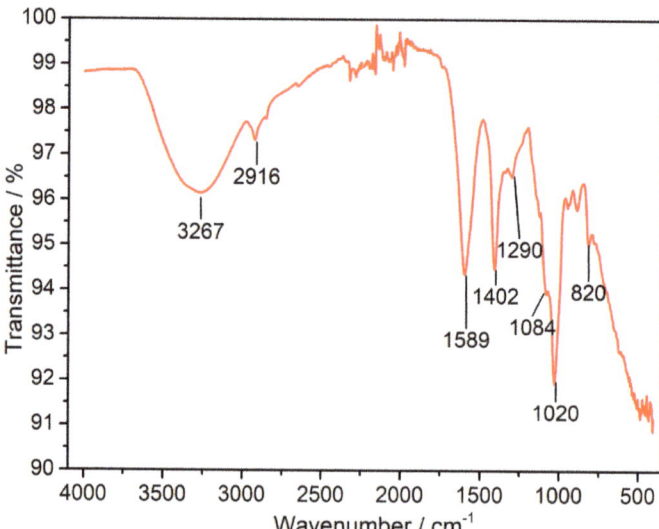

Figure 2. FTIR spectrum of Alg–Chi NPs. FTIR: Fourier transform infrared spectroscopy.

3.2. Preparation of the Ib–M1/Alg–Chi Bioconjugate

Figure 3A shows the reaction scheme for the immobilization of the Ib–M2 peptide on the surface of Alg–Chi NPs.

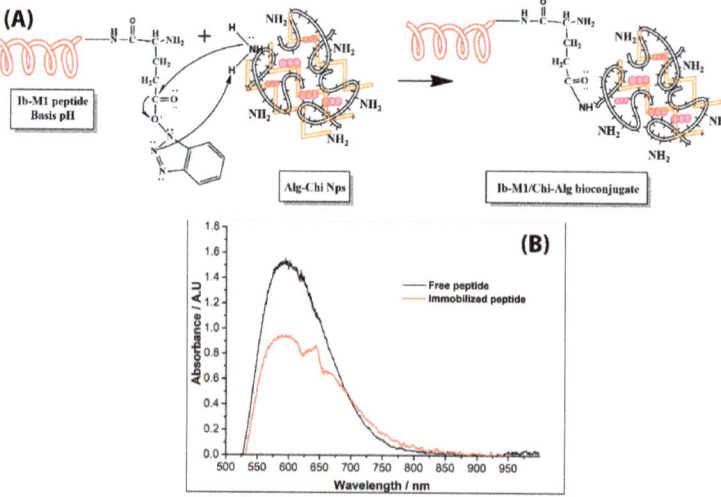

Figure 3. (**A**) Reaction scheme into amino groups of chitosan of Alg–Chi NPs and the carboxyl group of Glu residue of Ib–M1. (**B**) Monitoring of immobilization of the peptide by UV-Vis. Alg–Chit NPs: alginate-chitosan nanoparticles.

An amide bond was formed between the carboxyl group of Glu−1 of the peptide and the amino groups on the surface of the NPs. Immobilization was monitored by determining free amino groups in the suspension using the Bradford method. As shown in Figure 3B, after 2 h of reaction, 37% immobilization of the peptide on the Alg−Chi NPs was reached.

3.3. Antimicrobial Activity against E. coli

The antimicrobial activity of peptide Ib−M1 against *E. coli* 25922 was determined. To this end, the MIC was determined using the microdilution method, with streptomycin as the reference antibiotic. Figure 4 shows the growth kinetics of free Ib−M1 in the presence of streptomycin at 6.25 µM and peptide Ib−M1 at concentrations between 25 and 3.12 µM.

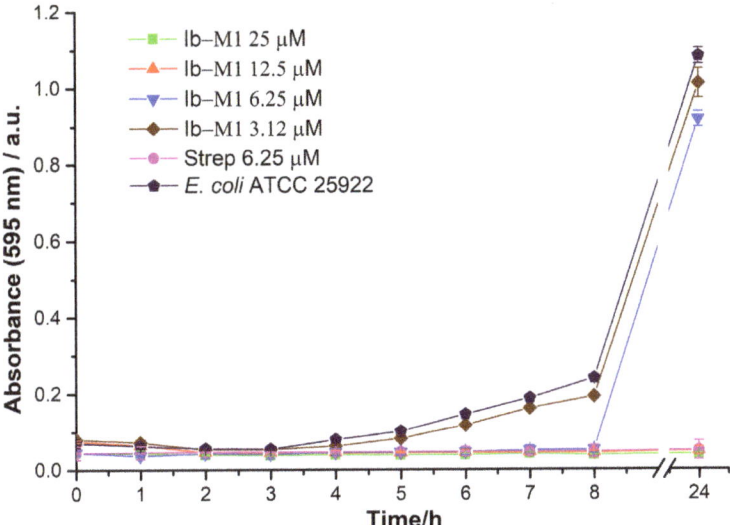

Figure 4. Growth kinetics of *E. coli* (ATCC 25922) in the presence of free Ib−M1 peptide at 25 µM (■), 12.5 µM (▲), 6.25 µM (▼), and 3.12 µM (♦), with streptomycin as reference antibiotic at 6.25 µM (●). Growth of *E. coli* without the addition of compounds (●) was included for comparison purposes. Each result was evaluated in triplicate in two independent experiments and expressed in terms of arithmetic average plus standard deviation. The Ib-M1 concentrations evaluated ranged from 100 µM to 0.78 µM; and are plotted in the Supplementary Figures (Figure S1A,B).

As shown in Figure 4, the minimum inhibitory concentration of Ib−M1 was 12.5 µM, which is concentration at which the growth of *E. coli* is inhibited for up to 24 h.

The MIC of the Ib−M1/Alg−Chi bioconjugate was determined by incubation with *E. coli* 25922 at an immobilized peptide concentration of between 12.5 and 0.78 µM. Figure 5 shows the growth kinetics of the bacteria in the presence of the Ib−M1/Alg−Chi bioconjugate and Alg−Chi NPs at a concentration of 0.4 mg/mL.

As shown in Figure 5, the peptide activity was maintained when it was immobilized on the nanoparticles. In addition, a synergistic effect was observed in the combination of the Alg−Chi NPs with the peptide. Figure 5 shows that Ib−M1 immobilized at a concentration of 6.25 µM inhibited approximately 35% of the growth of the microorganism after 24 h of incubation with the free peptide in the same concentration (Figure 4). On the other hand, the nanoparticles did not affect the growth of the bacteria. Therefore, Alg and chitosan did not show antimicrobial activity against *E. coli*.

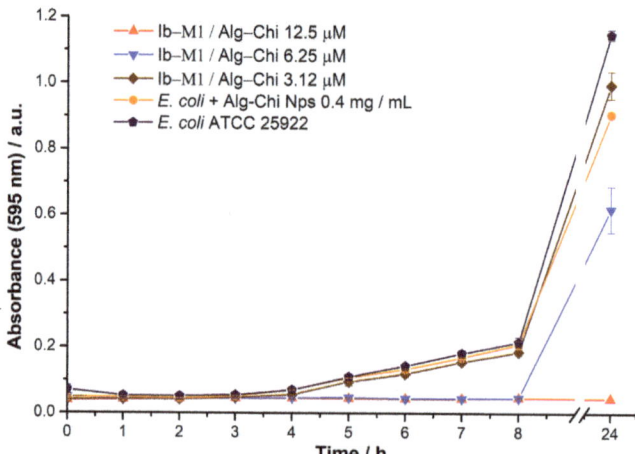

Figure 5. Growth kinetics of *E. coli* (ATCC 25922) in the presence of the Ib−M1/Alg−Chi bioconjugate at 12.5 µM (▲), 6.25 µM (▼), and 3.12 µM (♦). Results from *E. coli* with Alg−Chi NPs at 0.4 mg/mL (●) and *E. coli* without the addition of compounds (●) are included for comparison purposes. Each result was evaluated in triplicate in two independent experiments and expressed in terms of arithmetic average plus standard deviation.

3.4. Cytotoxicity

Figure 6 shows the percentage of cytotoxicity of Ib−M1 and Ib−M1/Alg-Chit at the MIC determined for *E. coli* strain 25922 (12.5 µM); the cytotoxicity of the Alg−Chi NPs (0.4 mg/mL) was also recorded.

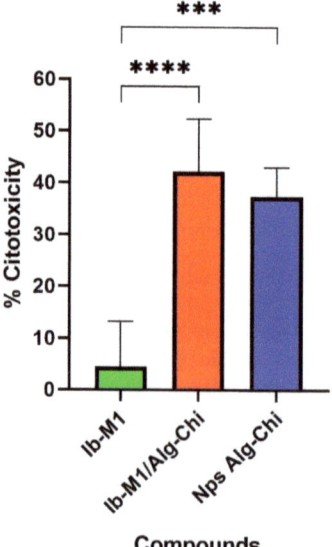

Figure 6. Cytotoxic activity of the free peptide Ib−M1 (12.5 µM), the bioconjugate Ib−M1/Alg−Chi (12.5 µM) and Alg−Chi NPs (0.4 mg/mL) in Vero cells. Each result was evaluated in triplicate in two independent experiments and expressed in terms of arithmetic average plus standard deviation. Cells maintained in culture medium without any treatment were used as a negative control (— Control; not shown). $p < 0.0001$: ****; $p \leq 0.0001$: ***.

The free peptide at a concentration of 12.5 µM (MIC) exhibited an approximate cytotoxicity of 5%, showing a statistically significant difference with respect to the Ib−M1/Alg−Chi bioconjugate, with a p value < 0.0001 (42.19 ± 10.13). The cytotoxicity value of the Alg−Chi NPs also showed a statistically significant difference ($p \leq 0.0001$) with respect to the free Ib−M1 peptide.

3.5. Stability of Ib−M1 and Ib−M1/Alg−Chit

3.5.1. pH Stability

Figure 7 shows the results of the antimicrobial activity with prior exposure to different pH conditions. As shown in Figure 5, the activity of the Ib−M1 peptide and the Ib−M1/Alg−Chi biconjugate was not affected under alkaline and acid conditions.

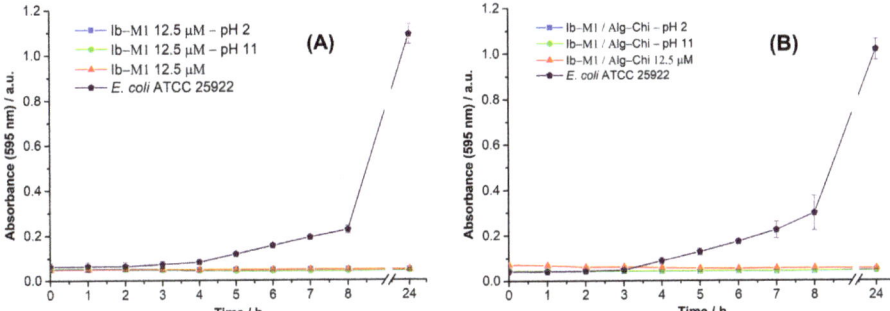

Figure 7. Effect of pH conditions on the antimicrobial activity of peptide Ib−M1 (**A**) and bioconjugate Ib−M1/Alg−Chi (**B**). Results of *E. coli* without the addition of compounds (●) are included for comparison purposes. Each result was evaluated in triplicate in two independent experiments and expressed in terms of arithmetic average plus standard deviation.

3.5.2. Thermal Stability

The influence of the temperatures used in the tests on Ib−M1 and Ib−M1/Alg−Chi are shown in Figure 8. There was no decrease in the activity of the compounds, which indicates good stability of the peptide and the bioconjugate during thermal treatments.

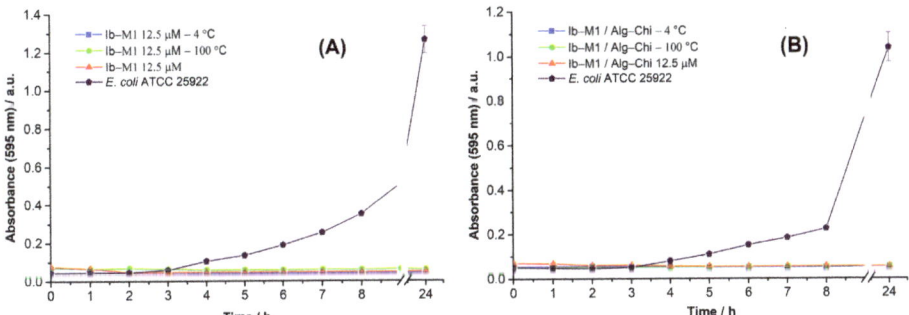

Figure 8. Effect of heat treatments on the antimicrobial activity of the Ib−M1 peptide (**A**) and the Ib−M1/Alg−Chi bioconjugate (**B**). Results of *E. coli* without the addition of compounds (●) are included for comparison purposes. Each result was evaluated in triplicate in two independent experiments and expressed in terms of arithmetic average plus standard deviation.

3.5.3. Proteolytic Stability

The effect of the proteolytic activity of trypsin and pepsin on the free peptide and the bioconjugate is shown Figure 9A,B, respectively.

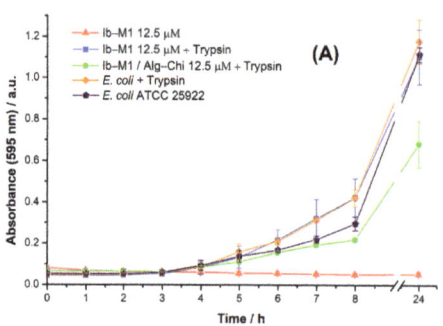 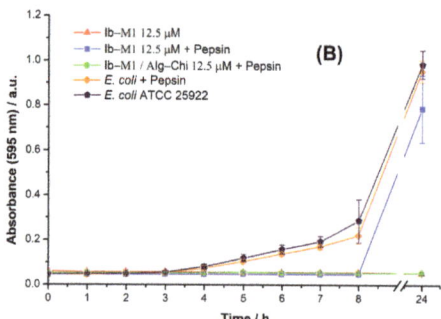

Figure 9. Antimicrobial activity of the peptide and the bioconjugate with prior exposure to trypsin (**A**) and pepsin (**B**). Results of *E. coli* without the addition of compounds (●) are included for comparison purposes. Each result was evaluated in triplicate in two independent experiments and expressed in terms of arithmetic mean plus standard deviation.

The free peptide Ib−M1, as well as the immobilized peptide Ib−M1/Alg-Chi, showed instability against the proteolytic action of trypsin, with proliferation of the microorganism at 24 h and optical densities like that of the growth control group. However, a decrease in the microbial population was observed at 24 h with the Ib−M1/Alg−Chi bioconjugate, and the growth of the microorganism slowed down within the first 8 h for the same wells, whereas the stability of the free peptide against pepsin was lost once it was exposed to this protease. On the contrary, the Ib−M1/Alg−Chi bioconjugate maintained its antimicrobial activity after being exposed to pepsin.

4. Discussion

Results obtained in this work are consistent with those reported by Bagre et al. [41]. Using SEM, they observed that alginate and chitosan nanoparticles obtained by ionic gelation were agglomerated. Furthermore, they observed particle sizes of 213 ± 3.8 nm, whereas in the present study, we obtained particle sizes of 150 ± 56 nm with a polydispersity value of 0.177, which indicates particles of homogeneous size. This aggregation is due to the tight crosslinking caused by the TPP with the chitosan molecules [41].

We prepared bioconjugates according to the methodology described by Ropero-Vega et al. [33], who used iron oxide nanoparticles coated with chitosan to immobilize the Ib-M2 peptide, obtaining a value of 30% of the immobilized peptide. In the present study, we achieved 37% immobilization, a value close to that reported by Ropero-Vega et al. The above shows that this method allows for the immobilization of Ib-M peptides on chitosan, which could expand the applications to obtaining antibacterial surfaces based on this polymer.

E. coli (ATCC 25922) was used as a reference microorganism for antimicrobial activity and was inhibited for the free peptide Ib−M1 at 12.5 μM. The result shows promising characteristics with respect to the clinical use of peptides to treat infectious diseases caused by microbial agents such as *E. coli*. Prada et al. [28] used the Ib−M1 peptide as an antimicrobial from different species of *E. coli* as ATCC reference strains and clinical isolates with MIC values of 4.7 and 1.6 μM, respectively. In addition, Flórez-Castillo et al. [30] showed the antimicrobial action of Ib−M1 against fourteen strains of pathogenic *E. coli*. The characteristics of the Ib-M peptide have been associated mainly with its positive charge and increased hydrophobicity as a result of the presence of arginine and tryptophan residues in the sequence [27].

The antimicrobial activity of the Ib−M1 peptide was not affected by immobilization in the Alg−Chi NPs. The above was evidenced by the MIC being maintained in the bioconjugate. These immobilization strategies have been used in various ways and have the advantage of preserving the properties of the compound of interest. Additionally, it was found that Alg−Chi NPs do not show antibacterial activity against *E. coli* at the evaluated concentrations, showing that the activity of the bioconjugate is due only to the presence of the peptide. This result is significant because it shows that the immobilization of the Ib−M1 peptide does not affect its activity. In addition, the activity of the peptide immobilized at $0.5 \times$ CMI (6.25 µM) prolonged the latent phase of the bacteria for at least 8 h and reduced its final concentration by up to 40% at 24 h relative to the growth control. These properties are similar to those reported in previous studies, wherein Ib-M2 was immobilized on iron oxide nanoparticles and its antibacterial activity was maintained against *E. coli* O157:H7 [33].

The cytotoxic effect of Ib−M1 and Ib−M1/Alg−Chi on Vero cells was evaluated by MTT assay. Our studies showed that Ib−M1 at the concentration of the CMI (12.5 µM) did not result in significant percentages of cytotoxicity, whereas the cytotoxicity generated by the bioconjugate showed a considerable increase, as well as the nanoparticles without the peptide. This result is consistent with that reported by Prada-Prada et al. [28], who observed that the Ib−M1 peptide presented cytotoxic concentration 50 (CC_{50}) in Vero cells at 395.2 µM. These results differ from those reported in previous studies with Alg−Chi polymers in the Vero cell line, as we observed 10% cytotoxicity (cell viability of 90%) [42]. Alginate and chitosan have been used in studies aimed at drug delivery and as healing polymers for skin wounds owing to their biocompatible properties, with low cytotoxicity in Vero cell lines and BHK-21 cells [43].

The stability tests at pH 2 and 11 during the 30 min of treatment and the thermal tests (4 and 100 °C) revealed that the Ib−M1 peptide and the Ib−M1/Alg−Chi bioconjugate maintained stable antimicrobial activity under these conditions. In contrast, Fahimirad et al. [44], evaluated the free antimicrobial peptide LL37 and observed that acidic pH (2) inhibits antibacterial activity. However, at neutral and basic pH, its activity is maintained, and when boiled for longer than 10 min, it loses stability, increasing the MIC. However, these characteristics improved once the LL37 peptide was immobilized in Chi NPs, maintaining its antimicrobial activity in terms of MIC under the evaluated pH and temperature conditions. These findings are also consistent with those reported by Yu et al., who showed higher stability of the antimicrobial peptide microcin J25 (MccJ25) when conjugated with chitosan nanoparticles [45]. Therefore, the stability of the Ib−M1 peptide is another characteristic that makes it promising for use as a new antimicrobial compound.

Immobilization of the peptide did not prevent the effect of trypsin; therefore, this protease can break peptide bonds, affecting the free and immobilized peptide activity. Trypsin is cleaved into amino acids adjacent to Arg and Lys [46]. The structure of the Ib−M1 peptide is rich in arginine; about 35% of the polypeptide chain is composed of this amino acid. The peptide is easily hydrolyzed upon exposure to trypsin, even when the peptide is immobilized in the NPs, generating oligopeptide chains with no activity against *E. coli*.

Pepsin is a protease that preferentially cleaves peptide bonds of amino acids, such as Leu, Ile, Phe, Val, and Trp [47]. Ib−M1 contains three Trp residues (W) susceptible to hydrolysis by pepsin. Therefore, the activity of the Ib−M1 peptide was affected in the presence of this protease. However, this effect was reduced in the Ib−M1/Alg−Chi bioconjugate. It is possible that the immobilization of the peptide makes it difficult for pepsin to access the Trp residues of the peptide.

5. Conclusions

Alg−Chi NPs was synthesized by the ionic gelation method. However, due to the properties of the polymers, agglomerates with average sizes of 150 nm were observed.

Bioconjugates were obtained by forming a bond between the amino group of chitosan and the carboxyl group of glutamic acid of Ib−M1.

Peptide Ib−M1 was found to be resistant to temperature and pH conditions. Furthermore, its immobilization in nanoparticles did not affect its activity against *E. coli*. On the contrary, it maintained its characteristics and achieved stability when the peptide was exposed to pepsin, maintaining its antimicrobial properties at MIC concentrations.

The Ib−M1 peptide exhibited low cytotoxic percentages against Vero cells, again showing this compound's favorable characteristics for biomedical applications. On the other hand, cytotoxicity percentages of 40% were observed for the Ib−M1/Alg−Chi bioconjugate. These values are attributed to Alg−Chi NPs and can be explained by their size. Therefore, further studies are necessary to reduce aggregation and obtain smaller particle sizes.

Supplementary Materials: The following supporting information can be downloaded at: https://www.mdpi.com/article/10.3390/polym14153149/s1. Figure S1A: Growth kinetics of *E. coli* ATTCC 25922 in the presence of peptide Ib−M1 at concentrations of 100 µM, 50 µM, 25 µM, and 12.5 µM, with streptomycin used as reference antibiotic at 6.25 µM and *E. coli* as growth control. Figure S1B. Growth kinetics of *E. coli* ATTCC 25922 in the presence of peptide Ib−M1 at concentrations of 6.25 µM, 3.12 µM, 1.56 µM, and 0.78 µM, with streptomycin used as reference antibiotic at 6.25 µM and *E. coli* as growth control.

Author Contributions: Conceptualization, J.M.F.-C.; data curation, C.E.O.-A. and J.L.R.-V.; formal analysis, C.E.O.-A., J.L.R.-V., A.E.F.-G. and J.M.F.-C.; funding acquisition, A.E.F.-G. and J.M.F.-C.; investigation, C.E.O.-A., J.L.R.-V and J.M.F.-C.; methodology, C.E.O.-A., J.L.R.-V. and J.M.F.-C.; project administration, A.E.F.-G.; resources, A.E.F.-G.; supervision, J.M.F.-C.; writing—original draft, C.E.O.-A.; writing—review and editing, J.L.R.-V., A.E.F.-G. and J.M.F.-C. All authors have read and agreed to the published version of the manuscript.

Funding: This research was funded by grant number 129980763392 CT-760-2018 of Ministerio de Ciencia, Tecnología e Innovación (MinCiencias), Call for projects on Science, Technology and Innovation in Health 807-2018, Colombia.

Institutional Review Board Statement: Not applicable.

Informed Consent Statement: Not applicable.

Data Availability Statement: Not applicable.

Conflicts of Interest: The authors declare no conflict of interest.

References

1. Puertas-Bartolomé, M.; Mora-Boza, A.; García-Fernández, L. Emerging Biofabrication Techniques: A Review on Natural Polymers for Biomedical Applications. *Polymers* **2021**, *13*, 1209. [CrossRef] [PubMed]
2. Liu, Q.; Li, Q.; Xu, S.; Zheng, Q.; Cao, X. Preparation and Properties of 3D Printed Alginate–Chitosan Polyion Complex Hydrogels for Tissue Engineering. *Polymers* **2018**, *10*, 664. [CrossRef]
3. Shi, D.; Shen, J.; Zhang, Z.; Shi, C.; Chen, M.; Gu, Y.; Liu, Y. Preparation and properties of dopamine-modified alginate/chitosan–hydroxyapatite scaffolds with gradient structure for bone tissue engineering. *J. Biomed. Mater. Res. Part A* **2019**, *107*, 1615–1627. [CrossRef]
4. Sibaja, B.; Culbertson, E.; Marshall, P.; Boy, R.; Broughton, R.M.; Solano, A.A.; Esquivel, M.; Parker, J.; De La Fuente, L.; Auad, M.L. Preparation of alginate–chitosan fibers with potential biomedical applications. *Carbohydr. Polym.* **2015**, *134*, 598–608. [CrossRef]
5. Kim, H.-J.; Lee, H.-C.; Oh, J.-S.; Shin, B.-A.; Oh, C.-S.; Park, R.-D.; Yang, K.-S.; Cho, C.-S. Polyelectrolyte complex composed of chitosan and sodium alginate for wound dressing application. *J. Biomater. Sci. Polym. Ed.* **1999**, *10*, 543–556. [CrossRef]
6. Bajas, D.; Vlase, G.; Mateescu, M.; Grad, O.A.; Bunoiu, M.; Vlase, T.; Avram, C. Formulation and Characterization of Alginate-Based Membranes for the Potential Transdermal Delivery of Methotrexate. *Polymers* **2021**, *13*, 161. [CrossRef]
7. Flynn, J.; Durack, E.; Collins, M.N.; Hudson, S.P. Tuning the strength and swelling of an injectable polysaccharide hydrogel and the subsequent release of a broad spectrum bacteriocin, nisin A. *J. Mater. Chem. B* **2020**, *8*, 4029–4038. [CrossRef]
8. Oliveira, D.M.L.; Rezende, P.S.; Barbosa, T.C.; Andrade, L.N.; Bani, C.; Tavares, D.S.; da Silva, C.F.; Chaud, M.V.; Padilha, F.; Cano, A.; et al. Double membrane based on lidocaine-coated polymyxin-alginate nanoparticles for wound healing: In vitro characterization and in vivo tissue repair. *Int. J. Pharm.* **2020**, *591*, 120001. [CrossRef] [PubMed]

9. Yoncheva, K.; Benbassat, N.; Zaharieva, M.M.; Dimitrova, L.; Kroumov, A.; Spassova, I.; Kovacheva, D.; Najdenski, H.M. Improvement of the Antimicrobial Activity of Oregano Oil by Encapsulation in Chitosan—Alginate Nanoparticles. *Molecules* **2021**, *26*, 7017. [CrossRef]
10. Li, J.; Wu, H.; Jiang, K.; Liu, Y.; Yang, L.; Park, H.J. Alginate Calcium Microbeads Containing Chitosan Nanoparticles for Controlled Insulin Release. *Appl. Biochem. Biotechnol.* **2021**, *193*, 463–478. [CrossRef] [PubMed]
11. Pant, A.; Negi, J.S. Novel controlled ionic gelation strategy for chitosan nanoparticles preparation using TPP-β-CD inclusion complex. *Eur. J. Pharm. Sci.* **2018**, *112*, 180–185. [CrossRef]
12. Choudhary, A.; Kant, V.; Jangir, B.L.; Joshi, V. Quercetin loaded chitosan tripolyphosphate nanoparticles accelerated cutaneous wound healing in Wistar rats. *Eur. J. Pharmacol.* **2020**, *880*, 173172. [CrossRef] [PubMed]
13. Reis, C.P.; Neufeld, R.J.; Ribeiro, A.J.; Veiga, F.; Nanoencapsulation, I. Methods for preparation of drug-loaded polymeric nanoparticles. *Nanomed. Nanotechnol. Biol. Med.* **2006**, *2*, 8–21. [CrossRef] [PubMed]
14. Zhang, Z.; Hao, G.; Liu, C.; Fu, J.; Hu, D.; Rong, J.; Yang, X. Recent progress in the preparation, chemical interactions and applications of biocompatible polysaccharide-protein nanogel carriers. *Food Res. Int.* **2021**, *147*, 110564. [CrossRef] [PubMed]
15. Coppi, G.; Bondi, M.; Coppi, A.; Rossi, T.; Sergi, S.; Iannuccelli, V. Toxicity and gut associated lymphoid tissue translocation of polymyxin B orally administered by alginate/chitosan microparticles in rats. *J. Pharm. Pharmacol.* **2010**, *60*, 21–26. [CrossRef]
16. Brown, K.L.; Hancock, R.E. Cationic host defense (antimicrobial) peptides. *Curr. Opin. Immunol.* **2006**, *18*, 24–30. [CrossRef]
17. Mookherjee, N.; Hancock, R.E.W. Cationic host defence peptides: Innate immune regulatory peptides as a novel approach for treating infections. *Cell. Mol. Life Sci.* **2007**, *64*, 922–933. [CrossRef]
18. Drucker, D.J. Advances in oral peptide therapeutics. *Nat. Rev. Drug Discov.* **2020**, *19*, 277–289. [CrossRef]
19. Chowdhury, R.; Ilyas, H.; Ghosh, A.; Ali, H.; Ghorai, A.; Midya, A.; Jana, N.R.; Das, S.; Bhunia, A. Multivalent gold nanoparticle–peptide conjugates for targeting intracellular bacterial infections. *Nanoscale* **2017**, *9*, 14074–14093. [CrossRef]
20. Liu, L.; Yang, J.; Xie, J.; Luo, Z.; Jiang, J.; Yang, Y.Y.; Liu, S. The potent antimicrobial properties of cell penetrating peptide-conjugated silver nanoparticles with excellent selectivity for Gram-positive bacteria over erythrocytes. *Nanoscale* **2013**, *5*, 3834–3840. [CrossRef]
21. Pal, I.; Brahmkhatri, V.P.; Bera, S.; Bhattacharyya, D.; Quirishi, Y.; Bhunia, A.; Atreya, H.S. Enhanced stability and activity of an antimicrobial peptide in conjugation with silver nanoparticle. *J. Colloid Interface Sci.* **2016**, *483*, 385–393. [CrossRef]
22. Poblete, H.; Agarwal, A.; Thomas, S.S.; Bohne, C.; Ravichandran, R.; Phopase, J.; Comer, J.; Alarcon, E.I. New Insights into Peptide–Silver Nanoparticle Interaction: Deciphering the Role of Cysteine and Lysine in the Peptide Sequence. *Langmuir* **2016**, *32*, 265–273. [CrossRef] [PubMed]
23. Bellich, B.; D'Agostino, I.; Semeraro, S.; Gamini, A.; Cesàro, A. "The Good, the Bad and the Ugly" of Chitosans. *Mar. Drugs* **2016**, *14*, 99. [CrossRef] [PubMed]
24. Azevedo, M.; Bourbon, A.I.; Vicente, A.; Cerqueira, M.A. Alginate/chitosan nanoparticles for encapsulation and controlled release of vitamin B2. *Int. J. Biol. Macromol.* **2014**, *71*, 141–146. [CrossRef] [PubMed]
25. Bekhit, M.; Sánchez-González, L.; Ben Messaoud, G.; Desobry, S. Encapsulation of Lactococcus lactis subsp. lactis on alginate/pectin composite microbeads: Effect of matrix composition on bacterial survival and nisin release. *J. Food Eng.* **2016**, *180*, 1–9. [CrossRef]
26. Poonguzhali, R.; Basha, S.K.; Kumari, V.S. Synthesis and characterization of chitosan/poly (vinylpyrrolidone) biocomposite for biomedical application. *Polym. Bull.* **2017**, *74*, 2185–2201. [CrossRef]
27. Flórez-Castillo, J.M.; Perullini, M.; Jobbágy, M.; Calle, H.D.J.C. Enhancing Antibacterial Activity Against Escherichia coli K-12 of Peptide Ib-AMP4 with Synthetic Analogues. *Int. J. Pept. Res. Ther.* **2014**, *20*, 365–369. [CrossRef]
28. Prada-Prada, S.; Flórez-Castillo, J.; García, A.E.F.; Guzmán, F.; Hernández-Peñaranda, I. Antimicrobial activity of Ib-M peptides against Escherichia coli O157: H7. *PLoS ONE* **2020**, *15*, e0229019. [CrossRef]
29. Flórez-Castillo, J.M.; Rondón-Villareal, P.; Ropero-Vega, J.L.; Mendoza-Espinel, S.Y.; Moreno-Amézquita, J.A.; Méndez-Jaimes, K.D.; Farfán-García, A.E.; Gómez-Rangel, S.Y.; Gómez-Duarte, O.G. Ib-M6 Antimicrobial Peptide: Antibacterial Activity against Clinical Isolates of Escherichia coli and Molecular Docking. *Antibiotics* **2020**, *9*, 79. [CrossRef]
30. Castillo, J.M.F.; Ropero-Vega, J.; Perullini, M.; Jobbágy, M. Biopolymeric pellets of polyvinyl alcohol and alginate for the encapsulation of Ib-M6 peptide and its antimicrobial activity against E. coli. *Heliyon* **2019**, *5*, e01872. [CrossRef]
31. Goycoolea, F.M.; Lollo, G.; Remuñán-López, C.; Quaglia, F.; Alonso, M.J. Chitosan-Alginate Blended Nanoparticles as Carriers for the Transmucosal Delivery of Macromolecules. *Biomacromolecules* **2009**, *10*, 1736–1743. [CrossRef]
32. Keawchaoon, L.; Yoksan, R. Preparation, characterization and in vitro release study of carvacrol-loaded chitosan nanoparticles. *Colloids Surf. B Biointerfaces* **2011**, *84*, 163–171. [CrossRef]
33. Ropero-Vega, J.; Ardila-Rosas, N.; Hernández, I.P.; Flórez-Castillo, J. Immobilization of Ib-M2 peptide on core@shell nanostructures based on SPION nanoparticles and their antibacterial activity against Escherichia coli O157:H7. *Appl. Surf. Sci.* **2020**, *515*, 146045. [CrossRef]
34. Bradford, M.M. A rapid and sensitive method for the quantitation of microgram quantities of protein utilizing the principle of protein-dye binding. *Anal. Biochem.* **1976**, *72*, 248–254. [CrossRef]
35. Wayne, P.A. Weinstein y Clinical and Laboratory Standards Institute. In *Performance Standards for Antimicrobial Susceptibility Testing: Supplement M100*, 30th ed.; Clinical and Laboratory Standards Institute: Wayne, PA, USA, 2020.

36. Fraser, P.; Nguyen, J.; Surewicz, W.; Kirschner, D. pH-dependent structural transitions of Alzheimer amyloid peptides. *Biophys. J.* **1991**, *60*, 1190–1201. [CrossRef]
37. Vishweshwaraiah, Y.L.; Acharya, A.; Hegde, V.; Prakash, B. Rational design of hyperstable antibacterial peptides for food preservation. *Npj Sci. Food* **2021**, *5*, 26. [CrossRef] [PubMed]
38. Lin, Y.; Pangloli, P.; Meng, X.; Dia, V.P. Effect of heating on the digestibility of isolated hempseed (*Cannabis sativa* L.) protein and bioactivity of its pepsin-pancreatin digests. *Food Chem.* **2020**, *314*, 126198. [CrossRef] [PubMed]
39. Lawrie, G.; Keen, I.; Drew, B.; Chandler-Temple, A.; Rintoul, L.; Fredericks, P.; Grøndahl, L. Interactions between Alginate and Chitosan Biopolymers Characterized Using FTIR and XPS. *Biomacromolecules* **2007**, *8*, 2533–2541. [CrossRef]
40. Tomaz, A.F.; de Carvalho, S.M.S.; Barbosa, R.C.; Silva, S.M.L.; Gutierrez, M.A.S.; de Lima, A.G.B.; Fook, M.V.L. Ionically Crosslinked Chitosan Membranes Used as Drug Carriers for Cancer Therapy Application. *Materials* **2018**, *11*, 2051. [CrossRef]
41. Bagre, A.P.; Jain, K.; Jain, N.K. Alginate coated chitosan core shell nanoparticles for oral delivery of enoxaparin: In vitro and in vivo assessment. *Int. J. Pharm.* **2013**, *456*, 31–40. [CrossRef]
42. Chalitangkoon, J.; Wongkittisin, M.; Monvisade, P. Silver loaded hydroxyethylacryl chitosan/sodium alginate hydrogel films for controlled drug release wound dressings. *Int. J. Biol. Macromol.* **2020**, *159*, 194–203. [CrossRef]
43. Rahman, A.; Islam, S.; Haque, P.; Khan, M.N.; Takafuji, M.; Begum, M.; Chowdhury, G.W.; Rahman, M.M. Calcium ion mediated rapid wound healing by nano-ZnO doped calcium phosphate-chitosan-alginate biocomposites. *Materialia* **2020**, *13*, 100839. [CrossRef]
44. Fahimirad, S.; Ghaznavi-Rad, E.; Abtahi, H.; Sarlak, N. Antimicrobial Activity, Stability and Wound Healing Performances of Chitosan Nanoparticles Loaded Recombinant LL37 Antimicrobial Peptide. *Int. J. Pept. Res. Ther.* **2021**, *27*, 2505–2515. [CrossRef]
45. Yu, H.; Ma, Z.; Meng, S.; Qiao, S.; Zeng, X.; Tong, Z.; Jeong, K.C. A novel nanohybrid antimicrobial based on chitosan nanoparticles and antimicrobial peptide microcin J25 with low toxicity. *Carbohydr. Polym.* **2021**, *253*, 117309. [CrossRef] [PubMed]
46. Dau, T.; Gupta, K.; Berger, I.; Rappsilber, J. Sequential Digestion with Trypsin and Elastase in Cross-Linking Mass Spectrometry. *Anal. Chem.* **2019**, *91*, 4472–4478. [CrossRef] [PubMed]
47. El-Zahar, K.; Sitohy, M.; Choiset, Y.; Métro, F.; Haertlé, T.; Chobert, J.-M. Peptic hydrolysis of ovine β-lactoglobulin and α-lactalbumin Exceptional susceptibility of native ovine β-lactoglobulin to pepsinolysis. *Int. Dairy J.* **2005**, *15*, 17–27. [CrossRef]

Article

Synergistics of Carboxymethyl Chitosan and Mangosteen Extract as Enhancing Moisturizing, Antioxidant, Antibacterial, and Deodorizing Properties in Emulsion Cream

Nareekan Chaiwong [1], Yuthana Phimolsiripol [1,2,3,*], Pimporn Leelapornpisid [4], Warintorn Ruksiriwanich [2,4], Kittisak Jantanasakulwong [1,2], Pornchai Rachtanapun [1,2], Phisit Seesuriyachan [1,2,3], Sarana Rose Sommano [5], Noppol Leksawasdi [1,2,3], Mario J. Simirgiotis [6], Francisco J. Barba [7] and Winita Punyodom [3]

1. Faculty of Agro-Industry, Chiang Mai University, Chiang Mai 50100, Thailand; meen.nareekan@gmail.com (N.C.); kittisak.jan@cmu.ac.th (K.J.); pornchai.r@cmu.ac.th (P.R.); phisit.s@cmu.ac.th (P.S.); noppol@hotmail.com (N.L.)
2. Cluster of Agro Bio-Circular-Green Industry, Chiang Mai University, Chiang Mai 50100, Thailand; warintorn.ruksiri@cmu.ac.th
3. Center of Excellence in Materials Science and Technology, Faculty of Science, Chiang Mai University, Chiang Mai 50200, Thailand; winita.punyodom@cmu.ac.th
4. Faculty of Pharmacy, Chiang Mai University, Chiang Mai 50200, Thailand; pimporn.lee@cmu.ac.th
5. Faculty of Agriculture, Chiang Mai University, Chiang Mai 50200, Thailand; sarana.s@cmu.ac.th
6. Faculty of Sciences, Institute of Pharmacy, Universidad Austral de Chile, Valdivia 509000, Chile; mario.simirgiotis@uach.cl
7. Department of Preventive Medicine and Public Health, Food Science, Toxicology and Forensic Medicine, Faculty of Pharmacy, Universitat de València, Avda. Vicent Andrés Estellés, 46100 Burjassot, València, Spain; francisco.barba@uv.es
* Correspondence: yuthana.p@cmu.ac.th; Tel.: +66-5-394-8236

Abstract: Carboxymethyl chitosan (CMCH) from native chitosan of high molecular weight (H, 310–375 kDa) was synthesized for improving water solubility. The water solubility of high-molecular-weight carboxymethyl chitosan (H-CMCH) was higher than that of native chitosan by 89%. The application of H-CMCH as enhancing the moisturizer in mangosteen extract deodorant cream was evaluated. Different concentrations of H-CMCH (0.5–2.5%) were investigated in physicochemical characteristics of creams, including appearance, phase separation, pH, and viscosity, by an accelerated stability test. The different degrees of skin moisturizing (DM) on pig skin after applying H-CMCH solution, compared with untreated skin, water, and propylene glycol for 15 and 30 min using a Corneometer®, were investigated. The results showed that the 0.5% H-CMCH provided the best DM after applying the solution on pig skin for 30 min. Trans-2-nonenal, as an unsatisfied odor component, was also evaluated against components of the mangosteen extract deodorant cream, which were compared to the standard, epigallocatechin gallate (EGCG). In addition, DPPH and ABTS radical scavenging activity, ferric reducing antioxidant power (FRAP), and antibacterial activities were examined for the mangosteen extract deodorant cream using 0.5% H-CMCH. Results indicated that the mangosteen extract synergized with H-CMCH, which had a good potential as an effective skin moisturizing agent enhancer, deodorizing activity on trans-2-nonenal odor, antioxidant properties, and antibacterial properties.

Keywords: carboxymethyl chitosan; mangosteen; deodorant; skin moisturizing; trans-2-nonenal; accelerated stability test

1. Introduction

Emulsions are the most common form of skin care products [1]. A variety of cosmetic emulsions are utilized for functional applications, such as sebum control, skin whitening, and UV protection [2]. Cosmetic emulsions are mainly classified as oil-in-water (O/W) [3], water-in-oil (W/O) [4], or water-in-silicone (W/S) emulsions [5,6]. Emulsions applied in cosmetic applications, both O/W and W/O types, need to satisfy several requirements, such as having the rheology for skin, feeling good on the skin, having good spreadability, and long-term physical stability under various conditions. The ingredients are safe and do not cause skin irritation or any harmful effects [1]; O/W emulsions are the most commonly used in the cosmetic industry. Many cosmetic industries produce a wide variety of beauty products to care for and to avoid excessive sweating and body odor [7]. Deodorants are one of the cosmetic preparations containing substances or ingredients able to eliminate or reduce body odor [8]. Body odor is caused by the growth of microorganisms, and odors are associated with perspiration and its breakdown by bacteria in the armpits, feet, or any other part of the body [9]. Many products claim to have skin benefits, such as anti-aging [10], skin tightening, and moisturizing activity [11]. In general, the preparation of deodorant is carried out using emulsion systems, where the active ingredients are mixed with waxes, oils, and silicones and produced in the desired form, such as deodorant cream, gel, roll-ons, and sticks [12].

Carboxymethyl chitosan (CMCH) is a polymer synthesized by introducing a carboxymethyl group into the main structure of chitosan, achieved by carboxymethylation of the hydroxyl and amine of chitosan, [13] and shows a potential application in cosmetics [14]. The water-soluble property of carboxymethyl chitosan provides conclusive insights into the utilization of its properties of biocompatibility, biodegradation, biological activity, and low toxicity [15]. CMCH in the cosmetics industry seems to be a promising avenue to boost its application as a multifunctional ingredient. Different aspects of CMCH have been applied in five major directions, as a moisture retention agent, antimicrobial agent, antioxidant agent, delivery system, and naturally derived emulsion stabilizer [14]. However, there are many herbs which have antimicrobial properties, this being a primary prerequisite for the development of deodorant formulations.

The use of mangosteen extract (ME) as a raw material to be used as an active ingredient for the preparation of cosmetic products has been evaluated by Ghasemzadeh et al. [16]. It contains the active ingredients xanthones, tannins, and proanthocyanins, which are predominant in mangosteen. In addition, ME has also demonstrated antibacterial properties, reduced acne inflammation, prevented acne, and also contains antioxidants which help to firm the skin and reduce melanin production in the skin, which whitens the skin [17]. ME is made from *Garcinia mangostana* with a standardized solvent extraction process; the product is easy to use and can be mixed with all types of cosmetics [16]. The mangosteen pericarp contains many compounds with outstanding antioxidant, anti-inflammatory and antibacterial properties, especially for the bacteria *Propionibacterium acnes* and *Staphylococcus epidermidis*. The solid waste obtained after extraction of these compounds from the pericarp is also ideal for the production of useful activated carbon [18], while wastewater generated from the extraction process can be efficiently treated with synergistic catalysts [19]. Pothitirat et al. [20] revealed that ME can inhibit the cause of acne and reduce acne rash. Deodorants are substances applied to the body in order to affect body odor caused by bacterial growth and the smell associated with bacterial breakdown of sweat in armpits, feet, and other areas of the body. Ham et al. [21] also reported properties that help in the elimination of body odor. ME is commonly used in various products, such as body cleansing products, soaps, shower creams, facial cleansers, acne treatment products, acne gel, serum, and facial cream [22]. Industries have also been promoted with specially developed synthetic cosmetics as the main ingredient. There are many herbs that have antimicrobial properties, which must be a key factor for improving the deodorant property. Again, herbal formulas require deodorizing properties with activity close to synthetic formulas. However, the synergistic effect of antibacterial activity and deodorizing and moisturizing performance has not been fully investigated. Therefore, this research aimed to determine the effect of CMCH with ME used as moisturizing, antioxidant, antibacterial, and deodorant agents by studying various parameters, then developing final products and evaluating the efficacy of the cream emulsion system.

2. Materials and Methods

2.1. Materials

High-molecular-weight native chitosan (H, 310–375 kDa) with a degree of deacetylation above 90% was purchased from Kritnarong Limited Partnership, Phitsanulok, Thailand. Ethanol, methanol, isopropanol, sodium hydroxide, and glacial acetic acid were purchased from RCI Labscan (Bangkok, Thailand). Monochloroacetic acid and trans-2-nonenal were obtained from Merck KGaA (Darmstadt, Germany). Aluminium chlorohydrate, ceteareth-25, glyceryl monostearate, propylene glycol, and stearyl alcohol were bought from Thai Poly Chemicals (Samutsakhon, Thailand). All other reagents were of analytical grade. Six species of bacteria: *Corynebacterium* spp. (TISTR 1259), *Staphylococcus epidermidis* (TISTR 518), *Staphylococcus aureus* (ATCC 25923), *Bacillus subtilis* (DMST 15896), *Pseudomonas aeruginosa* (TISTR 781), and *Escherichia coli* (ATCC 25922) were obtained from Thailand Institute of Scientific and Technological Research (Pathum Thani, Thailand).

2.2. Synthesis of CMCH

H-CMCH was synthesized following the method of Chaiwong et al. [23]. Chitosan flake was ground and sieved to obtain a particle size under 60 mesh (Endecotts, UK). The chitosan (25 g) was suspended in 50% (w/v) sodium hydroxide solution (400 mL), and 100 mL of isopropanol was added and mixed well at 50 °C for 1 h. Monochloroacetic acid (50 g) was dissolved in isopropanol (50 mL) and gradually dropped into the reaction for 30 min, and the system was left to stand in reaction at 50 °C for 4 h. The reaction was stopped by adding 70% (v/v) methanol. The pH of the sample was later adjusted to 7.0 by 1% (v/v) glacial acetic acid. From that point, the solid was separated and washed with 70% (v/v) ethanol 5 times, 250 mL each time, and finally washed with 250 mL of 95% (v/v) ethanol for desalting and dried in a hot air oven (ED56, Binder GmbH, Tuttlingen, Germany) at 80 °C for 12 h. The functional groups of high-MW native chitosan and H-CMCH were measured using a Fourier transform infrared spectrometer (Frontier, PerkinElmer, Waltham, MA, USA) in the range of 500–4000 cm^{-1} as shown in Figure S1. Chitosan was converted to CMCH. The –COO groups enhanced the hydrophilic properties of the CMCH. A certain amount (10 g) of H-CMCH powder was dissolved in 20 mL of deionized water. The suspension was mechanically stirred at 50 °C for 10 min by following the method of Rachtanapun et al. [15] for use as a moisturizing agent.

2.3. ME Preparation

Mangosteen fruits were purchased from a local market in Fresh Fruits Market, Chiang Mai, Thailand. Mangosteens with reddish purple skin were selected at the fifth color level according to the Thai Agricultural Standard for Mangosteen (TAS 2-2013) color index [24]. The mangosteen fruits were rinsed with distilled water to remove impurities such as dust before the pericarp was separated from the fruits manually. The mangosteen pericarp was chopped into small pieces and dried at 60 °C. The dried mangosteen pericarp was ground into powder using a blending machine. The powdered mangosteen pericarp (5 g) was thoroughly extracted using a sonication bath (SB25-12 DTD, Ningbo SCIENTZ Biotechnology Co., Ltd., Zhejiang, China) with 50% ethanol (200 mL). The filtrates were concentrated by a rotary evaporator (R-250, Buchi, Flawil, Switzerland) at 50 °C to give a crude extract (200 mg). The ME was kept in air-tight amber bottles and stored at 4 °C until use [20].

2.4. Deodorant Cream (O/W) Preparation

Cream base was prepared using an emulsification technique according to the method of Kassakul et al. [25]. First, all ingredients were weighed accurately by a calibrated analytical balance as shown in Table 1. The oil phase was prepared by mixing aluminium chlorohydrate, stearyl alcohol, ceteareth-25, glyceryl monostearate, and mineral oil and then heating to 70 °C. The aqueous phase was prepared by dissolving glycerin and propylene glycol in distilled water in a beaker and heating to 75°C. Both phases (oil and aqueous) were heated up to the same temperature (45 °C) in a water bath; the aqueous phase was added into the oil phase gradually with stirring. Then, ME was added, and different concentrations of H-CMCH, including 0.5, 1.0, 1.5, 2.0, and 2.5% (H1, H2, H3, H4, and H5) were added into the cream base. Finally, the weight of cream was 100 g, and 0.01 mL of perfume was added into the mixture with continuous stirring for 20 min until the cream cooled to 25 °C.

Table 1. Deodorant cream formulas with different content of H-CMCH (0.5, 1.0, 1.5, 2.0, and 2.5% (w/v)).

Ingredients	(% w/v)				
	H1 (0.5)	H2 (1.0)	H3 (1.5)	H4 (2.0)	H5 (2.5)
Phase A (oil phase)					
Aluminium chlorohydrate	40.0	40.0	40.0	40.0	40.0
Stearyl alcohol	2.0	2.0	2.0	2.0	2.0
Ceteareth-25	2.0	2.0	2.0	2.0	2.0
Glyceryl monostearate	2.0	2.0	2.0	2.0	2.0
Mineral oil	5.0	5.0	5.0	5.0	5.0
Phase B (aqueous phase)					
Glycerin	2.0	2.0	2.0	2.0	2.0
Propylene glycol	5.0	5.0	5.0	5.0	5.0
Distilled water	41.4	40.9	40.4	39.9	39.4
Phase C					
ME	0.1	0.1	0.1	0.1	0.1
Phase D					
H-CMCH	0.5	1.0	1.5	2.0	2.5
Phase E					
Perfume	0.01	0.01	0.01	0.01	0.01

2.5. Degree of Skin Moisturizing (DM)

The degree of skin moisturizing of the deodorant cream was examined with 0.5, 1.0, 1.5, 2.0, and 2.5% (w/v) of H-CMCH on pig skin and compared with untreated skin, water, and propylene glycol. The pig skins were prepared from the back side of the pig ear obtained from local market sources (Chiang Mai, Thailand). The samples were washed and cleaned, with removal of the fat layer, prior to cutting into 3 × 3 cm pieces. Each sample (100 µL) was applied to the skin surface. The skin without any substance was used as a control. Skin moisturizing was measured before application to samples and after application at 0, 15, and 30 min intervals using a Corneometer® (Courage-Khazaka Electronic GmbH, Cologne, Germany). Before applying the sample and recording the parameter, the pig skins were kept at 25 °C for 30 min. This method was adapted from Kassakul et al. [25]. The degree of skin moisturizing (%) was tested in triplicate to detect random error and calculated using Equation (1).

$$DM\ (\%) = \frac{M_i - M_a}{M_i} \times 100 \quad (1)$$

where M_i is the initial moisturizing content before the sample was applied to the skin and M_a is moisturizing content after the sample was applied to the skin.

2.6. Accelerated Stability Study

The deodorant creams with different concentrations of H-CMCH (0.5, 1.0, 1.5, 2.0, and 2.5% (w/v)) were centrifuged (Universal 320R, Hettich, Tuttlingen, Germany) at 6000× g for 20 min. Accelerated stability tests were performed at both 4 °C and 45 °C for 24 h in an Incucell incubator (MMM Medcenter Einrichtungen GmbH, München, Germany) for 6 cycles. The physicochemical characteristics of the creams—including visual appearance, pH using a pH meter (FiveEasy F20, Mettler Toledo, Greifensee, Switzerland), viscosity using a Brookfield viscometer (DV-II+, Brookfield, Middleboro, MA, USA), and color L*, a*, b* using a colorimeter (CR-410, Konica-Minolta, Tokyo, Japan)—were monitored at every cycle. Total color difference (ΔE) was calculated for each sample and each cycle following the strategy of Tkacz et al. [26].

2.7. Deodorizing Activity

The deodorizing activity of the developed deodorant cream was evaluated against the odor component trans-2-nonenal, following the method of Ham et al. [21] by using solid-phase microextraction and gas chromatography. The dilute deodorant cream solutions of different concentrations (1–100 mg/mL) were prepared by dissolving the extract in 0.2 M potassium phosphate buffer solution (pH 7.4). An aliquot (1 mL) of the dilute deodorant cream solutions was mixed with aqueous solution (100 µL) containing the odor compound: 10 ppm of trans-2-nonenal or solution. The mixture of deodorant cream and odor substance was placed in a vial (20 mL), which was tightly sealed with a cap furnished with PTFE/silicone septa (Supelco, Bellefonte, PA, USA). The sample vial was then placed in a stirring water bath at 35 °C for 10 min to achieve phase equilibrium, and then the odor substance in the headspace of the vial was taken by a SPME fiber during additional stirring for 5 min at 37 °C. Carboxen/polydimethylsiloxane (Carboxen/PDMS; 75 µm film thickness) was used for detecting trans-2-nonenal. After the adsorption of the odor substance, the fiber was removed from the vials and immediately inserted into the injector of a gas chromatography system for quantitative analysis. The odor compounds were desorbed from the fiber by heating at 250 °C for 2 min in the gas chromatography system.

Gas chromatography was carried out using a gas chromatography flame ionization detector (GC-2010 Series, Shimadzu, Santa Clara, CA, USA) equipped with a flame ionization detection system. The oven temperature for trans-2-nonenal analysis was programmed at 50 °C for 2 min, from 50 °C to 200 °C at a heating rate of 8 °C/min, 200 °C for 2 min, and finally 250 °C hold for 1 min. Injector and detector temperatures for the analysis of trans-2-nonenal were also 250 °C. The samples were injected in a spitless mode using nitrogen as the carrier gas (1 mL/min) at a volume of 1.0 µL. Deodorizing activity (%) was calculated by Equation (2).

$$\text{Deodorizing activity (\%)} = \frac{H_n - H_c}{H_n} \times 100 \qquad (2)$$

where H_n is the headspace amount of the odor substance (trans-2-nonenal) and H_c is the headspace amount of the deodorant cream.

2.8. Antioxidant Properties

The optimal cream formula was selected based on skin moisturizing and deodorizing properties and a stability test for antioxidant activity. These properties were compared in the prototype cream formula (no ME and H-CMCH). Solutions of the developed deodorant cream (stock: 5 mg/mL in distilled water) at different concentrations of 1, 2, 3, 4, and 5 mg/mL were prepared and used for DPPH, ABTS, and FRAP assays.

2.8.1. DPPH Radical Scavenging Activity

The ability of antioxidants to scavenge the 2,2-diphenyl-1-picrylhydrazyl (DPPH) free radical was tested by using the modified method of Surin et al. [27] and Phimolsiripol et al. [28]. After that, 100 µL of the stock samples (as described above, concentrations: 1, 2, 3, 4, and 5 mg/mL) were blended with 100 µL of 0.2 mM DPPH reagent (Sigma-Aldrich, Singapore) and incubated at 25 °C for 30 min in the dark. Absorbance was measured at 517 nm in a 96-well microplate reader (SpectraMax® i3x, Molecular Devices, San Jose, CA, USA). The radical scavenging activity of the sample was calculated based on gallic acid (Sigma-Aldrich, Darmstadt, Germany). Results were expressed as milligram gallic equivalent per gram of sample (mgGAE/g sample). The percentage of DPPH radical scavenging activity can be calculated as shown in Equation (3) before plotting of IC_{50} against respective concentration.

$$\% \text{ DPPH radical inhibition} = \frac{Ac - As}{Ac} \times 100 \quad (3)$$

where A_c is absorbance of the DPPH solution and A_s is absorbance of different concentrations of samples.

2.8.2. ABTS Radical Scavenging Activity

2,2-azino-bis-(3-ethylbenzothiazoline-6-sulfonic acid) (ABTS) radical scavenging activity was tested according to the method described by Surin et al. [27,29] and Ruksiriwanich et al. [30]. ABTS (Sigma-Aldrich, Singapore) reagent solution was freshly prepared by mixing 7 mM of ABTS solution with 2.45 mM of potassium persulfate (Sigma-Aldrich, Singapore). ABTS powder and potassium persulfate powder were individually dissolved in water to the required concentration and then combined in a bottle. After 16 h of incubation in the dark at 25 °C, the resultant dark blue color of the ABTS reagent solution was diluted with ethanol until the absorbance reading reached 0.7 ± 0.2. The solutions of H-CMCH were prepared as described previously in 2.6. Each sample solution (0.5 mL at concentrations: 1, 2, 3, 4, and 5 mg/mL) was mixed with 1.0 mL of ABTS stock solution and incubated for 6 min in the dark. Absorbance was measured at 734 nm in the 96-well microplate reader. The ABTS radical scavenging activity was expressed as milligram gallic equivalent per gram of sample (mgGAE/g sample). The percentage of ABTS radical scavenging activity can be calculated as shown in Equation (4) with plotting of IC_{50} against respective concentration.

$$\% \text{ ABTS radical inhibition} = \frac{Ac - As}{Ac} \times 100 \quad (4)$$

where A_c is absorbance of the ABTS solution and A_s is absorbance of different concentrations of samples.

2.8.3. Ferric Reducing Antioxidant Power (FRAP)

The ferric reducing antioxidant power (FRAP) assay was carried out according to the technique of Surin et al. [31]. The FRAP reagent was prepared by mixing 25 mL of 0.3 M acetate buffer (pH 3.6), 2.5 mL of 4,6-tripyridyl-s-triazine (TPTZ) (Sigma-Aldrich, Darmstadt, Germany) solution in 40 mM HCl (RCI Labscan, Bangkok, Thailand), and 2.5 mL of 20 mM ferrous sulphate (Loba Chemie, Mumbai, India). Then, 50 µL of sample (5 mg/mL) was mixed with 950 µL of FRAP reagent and incubated in the dark for 30 min. Absorbance was measured at 593 nm in a 96-well microplate. The ferric reducing antioxidant power of the sample was determined based on ferrous sulphate (Merck KGaA, Darmstadt, Germany). Results were expressed as ferrous sulphate equivalent antioxidant capacity (FEAC) with µmol Fe^{2+}/g sample.

2.9. Antibacterial Properties

The antibacterial properties on six species of bacteria—*Corynebacterium* spp., *S. epidermidis*, *S. aureus*, *B. subtilis*, *P. aeruginosa*, and *E. coli*—were tested using the agar well diffusion method by Bai-Ngew et al. [32]. For this step, a good representative formula was compared with the prototype cream formula (no ME and H-CMCH). The bacterial culture was swabbed on sterile nutrient agar plates. Subsequently, filter paper discs (6 mm in diameter) were dipped into the prototype deodorant cream, developed deodorant cream (stock 5 mg/mL in distilled water), and positive control (10 mg/mL Streptomycin). The plates were incubated at 37 °C for 18–24 h in an upright position. The experiment was carried out in triplicate, and the inhibition zone was recorded and expressed in millimeters.

2.10. Statistical Analysis

All data were analyzed by one-way ANOVA. Mean comparison was performed by Duncan's multiple range tests with significance level $p < 0.05$. Statistical analyses were performed with the SPSS 17.0 (SPSS Inc.; IBM Corp.; Chicago, IL, USA).

3. Results and Discussion

3.1. Effect of H-CMCH Synthesis

The yield, moisture content, water solubility, viscosity, and pH of H-CMCH were 45.36%, 5.56%, 89.5%, 360 cP, and 7.33, respectively. Solubility is a significant property of CMCH that measures its resistance to water. The H-CMCH improved water solubility by about 89% when compared to chitosan. The solubility and conformation of CMCH are a result of the deacetylation, pH, and MW of native chitosan. The solubilization process of CMCH is related to functionalized polymers and different types of chemical and physical interactions, such as hydrogen bonds, hydrophobic interactions, and van der Waals forces. High water solubility suggests that CMCH is moisture absorbent and has a greater ability to bind with water than chitosan. However, the solubility of chitosan was relatively poor in water and organic solvents, which resulted in limitation in its uses. At pH < 6.0, chitosan is positively charged ($-NH_3^+$), with increased solubility in water. As the pH increases, chitosan loses its charges, due to protonation, while the amino groups decrease as the solutes begin to precipitate. Chemical modification of amino groups and hydroxyl groups using a carboxymethylation reaction results in a large number of water-soluble chitosan derivatives [14]. This causes the increase of hydrated water molecules around the chains of CMCH than surround the chitosan chains, resulting in higher water solubility. The results are also consistent with the report of Siahaan et al. [33], who found that temperature and NaOH concentration affected CMCH synthesis, and Rachtanapun et al. [34] in carboxymethyl bacterial cellulose. The interactions between NaOH and monochloroacetic acid resulted in reduced CMCH forming and lower solubility. The mitigation in solubility might stem from the loss of free amino functional groups that enhance the hydrophobic nature of the compounds [35]. The greater solubility of L-CMCH and M-CMCH resulted in a decrease in viscosity, but H-CMCH showed higher viscosity. This could be explained as follows, CMCHs with longer chains or higher MW were contributing to the gel. H-CMCH is an effective water-soluble polymer with high viscosity which could be successfully utilized in pharmaceuticals and cosmetics as an emulsion stabilizer and thickening agent. Thanakkasaranee et al. [36] also reported that the yield of CMCH was also dependent on the concentration of NaOH, MW of chitosan, solvent, and reaction temperature. In addition, the solvent ratio and the processing also affect the yield and antioxidant activities [37].

3.2. Degree of Skin Moisturizing (% DM)

The degree of skin moisturizing indicates the water-holding capacity of the skin, which can be tested by the Corneometer® method. The Corneometer® measures the changes in electrical capacitance related to the moisture content of the skin before and after applying the solutions [25]. The degree of skin moisturizing of the deodorant cream with different concentrations of H-CMCH (0.5, 1.0, 1.5, 2.0, and 2.5% (w/v)) was examined on pig skin and compared with untreated skin, water, and propylene glycol at 15 and 30 min as presented in Figure 1.

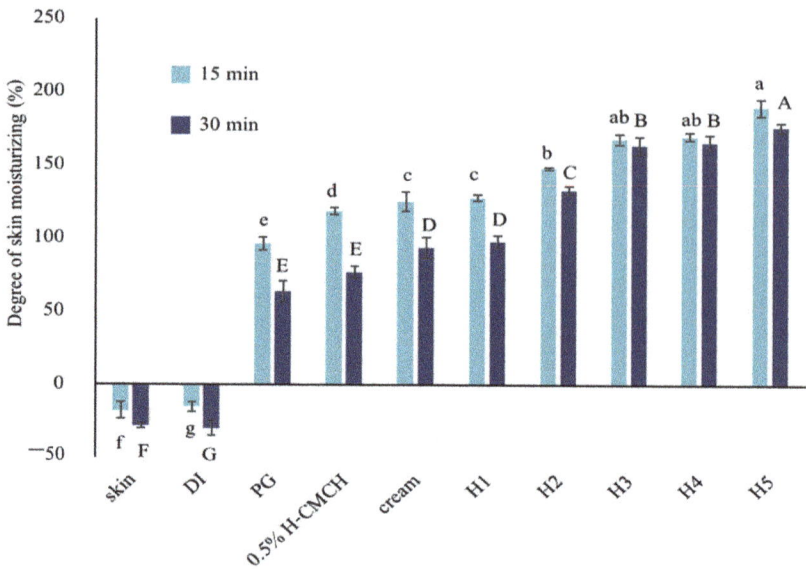

Figure 1. DM (%) as affected by time (15 and 30 min) and different treatments (skin, DI, PG, H1, H2, H3, H4, and H5) on pig skin. Different lowercase letters (a,b,c...) indicate significant differences between solutions at 15 min and different uppercase letters (A,B,C...) indicate significant differences between solutions at 30 min.

The degree of moisturizing at time 15 and 30 min showed that the degree of skin moisturizing of the solutions decreased with increasing time after applying solutions. At the same time, the degree of skin moisturizing of all treatments showed a significant difference after applying between 15 and 30 min ($p < 0.05$). Applying H-CMCH solution for 15 min gave a higher degree of skin moisturizing than 30 min, showing the degree of skin moisturizing of untreated skin, water, propylene glycol, H1, H2, H3, H4, and H5 solutions applied on pig skin for 30 min were significantly decreased compared to 15 min. This confirms that the H-CMCH solution provided a good moisture absorption. In fact, the skin moisturizing effects appeared to decrease with increasing time due to a lack of mechanisms to maintain skin moisturizing and the dryness of pig skin cells [38]. The higher MW CMCH also had superior moisture retention capacity. Kassakul et al. [25] found that 0.2% *Hibiscus rosa-sinensis* mucilage as a natural ingredient provided good results for skin moisturizing after applying for 30 min, improving by about 130%. The results showed that moisturizing products could increase the water content of the skin while maintaining softness and smoothness [39]. Chaiwong et al. [23] reported, after applying solutions containing different MW of water-soluble CMCH (L-CMCH, M-CMCH, H-CMCH), that the moisture content of the skin increased. The mechanism of the moisturizing effect is based on the formation of a water film on the skin surface after dissolution of CMCH, and a subsequent stage of water evaporation could further prevent water evaporation from the skin [40]. Positive electrical charges and relatively high MW facilitate prolonged skin adherence [14]. Our results also showed that H-CMCH decreased the loss of water while elevating skin humidity. The higher apparent viscosity of H-CMCH can improve stability and enhance skin hydration. In fact, H-CMCH was superior to untreated skin, water, and propylene glycol in terms of degree of skin moisturizing effect. The higher concentrations of H-CMCH also indicated potential for film forming and multilayer coating of the skin. Subsequently, it could be used in cosmetic preparations, with further studies suggested to test skin irritation in human subjects.

3.3. Effect of Accelerated Stability Study

3.3.1. Visual Appearance

A heating/cooling cycle test was performed while the formulations were stored in an incubator. The temperature was alternated between 4 °C and 40 °C every 24 h during any period of time. This method is commonly used during the initial stage of developmental screening. Useful information relating to stability could be obtained from such tests [1]. The accelerated stability test of the deodorant cream with different concentrations of H-CMCH (0.5, 1.0, 1.5, 2.0, and 2.5% (w/v)) was performed at 4 °C in a refrigerator and 45 °C in a hot oven and a heating/cooling cycle for up to 6 cycles. The result of each formulation was randomly checked every 1 cycle by centrifugation at 6000× g for 20 min at 25 °C. Deodorant creams with 0.5 and 1.0% H-CMCH indicated acceptable stability after 6 cycles by less phase separation. For the cream containing 1.5–2.5% H-CMCH, higher phase separation and greater changes in color are evident, as shown in Figures 2–6.

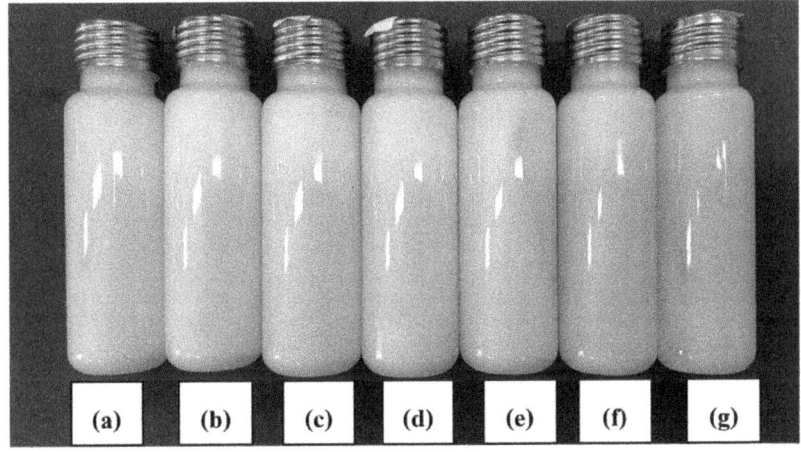

Figure 2. Deodorant cream with 0.5% H-CMCH by heating/cooling cycle; (**a**) cycle 0, (**b**) cycle 1, (**c**) cycle 2, (**d**) cycle 3, (**e**) cycle 4, (**f**) cycle 5, and (**g**) cycle 6.

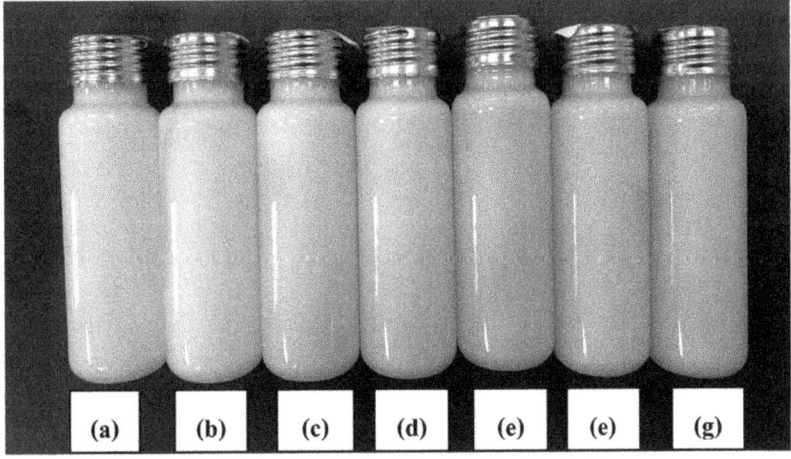

Figure 3. Deodorant cream with 1.0% H-CMCH by heating/cooling cycle; (**a**) cycle 0, (**b**) cycle 1, (**c**) cycle 2, (**d**) cycle 3, (**e**) cycle 4, (**f**) cycle 5, and (**g**) cycle 6.

Figure 4. Deodorant cream with 1.5% H-CMCH by heating/cooling cycle; (**a**) cycle 0, (**b**) cycle 1, (**c**) cycle 2, (**d**) cycle 3, (**e**) cycle 4, (**f**) cycle 5, and (**g**) cycle 6.

Figure 5. Deodorant cream with 2.0% H-CMCH by heating/cooling cycle; (**a**) cycle 0, (**b**) cycle 1, (**c**) cycle 2, (**d**) cycle 3, (**e**) cycle 4, (**f**) cycle 5, and (**g**) cycle 6.

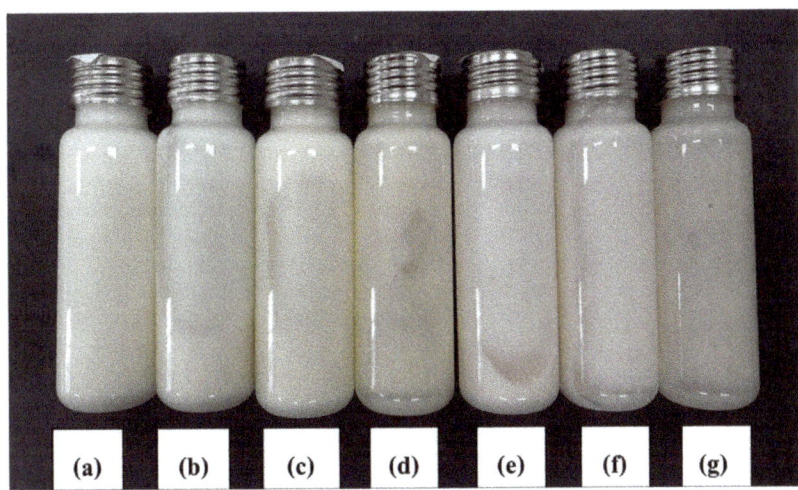

Figure 6. Deodorant cream with 2.5% H-CMCH by heating/cooling cycle; (**a**) cycle 0, (**b**) cycle 1, (**c**) cycle 2, (**d**) cycle 3, (**e**) cycle 4, (**f**) cycle 5, and (**g**) cycle 6.

3.3.2. pH Value

The measurement of pH for all five deodorant formulas in the accelerated state with the heating/cooling cycle method, at the 0 cycle of the stability study the developed deodorant cream samples had pH in the range of 6.32–6.41. After the end of the test, the pH values of all five formulas of deodorant products were found to be in the range of 6.26 to 6.37. This was similar to the pH of human skin and thus suitable for application to the skin with a good stability; for example, the cream texture was a fine opaque white. While applying, the consistency of the cream spreads well; it is easy to blend and can be absorbed into the skin and enhance the moisture of skin for a long time. Moreover, an effective cream should have a pH of about 6.55 [40]. After acceleration to cycle 6, it was found that the pH value decreased significantly ($p < 0.05$), with the pH value in the range of 6.34–6.37 as shown in Figure S2. The pH range should not be too acidic or alkaline, because irritation to the skin might be unavoidable. The pH mitigation is probably due to separation of the cream emulsion from its matrix and ionization, resulting in net negative charge and causing the rise in acidity [41,42]. To avoid exceedingly high pH values beyond the skin's physiological range, a preservative solution was added to the formulations. Implementation of sodium benzoate or sodium salt of benzoic acid also facilitates stabilization of the skin's pH. In general, sodium benzoate is commonly used in combination with antiseptics as food preservatives, cosmetics, and medicines [42]. The oxidative/reductive mechanisms, deactivating properties, and safety assessment of benzoic acid or related compounds in biological systems are well documented in the literature [43–45]. Considering the kinetic rate (k) of pH, it was found that increasing CMCH content (0.5–2.5%) in the formula affected the decrease of the pH's k-value when compared to the formula with 0.5% (w/v) CMCH. It was evident that cream containing 0.5% (w/v) H-CMCH showed a good pH stability, which plateaued after six cycles. For other samples, the pH decrease was still ongoing in a linear trend after six cycles. This is probably due to the implementation of higher H-CMCH concentration.

3.3.3. Viscosity

Viscosity is one of key parameters indicating cream quality. The forecasting of this parameter is commonly performed in accelerated stability testing [1]. From the viscosity stability of the deodorant cream after passing the accelerated state for up to six cycles, the results showed that H1 and H2 were not separated and precipitated; the cream texture had a smooth appearance. The viscosity was not significantly changed ($p \geq 0.05$), being in the ranges 234–300 and 231–313 cP, respectively. H3, H4, and H5 started to separate. The viscosity was significantly decreased ($p < 0.05$), with values in the ranges 154–363, 125–464, and 114–500 cP, respectively, as shown in Figure S3. As for the viscosity changing rate (k), as the CMCH content increased from 0.5% to 2.5% (w/v), the k-value increased (from 9.9 to 71.6 cP/cycle). The increasing CMCH content could have affected the decrease in the viscosity value and stability under changing temperature. CMCH is an amphiprotic ether of chitosan derivative. The functional groups include active hydroxyl (–OH), carboxyl (–COOH), and amine (–NH_2) in the molecule. CMCH is soluble in water at neutral pH (pH = 7). It also exhibits high viscosity as well as film and gel forming capability, which encourages its use in foods and cosmetics [14]. These are excellent properties for work as stabilizers in emulsion preparation [14]. Chaiwong et al. [23] have reported that the greater solubility also corresponded to the decrease in viscosity of the low- and medium-molecular-weight CMCHs, which are slightly different, but for the high-MW CMCH, it required significantly higher viscosity. This could be explained by the fact that CMCHs with chains longer or higher in MW were contributing to the gel. Moreover, it also has been pointed out by Tzaneva et al. [46] that with increasing temperature of emulsions, viscosity and shear stress decreased with different gradients. Using CMCH as a stabilizing agent indicates the ability of its rheological characteristics. After measurement of thermophysical properties by TGA/DTA analysis, it can be concluded that CMCH is suitable to work in the heating process and sterilization at temperatures up to 220 °C without changing the quality of components. The emulsions containing 0.3–0.5% (w/v) of CMCH could be applied in terms of pharmaceutical and cosmetic oil/water emulsions.3.3.4. Color L*, a*, b* and ΔE.

Color measurements with the colorimeter of deodorant creams with concentrations of 0.5, 1.0, 1.5, 2.0, and 2.5% (w/v) H-CMCH (H1, H2, H3, H4, and H5) were carried out during the accelerated stability test under 4 °C in a refrigerator and 45 °C in an incubator by heating/cooling cycle (4 °C, 24 h and 45 °C, 24 h) for six cycles. Defined by the Commission Internationale de l'Eclairage (CIE), the L*, a*, and b* color space was modeled after a color-opponent theory. As L* indicates lightness, a* is the red/green coordinate, and b* is the yellow/blue coordinate. The results showed that all five formulas of deodorant cream have an initial L* value (cycle 0) in the range of 79.63–80.07. Moreover, it was found that as the number of cycles of the acceleration test increased, the brightness of five deodorant formulas was significantly reduced ($p < 0.05$), as shown in Figure S4. The H4 and H5 deodorant creams had the lowest L* values compared with H1, H2, and H3 at accelerated cycles 3–6. The separation is caused by high-speed centrifugation, which can accelerate the emulsion precipitation. A good emulsion must withstand a long centrifugal force of 5000–10,000× g for 30 min without separation. Shaking or stirring causes more particles in the emulsion to be more mixed. Moreover, reducing the viscosity accelerates the integration of the internal or dispersed phase. This acceleration is achieved by continuously centrifuging the emulsion. Normally, the emulsion stability limited by agglomeration, sedimentation, viscosity of the aqueous phase and rheological properties of the emulsion [47]. This is a result of disintegration or changes in the structure of important substances in the ME and H-CMCH.

For a* (red/green coordinate) values, the result is shown in Figure S5 when considering each formula of deodorant cream during the accelerated stability test for six cycles. It was found that the a* tended to increase in cycle 2 and tended to decrease in cycle 3 until the end of storage. For each cycle in accelerated storage, the results showed that the a* value of the cream deodorant formulas H1, H2, and H3 in cycles 3–6 were in steady decline ($p \geq 0.05$), ranging from 1.41 to 1.43. Due to the instability of deodorant cream with poor emulsion and lower smoothness, it was clearly seen that the consistency of the cream changed as the number of stability tests increased.

For b* (yellow/blue coordinate) values, the result is shown in Figure S6 when considering each formula of deodorant cream during the accelerated stability test for six cycles. It was found that there was a tendency of the b* value of the deodorant creams to increase from the initial cycle (cycle 0) in the range of 2.1–2.5. For H4 and H5, the b* increased from cycle 1 until the end; the values remained in the ranges 3.26–4.34 and 3.49–4.34, respectively. Meanwhile, b* values for H2 and H3 tended to increase in cycle 2 and gradually remained constant until cycle 2–6 retention, ranging from 2.53–2.93. and 2.54–3.93. For H1, the b* value changed at accelerated cycle 3 until the end of storage (cycles 2–6), being in the range of 2.63–3.43. This showed that increasing storage time had the effect of increasing the b* value of deodorant creams ($p < 0.05$).

The total color difference (ΔE) describes incorporated changes in the qualities of L*, a*, and b* through the square root of the sum of square differences between two sets of complete color values [48]. The ΔE of deodorant creams with different concentrations of H-CMCH (0.5–2.5%) was measured during an accelerated stability test performed at 4 °C in a refrigerator and 45 °C in an incubator by heating/cooling cycle (4 °C and 45 °C for 24 h) for six cycles and compared with the basic formula deodorant cream, measuring with the CIE system colorimeter and calculating in the form of ΔE as in Figure 7. In comparison with the white basic deodorant cream, an effect on ΔE resulted. As the development of the deodorant cream involved adding mangosteen extract for the deodorizing agent, the initial color of all five deodorant formulas was white and pale yellow. This could be clearly observed with ΔE in the range of 2.65–3.06. However, as the retention period increased, the results showed that the stability of the cream changed, with visible separation occurring and unstable color, affecting the ΔE, which tended to increase significantly ($p < 0.05$). However, their colors were still acceptable by consumers if the ΔE values were less than 5 [49].

3.4. Deodorizing Activity

Trans-2-nonenal is an unsaturated aldehyde produced from lipid oxidation, which generates an unpleasant greasy odor. It is known to be a major odor component detected from the bodies of old people [20]. Different concentrations (1, 10, and 100 mg/mL) of each sample—(a) ME, (b) standard EGCG, (c) prototype cream, (d) developed deodorant cream mixed ME and 1.0% H-CMCH (H2), and (e) prototype cream mixed with EGCG standard and 1.0% H-CMCH—were used for deodorizing activity against trans-2-nonenal as shown in Figure 8. It was found that the basic deodorant had the lowest deodorizing activity (18–37%). The deodorizing activity was significantly increased ($p < 0.05$) when ME and EGCG were added to the basic formula deodorant. However, the samples of deodorant cream with ME added and formula with EGCG added at a concentration of 1–100 mg/mL. The results showed that the deodorizing activities were in the range of 27–70% and 21–68%, respectively, which was slightly lower than ME and EGCG standards. The basic formula deodorant contains waxes and fatty acids (fatty acids or fatty alcohol), which are of high MW, high viscosity, non-volatile, and have skin moisturizing properties (by reducing the evaporation of water), but it has no deodorizing properties [50]. Therefore, for some types of deodorant creams or cosmetics, it is imperative to add an active substance to the product in order to increase its antioxidant properties and deodorizing activity.

3.5. Antioxidant Properties

The developed formula (ME + 1% (w/v) H-CMCH) was selected from former experiments in order to compare the antioxidant activities to the prototype formula (no ME and H-CMCH) as presented in Table 2. The developed formula had strong antioxidant activity. Although the DPPH values of the two formulas were not statistically different ($p \geq 0.05$), the developed deodorant cream showed the greater ABTS values and had higher ferric ion reducing antioxidant power than the prototype formula.

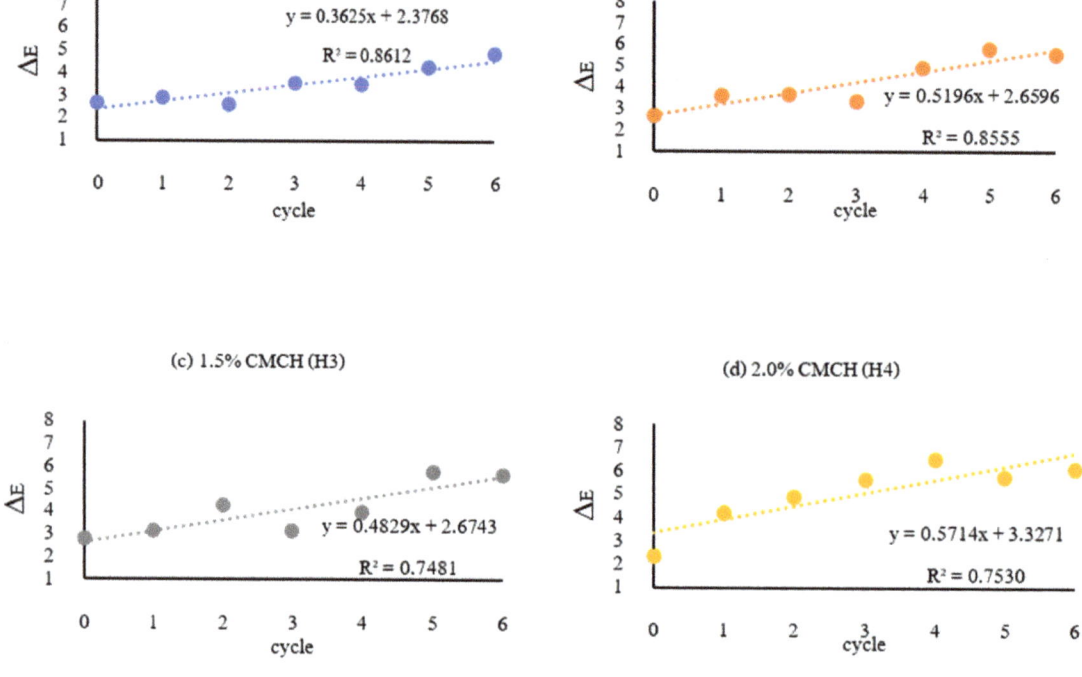
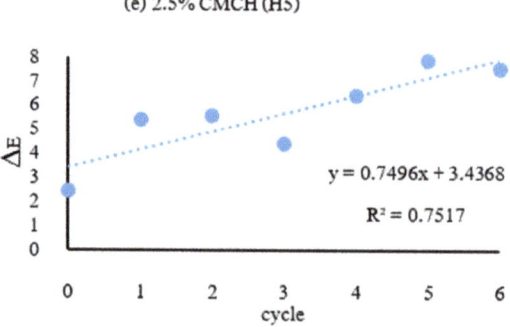

Figure 7. Total color difference (ΔE) of deodorant cream adding (**a**) 0.5%, (**b**) 1.0 %, (**c**) 1.5%, (**d**) 2.0% and (**e**) 2.5% (w/v) H-CMCH; heating/cooling cycle for up to 6 cycles.

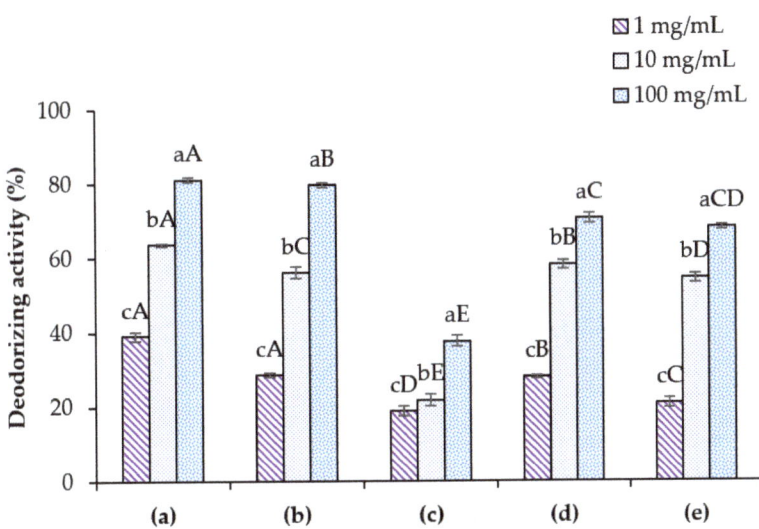

Figure 8. Deodorizing activity of (**a**) ME, (**b**) EGCG, (**c**) prototype cream, (**d**) developed deodorant cream mixed with ME and 1.0% (w/v) H-CMCH (H2), and (**e**) prototype cream mixed with EGCG and 1.0% (w/v) H-CMCH at different concentrations (1, 10, and 100 mg/mL). Different lowercase letters (a,b,c...) indicate significant differences between concentrations in the same formula, and different uppercase letters (A,B,C...) indicate significant differences between formulas at the same concentration.

Table 2. Antioxidant properties of the developed deodorant cream compared to prototype formula.

	Samples	IC_{50} DPPH^{ns} (µg/mL)	IC_{50} ABTS (µg/mL)	FRAP (µmoL Fe^{2+}/g Sample)
(1)	Deodorant cream prototype (no ME and H-CMCH)	11.4 ± 2.8	12.7 [a] ± 1.0	47.5 [b] ± 0.4
(2)	Developed deodorant cream (ME + 1% H-CMCH)	13.7 ± 3.0	7.7 [b] ± 2.8	51.8 [a] ± 0.6

Different letters indicate significant different between columns ($p < 0.05$).

3.6. Antibacterial Properties

For the antibacterial properties, the deodorant cream with the mixture of ME and 1% (w/v) H-CMCH was compared with a basic formula deodorant cream and streptomycin. It was found that the developed deodorant cream could inhibit all six types of bacteria, including *S. aureus*, *S. epidermidis*, *Corynebacterium* spp., *B. subtilis*, *P. aeruginosa*, and *E. coli*, and it was more effective in antibacterial activity than the basic formula (without ME and H-CMCH), as reflected by a greater inhibition zone (Table 3). Table 3 showed that the incorporation of ME and H-CMCH improved the antimicrobial properties of the deodorant cream. Janardhanan et al. [51] reported that mangosteen pericarp extract is known for its antibacterial activity against several pathogens that cause skin infection and acne. Moreover, He et al. [52] prepared the CMCH/lincomycin hydrogels for investigation into antibacterial properties. The antibacterial activities of the hydrogels were tested against Gram-negative (*E. coli*) and Gram-positive (*S. aureus*) bacteria. The result showed that the CMCH/lincomycin hydrogel was expected to be used as an antibacterial agent. Mohamed and Sabaa [53] studied CMCH/silver nanoparticle (Ag) hydrogels with high antibacterial activity against three Gram-positive bacteria (*S. aureus*, *B. subtilis*, and *Streptococcus faecalis*), three Gram-negative bacteria (*E. coli*, *P. aeruginosa*, and *Neisseria gonorrhoeae*), and *Candida albicans* fungus. The hydrophobicity and antibacterial properties of the solid surface are closely correlated with adhesion forces [54].

Table 3. Inhibition zone of prototype deodorant cream and developed deodorant cream.

Samples (10 mg/mL)	Inhibition Zone (mm)					
	S. aureus	*S. epidermidis*	*Corynebacterium* spp.	*B. subtilis*	*P. aeruginosa*	*E. coli*
(1) Deodorant cream prototype (no ME and H-CMCH)	6.6 [c] ± 0.3	6.6 [c] ± 0.8	7.2 [c] ± 0.9	6.9 [c] ± 1.1	6.5 [c] ± 0.5	6.0 [c] ± 0.7
(2) Developed deodorant cream (ME + 1% H-CMCH)	13.3 [b] ± 0.6	19.2 [b] ± 1.2	25.3 [b] ± 0.5	21.7 [b] ± 2.1	7.9 [b] ± 0.8	12.4 [b] ± 1.2
(3) Streptomycin	19.3 [a] ± 0.5	27.2 [a] ± 0.4	36.8 [a] ± 1.1	30.0 [a] ± 1.9	10.6 [a] ± 0.5	19.5 [a] ± 0.5

Different letters indicate significant different between columns ($p < 0.05$).

4. Conclusions

H-CMCH showed to be an effective polymer in retaining skin moisture for longer than untreated skin, water, propylene glycol, and native chitosan. Additionally, from the mangosteen extract deodorant creams with different H-CMCH concentrations (0.5–2.5% w/v), the appropriate H-CMCH content was selected from an accelerated stability test with six heating/cooling cycles. For the developed deodorant cream with 1.0% (w/v) H-CMCH, the viscosity and pH were unchanged after storage in the accelerated state, while the a* and b* values of the other formulas were slightly increased and the L* values was moderately decreased. Therefore, in deodorant cream development, 1.0%(w/v) H-CMCH was used for the optimal formula. Results indicated that the synergistic activity of ME and H-CMCH in emulsion creams had good potential as an effective skin moisturizing agent enhancer and good deodorizing activity against trans-2-nonenal odor, antioxidant properties, and antibacterial properties. Future studies may include investigation on modeling and numerical simulation of product stability. In addition, the engineering rheological properties of CMCH and creams should also be subsequently investigated.

Supplementary Materials: The following are available online at https://www.mdpi.com/article/10.3390/polym14010178/s1. Figure S1. FT-IR spectra of (**a**) high Mw native chitosan; (**b**) H-CMCH; Figure S2. pH of deodorant creams adding H-CMCH by heating-cooling cycle for up to 6 cycles; Figure S3. Viscosity of deodorant creams adding H-CMCH by heating-cooling cycle for up to 6 cycles; Figure S4. Lightness (L*) of deodorant creams adding H-CMCH by heating-cooling cycle for up to 6 cycles; Figure S5. Red/green coordinate (a*) of deodorant creams adding H-CMCH by heating-cooling cycle for up to 6 cycles; Figure S6. Yellow/blue coordinate (b*) of deodorant creams adding H-CMCH by heating-cooling cycle for up to 6 cycles.

Author Contributions: Conceptualization, N.C., P.L. and Y.P.; methodology, N.C., P.R., P.L. and W.R.; formal analysis, N.C., M.J.S. and Y.P.; investigation, N.C. and Y.P.; resources, P.L. and S.R.S.; writing—original draft preparation, N.C. and Y.P.; writing—review and editing, K.J., P.S., N.L. and F.J.B.; supervision, Y.P. and P.L.; funding acquisition, Y.P. and W.P. All authors have read and agreed to the published version of the manuscript.

Funding: This research was funded by the National Research Council of Thailand (NRCT), grant number 256106A1040012, and the APC was funded by Chiang Mai University.

Institutional Review Board Statement: Not applicable.

Informed Consent Statement: Not applicable.

Data Availability Statement: Not applicable.

Acknowledgments: We wish to thank Center of Excellence in Materials Science and Technology, Chiang Mai University for financial support under the administration of Materials Science Research Center, Faculty of Science, Chiang Mai University. This research work was also partially supported by Chiang Mai University under the Cluster of Agro Bio-Circular-Green.

Conflicts of Interest: The authors declare no conflict of interest associated with this research.

References

1. Estanqueiro, M.; Conceição, J.; Amaral, M.H.; Santos, D.; Silva, J.B.; Lobo, J.M.S. Characterization and stability studies of emulsion systems containing pumice. *Braz. J. Pharm. Sci.* **2014**, *50*, 361–369. [CrossRef]
2. Kim, K.-M.; Oh, H.M.; Lee, J.H. Controlling the emulsion stability of cosmetics through shear mixing process. *Korea Aust. Rheol. J.* **2020**, *32*, 243–249. [CrossRef]
3. Gilbert, L.; Picard, C.; Savary, G.; Grisel, M. Rheological and textural characterization of cosmetic emulsions containing natural and synthetic polymers: Relationships between both data. *Colloids Surf. Physicochem. Eng. Asp.* **2013**, *421*, 150–163. [CrossRef]
4. Colucci, G.; Santamaria-Echart, A.; Silva, S.C.; Fernandes, I.P.; Sipoli, C.C.; Barreiro, M.F. Development of water-in-oil emulsions as delivery vehicles and testing with a natural antimicrobial extract. *Molecules* **2020**, *25*, 2105. [CrossRef]
5. Zelisko, P.M.; Flora, K.K.; Brennan, J.D.; Brook, M.A. Water-in-silicone oil emulsion stabilizing surfactants formed from native albumin and α, ω-triethoxysilylpropyl-polydimethylsiloxane. *Biomacromolecules* **2008**, *9*, 2153–2161. [CrossRef]
6. Nazir, H.; Zhang, W.; Liu, Y.; Chen, X.; Wang, L.; Naseer, M.; Ma, G. Silicone oil emulsions: Strategies to improve their stability and applications in hair care products. *Int. J. Cosmet. Sci.* **2014**, *36*, 124–133. [CrossRef]
7. Chaiyasut, C.; Kesika, P.; Sakdakampanat, P.; Peerajan, S.; Sivamaruthi, B.S. Formulation and evaluation of stability of Thai purple rice bran-based cosmetic products. *Asian J. Pharm. Clin. Res.* **2018**, *11*, 99–104. [CrossRef]
8. Lestari, U.; Farid, F.; Fudholi, A. Formulation and effectivity test of deodorant from activated charcoal of palm shell as excessive sweat adsorbent on body. *Asian J. Pharm. Clin. Res.* **2019**, *12*, 193–196. [CrossRef]
9. Kanlayavattanakul, M.; Lourith, N. Body malodours and their topical treatment agents. *Int. J. Cosmet. Sci.* **2011**, *33*, 298–311. [CrossRef]
10. Ganceviciene, R.; Liakou, A.I.; Theodoridis, A.; Makrantonaki, E.; Zouboulis, C.C. Skin anti-aging strategies. *Derm. Endocrinol.* **2012**, *4*, 308–319. [CrossRef]
11. Lintner, K. Benefits of anti-aging actives in sunscreens. *Cosmetics* **2017**, *4*, 7. [CrossRef]
12. Debnath, S.; Babu, M.N.; Kusuma, G. Formulation and evaluation of herbal antimicrobial deodorant stick. *Res. J. Top. Cosmet. Sci.* **2011**, *2*, 21–24.
13. Bose, A.; Wong, T.W. Oral colon cancer targeting by chitosan nanocomposites. In *Applications of Nanocomposite Materials in Drug Delivery*; Elsevier: Amsterdam, The Netherlands, 2018; pp. 409–429.
14. Jimtaisong, A.; Saewan, N. Utilization of carboxymethyl chitosan in cosmetics. *Int. J. Cosmet. Sci.* **2014**, *36*, 12–21. [CrossRef] [PubMed]
15. Rachtanapun, P.; Simasatitkul, P.; Chaiwan, W.; Watthanaworasakun, Y. Effect of sodium hydroxide concentration on properties of carboxymethyl rice starch. *Int. Food Res. J.* **2012**, *19*, 923.
16. Ghasemzadeh, A.; Jaafar, H.Z.; Baghdadi, A.; Tayebi-Meigooni, A. Alpha-mangostin-rich extracts from mangosteen pericarp: Optimization of green extraction protocol and evaluation of biological activity. *Molecules* **2018**, *23*, 1852. [CrossRef]
17. Moosophin, K.; Wetthaisong, T.; Seeratchakot, L.; Kokluecha, W. Tannin extraction from mangosteen peel for protein precipitation in wine. *Asia Pac. J. Sci. Technol.* **2010**, *15*, 377–385.
18. Aizat, W.M.; Ahmad-Hashim, F.H.; Jaafar, S.N.S. Valorization of mangosteen, "The Queen of Fruits," and new advances in postharvest and in food and engineering applications: A review. *J. Adv. Res.* **2019**, *20*, 61–70. [CrossRef] [PubMed]
19. Seesuriyachan, P.; Kuntiya, A.; Kawee-ai, A.; Techapun, C.; Chaiyaso, T.; Leksawasdi, N. Improvement in efficiency of lignin degradation by Fenton reaction using synergistic catalytic action. *Ecol. Eng.* **2015**, *85*, 283–287. [CrossRef]
20. Pothitirat, W.; Chomnawang, M.T.; Supabphol, R.; Gritsanapan, W. Free radical scavenging and anti-acne activities of mangosteen fruit rind extracts prepared by different extraction methods. *Pharm. Biol.* **2010**, *48*, 182–186. [CrossRef]
21. Ham, J.-S.; Kim, H.-Y.; Lim, S.-T. Antioxidant and deodorizing activities of phenolic components in chestnut inner shell extracts. *Ind. Crops Prod.* **2015**, *73*, 99–105. [CrossRef]
22. Pothitirat, W.; Chomnawang, M.T.; Gritsanapan, W. Anti-acne-inducing bacterial activity of mangosteen fruit rind extracts. *Med. Princ. Pract.* **2010**, *19*, 281–286. [CrossRef] [PubMed]
23. Chaiwong, N.; Leelapornpisid, P.; Jantanasakulwong, K.; Rachtanapun, P.; Seesuriyachan, P.; Sakdatorn, V.; Leksawasdi, N.; Phimolsiripol, Y. Antioxidant and moisturizing properties of carboxymethyl chitosan with different molecular weights. *Polymers* **2020**, *12*, 1445. [CrossRef]
24. Standard, T.A. *Mangosteen*; National Bureau of Agricultural Commodity and Food Standards Ministry of Agriculture and Cooperatives: Bangkok, Thailand, 2013.

25. Kassakul, W.; Praznik, W.; Viernstein, H.; Hongwiset, D.; Phrutivorapongkul, A.; Leelapornpisid, P. Characterization of the mucilages extracted from *Hibiscus rosasinensis* linn and *Hibiscus mutabilis* linn and their skin moisturizing effect. *Int. J. Pharm. Sci. Res.* **2014**, *6*, 453–457.
26. Tkacz, K.; Modzelewska-Kapituła, M.; Wiek, A.; Nogalski, Z. The applicability of total color difference hE for determining the blooming time in *Longissimus lumborum* and *Semimembranosus* muscles from Holstein-Friesian bulls at different ageing times. *Appl. Sci.* **2020**, *10*, 8215. [CrossRef]
27. Surin, S.; You, S.G.; Seesuriyachan, P.; Muangrat, R.; Wangtueai, S.; Jambrak, A.R.; Phongthai, S.; Jantanasakulwong, K.; Chaiyaso, T.; Phimolsiripol, Y. Optimization of ultrasonic-assisted extraction of polysaccharides from purple glutinous rice bran (*Oryza sativa* L.) and their antioxidant activities. *Sci. Rep.* **2020**, *10*, 10410. [CrossRef]
28. Phimolsiripol, Y.; Buadoktoom, S.; Leelapornpisid, P.; Jantanasakulwong, K.; Seesuriyachan, P.; Chaiyaso, T.; Leksawasdi, N.; Rachtanapun, P.; Chaiwong, N.; Sommano, S.R.; et al. Shelf life extension of chilled pork by optimal ultrasonicated Ceylon Spinach (Basella alba) extracts: Physicochemical and microbial properties. *Foods* **2021**, *10*, 1241. [CrossRef]
29. Surin, S.; Surayot, U.; Seesuriyachan, P.; You, S.G.; Phimolsiripol, Y. Antioxidant and immunomodulatory activities of sulphated polysaccharides from purple glutinous rice bran (*Oryza sativa* L.). *Int. J. Food Sci. Tech.* **2018**, *53*, 994–1004. [CrossRef]
30. Ruksiriwanich, W.; Khantham, C.; Linsaenkart, P.; Chaitep, T.; Rachtanapun, P.; Jantanasakulwong, K.; Phimolsiripol, Y.; Jambrak, A.R.; Nazir, Y.; Yooin, W.; et al. Anti-inflammation of bioactive compounds from ethanolic extracts of edible mushroom (*Dictyophora indusiata*) as functional health promoting food ingredients. *Int. J. Food Sci. Tech.* **2021**, 1–13. [CrossRef]
31. Surin, S.; Seesuriyachan, P.; Thakeow, P.; You, S.G.; Phimolsiripol, Y. Antioxidant and antimicrobial properties of polysaccharides from rice brans. *Chiang Mai J. Sci.* **2018**, *45*, 1372–1382.
32. Bai-Ngew, S.; Chuensun, T.; Wangtueai, S.; Phongthai, S.; Jantanasakulwong, K.; Rachtanapun, P.; Sakdatorn, V.; Klunklin, W.; Regenstein, J.M.; Phimolsiripol, Y. Antimicrobial activity of a crude peptide extract from lablab bean (*Dolichos lablab*) with semi-dried rice noodles shelf-life. *Qual. Assur. Saf. Crop.* **2021**, *13*, 25–33. [CrossRef]
33. Siahaan, P.; Mentari, N.C.; Wiedyanto, U.O.; Hudiyanti, D.; Hildayani, S.Z.; Laksitorini, M.D. The optimum conditions of carboxymethyl chitosan synthesis on drug delivery application and its release of kinetics study. *Indones. J. Chem.* **2017**, *17*, 291–300. [CrossRef]
34. Rachtanapun, P.; Jantrawut, P.; Klunklin, W.; Jantanasakulwong, K.; Phimolsiripol, Y.; Leksawasdi, N.; Seesuriyachan, P.; Chaiyaso, T.; Insomphun, C.; Phongthai, S.; et al. Carboxymethyl bacterial cellulose from Nata de coco: Effects of NaOH. *Polymers* **2021**, *13*, 348. [CrossRef]
35. Samar, M.M.; El-Kalyoubi, M.; Khalaf, M.; Abd El-Razik, M. Physicochemical, functional, antioxidant and antibacterial properties of chitosan extracted from shrimp wastes by microwave technique. *Ann. Agric. Sci.* **2013**, *58*, 33–41. [CrossRef]
36. Thanakkasaranee, S.; Jantanasakulwong, K.; Phimolsiripol, Y.; Leksawasdi, N.; Seesuriyachan, P.; Chaiyaso, T.; Jantrawut, P.; Ruksiriwanich, W.; Rose Sommano, S.; Punyodom, W.; et al. High substitution synthesis of carboxymethyl chitosan for properties improvement of carboxymethyl chitosan films depending on particle sizes. *Molecules* **2021**, *26*, 6013. [CrossRef] [PubMed]
37. Martí-Quijal, F.J.; Ramon-Mascarell, F.; Pallarés, N.; Ferrer, E.; Berrada, H.; Phimolsiripol, Y.; Barba, F.J. Extraction of antioxidant compounds and pigments from spirulina (*Arthrospira platensis*) assisted by pulsed electric fields and the binary mixture of organic solvents and water. *Appl. Sci.* **2021**, *11*, 7629. [CrossRef]
38. Tamer, T.M.; Hassan, M.A.; Omer, A.M.; Baset, W.M.; Hassan, M.E.; El-Shafeey, M.; Eldin, M.M. Synthesis, characterization and antibacterial evaluation of two aromatic chitosan Schiff base derivatives. *Process Biochem.* **2016**, *51*, 1721–1730. [CrossRef]
39. Aranaz, I.; Acosta, N.; Civera, C.; Elorza, B.; Mingo, J.; Castro, C.; Gandía, M.D.L.L.; Caballero, A.H. Cosmetics and cosmeceutical applications of chitin, chitosan and their derivatives. *Polymers* **2018**, *10*, 213. [CrossRef]
40. Szymańska, E.; Winnicka, K. Stability of chitosan—A challenge for pharmaceutical and biomedical applications. *Mar. Drugs* **2015**, *13*, 1819–1846. [CrossRef]
41. Kartini, K.; Winarjo, B.M.; Fitriani, E.W.; Islamie, R. Formulation and pH-physical stability evaluation of gel and cream of Plantago major leaves extract. *MPI (Media Pharm. Indones.)* **2017**, *1*, 174–180. [CrossRef]
42. Navarro-Pérez, Y.M.; Cedeño-Linares, E.; Norman-Montenegro, O.; Ruz-Sanjuan, V.; Mondeja-Rivera, Y.; Hernández-Monzón, A.M.; González-Bedia, M.M. Prediction of the physical stability and quality of O/W cosmetic emulsions using full factorial design. *J. Pharm. Pharmacogn. Res.* **2021**, *9*, 98–112.
43. Sankar, M.; Nowicka, E.; Carter, E.; Murphy, D.M.; Knight, D.W.; Bethell, D.; Hutchings, G.J. The benzaldehyde oxidation paradox explained by the interception of peroxy radical by benzyl alcohol. *Nat. Commun.* **2014**, *5*, 3332. [CrossRef] [PubMed]
44. Leksawasdi, N.; Breuer, M.; Hauer, B.; Rosche, B.; Rogers, P.L. Kinetics of pyruvate decarboxylase deactivation by benzaldehyde. *Biocatal. Biotrans.* **2003**, *21*, 315–320. [CrossRef]
45. Anderson, F. Final report on the safety assessment of benzaldehyde. *Int. J. Toxicol.* **2006**, *25* (Suppl. S1), 11–27.
46. Tzaneva, D.; Simitchiev, A.; Petkova, N.; Nenov, V.; Stoyanova, A.; Denev, P. Synthesis of carboxymethyl chitosan and its rheological behaviour in pharmaceutical and cosmetic emulsions. *J. Appl. Pharm. Sci.* **2017**, *7*, 70–80.
47. Wang, B.; Tian, H.; Xiang, D. Stabilizing the oil-in-water emulsions using the mixtures of *Dendrobium officinale* polysaccharides and gum arabic or propylene glycol alginate. *Molecules* **2020**, *25*, 759. [CrossRef]
48. Yan, Y.; Lee, J.; Hong, J.; Suk, H.J. Measuring and describing the discoloration of liquid foundation. *Color Res. Appl.* **2021**, *46*, 362–375. [CrossRef]
49. Mokrzycki, W.; Tatol, M. Colour difference ΔE-A survey. *Mach. Graph. Vis.* **2011**, *20*, 383–411.

50. Ilievska, J.; Cicimov, V.; Antova, E.; Gjorgoski, I.; Hadzy-Petrushev, N.; Mladenov, M. Heat-induced oxidative stress and inflammation in rats in relation to age. *Res. Phys. Educ. Sport Health* **2016**, *5*, 123–130.
51. Janardhanan, S.; Mahendra, J.; Girija, A.S.; Mahendra, L.; Priyadharsini, V. Antimicrobial effects of *Garcinia mangostana* on cariogenic microorganisms. *J. Clin. Diagn. Res.* **2017**, *11*, 19–22. [CrossRef]
52. He, G.; Chen, X.; Yin, Y.; Cai, W.; Ke, W.; Kong, Y.; Zheng, H. Preparation and antibacterial properties of O-carboxymethyl chitosan/lincomycin hydrogels. *J. Biomater. Sci. Polym. Ed.* **2016**, *27*, 370–384. [CrossRef] [PubMed]
53. Mohamed, R.R.; Sabaa, M.W. Synthesis and characterization of antimicrobial crosslinked carboxymethyl chitosan nanoparticles loaded with silver. *Int. J. Biol. Macromol.* **2014**, *69*, 95–99. [CrossRef] [PubMed]
54. Liu, J.-L.; Xia, R. A unified analysis of a micro-beam, droplet and CNT ring adhered on a substrate: Calculation of variation with movable boundaries. *Acta Mech. Sin.* **2013**, *29*, 62–72. [CrossRef]

Article

Accelerating Payload Release from Complex Coacervates through Mechanical Stimulation

Wesam A. Hatem and Yakov Lapitsky *

Department of Chemical Engineering, University of Toledo, Toledo, OH 43606, USA
* Correspondence: yakov.lapitsky@utoledo.edu; Tel.: +1-419-530-8254

Abstract: Complex coacervates formed through the association of charged polymers with oppositely charged species are often investigated for controlled release applications and can provide highly sustained (multi-day, -week or -month) release of both small-molecule and macromolecular actives. This release, however, can sometimes be too slow to deliver the active molecules in the doses needed to achieve the desired effect. Here, we explore how the slow release of small molecules from coacervate matrices can be accelerated through mechanical stimulation. Using coacervates formed through the association of poly(allylamine hydrochloride) (PAH) with pentavalent tripolyphosphate (TPP) ions and Rhodamine B dye as the model coacervate and payload, we demonstrate that slow payload release from complex coacervates can be accelerated severalfold through mechanical stimulation (akin to flavor release from a chewed piece of gum). The stimulation leading to this effect can be readily achieved through either perforation (with needles) or compression of the coacervates and, besides accelerating the release, can result in a deswelling of the coacervate phases. The mechanical activation effect evidently reflects the rupture and collapse of solvent-filled pores, which form due to osmotic swelling of the solute-charged coacervate pellets and is most pronounced in release media that favor swelling. This stimulation effect is therefore strong in deionized water (where the swelling is substantial) and only subtle and shorter-lived in phosphate buffered saline (where the PAH/TPP coacervate swelling is inhibited). Taken together, these findings suggest that mechanical activation could be useful in extending the complex coacervate matrix efficacy in highly sustained release applications where the slowly releasing coacervate-based sustained release vehicles undergo significant osmotic swelling.

Keywords: polyelectrolyte; complex coacervate; polyamine; stimulus-responsive materials; controlled release

Citation: Hatem, W.A.; Lapitsky, Y. Accelerating Payload Release from Complex Coacervates through Mechanical Stimulation. *Polymers* **2023**, *15*, 586. https://doi.org/10.3390/polym15030586

Academic Editors: Lorenzo Antonio Picos Corrales, Angel Licea-Claverie and Grégorio Crini

Received: 5 January 2023
Revised: 17 January 2023
Accepted: 17 January 2023
Published: 23 January 2023

Copyright: © 2023 by the authors. Licensee MDPI, Basel, Switzerland. This article is an open access article distributed under the terms and conditions of the Creative Commons Attribution (CC BY) license (https://creativecommons.org/licenses/by/4.0/).

1. Introduction

Complex coacervation is a liquid-liquid phase separation that occurs through the complexation of colloidal (or macromolecular) solutes with other solute species [1–3]. This phase separation generates a solute-rich coacervate phase, which is rich in both the associating solution species and in equilibrium with a dilute, solvent-rich supernatant phase. The colloid-rich coacervate phase typically has viscoelastic fluid- or gel-like properties [4–6], and offers numerous benefits: easy formation under mild, aqueous conditions [7,8]; low toxicity [9,10]; and an ability to form, transform their properties, and dissolve in response to external stimuli [1,11–13]. Among their many potential applications (which range from drug delivery [8,14,15] to separation processes [16,17], foods [18], and adhesives [11,19,20]), complex coacervates are frequently used in the controlled release of various active compounds [8,14,15,21–23].

One aspect of complex coacervates that makes them potentially effective for sustained release applications is their polymer-rich composition, which—in contrast to most hydrogels [24]—makes them highly effective diffusion barriers [9,15,25,26]. This barrier property enables them to sustain the release of small water-soluble molecules (which tend to elute rapidly from gels) over highly extended timescales. Complex coacervates formed from poly(allylamine hydrochloride) (PAH) complexed with the multivalent anion tripolyphosphate (TPP), for instance, have recently been shown to sustain the release of diverse small molecules (including drugs and disinfectants) over multiple months [9,15,21,27].

While the highly sustained release enabled by these viscoelastic materials could be useful in an array of biomedical, household, and industrial technologies, the duration of the benefits derived from such release can sometimes be limited by the payload delivery ultimately becoming too slow to be efficacious (once the most-accessible portion of their payload closest to the surface is released) [21,27]. A recent study, for instance, revealed that, while the long-term bactericide release from PAH/TPP coacervates can provide antibacterial benefits over multiple weeks, this activity is ultimately lost (even though only a fraction of the bactericidal payload is released) [21]. The evident reason for this ultimate activity loss is that the release rate declines with time and, after a few weeks, becomes too slow to be effective. Indeed, with certain coacervate/payload molecule combinations, the slower-than-desired release can impose even greater limitations on the coacervate's applicability (e.g., to situations/applications where the volumes of the release media relative to the coacervate are very low, such that the released actives are not diluted below the minimal concentration needed for their efficacious use) [27].

To overcome these limitations of insufficient release rates or early activity loss, here we explore the use of mechanical stimulation—namely, periodic perforation and compression—as a method for stimulating or reactivating small molecule release from coacervate matrices. Mechanical crushing of coacervate-based microcapsules (e.g., in ink, fragrance, or food formulations) is a well-known approach to achieving rapid stimulus-responsive release of hydrophobic payloads such as oil droplets [28–31]. However, we are not aware of any reports demonstrating mechanical stimulus use for controlling (1) long-term (multiday) release from complex coacervates, (2) the release of hydrophilic/water-soluble actives from complex coacervates, or (3) the release from continuous coacervate phases (i.e., macroscopic matrices) rather than dispersed coacervate microcapsules. To this end, using macroscopic PAH/TPP coacervates and the Rhodamine B (RhB) dye as a model complex coacervate and small, hydrophilic payload system, we analyze the effect of mechanical stimulation (using UV-vis spectroscopy) on long-term small molecule release. To gain further insight into these effects, the impacts of mechanical stimulation on coacervate swelling are also analyzed (through gravimetry and digital photography) and related to changes in the release profiles. Finally, we discuss the opportunities offered by (and limitations of) this approach in potential applications of complex coacervates.

2. Materials and Methods

2.1. Materials

All experiments were conducted using Millipore Direct-Q 3 deionized water (18.2 MΩ cm). The PAH (nominal molecular weight of 150 kDa; supplied as a 40 wt% aqueous solution) was purchased from Nittobo Medical Co. Ltd. (Tokyo, Japan). TPP ($\geq$98% pure) and RhB ($\geq$95% pure) were purchased from Sigma-Aldrich (St. Louis, MO, USA). Phosphate buffered saline (PBS) powder, HCl (12 M), and NaOH pellets ($\geq$97% pure) were purchased from Fischer Scientific (Nazareth, PA, Fair Lawn, NJ and Hampton, NH, respectively). All materials were used as received.

2.2. Coacervate Preparation

The coacervates were prepared in 2.0 mL microcentrifuge tubes at room temperature using 0.33 mL of 10 wt% PAH and 0.35 mL of 7.5 wt% TPP, both adjusted to pH 7.0 with 6 M HCl and 6 M NaOH solutions. Small (0.097 mL) aliquots of 16 mg/mL aqueous RhB solution were added to the PAH before adding TPP, whereupon the PAH/RhB solutions were mixed for 10–15 s on a vortex mixer. TPP was then immediately added (to generate a 0.20:1 TPP:PAH amine group molar ratio), shaking vigorously for ~10 s by hand after the single-shot addition. The phase-separating PAH/TPP/RhB mixtures were then centrifuged at 15,000 rpm for 90 min, which yielded single, macroscopic coacervate pellets (3–4 mm in height) at the bottoms of the microcentrifuge tubes with dilute/solvent-rich supernatant phases on the top.

2.3. Release Experiments

Upon forming the macroscopic coacervate matrix pellets, the supernatant phases were removed, after which the coacervates were weighed, submerged in 1 mL of release media—either deionized water or 1 × PBS (pH 7.4)—and agitated at 400 rpm and 37 °C using a Benchmark Scientific Multitherm shaker (South Plainfield, NJ, USA). To maintain sink conditions and determine the RhB amounts released, the release medium was collected (after 1 h and every 24 h thereafter) and replaced with fresh medium after rinsing the coacervate and tube with 1 mL of fresh, RhB-free release medium for 3–5 s. The released RhB was then quantified by UV-vis spectroscopy, employing a Varian Cary 50 spectrophotometer (λ = 555 nm; ε = 143 mL mg^{-1} cm^{-1} in deionized water and 192 mL mg^{-1} cm^{-1} in PBS).

To mechanically stimulate the release, two methods were used: perforation and compression. In the first release experiment, this agitation was performed every 3 d, with deionized water as the release medium and the first treatment occurring 3 d into the release process. Here, the perforation procedure was performed manually with a needle (a size 7 cotton darner), which was 47 mm long and 0.69 mm in diameter. During each mechanical stimulation step, the perforated samples were subjected to 5 needle jabs with locations resembling the Number 5 face of a game die (Figure 1a). Conversely, stimulation through coacervate compression was performed by wedging a 6.5 mm wide and 0.9 mm thick spatula (with a U-shaped tip) between the coacervate and centrifuge tube. These spatula insertions were performed one per treatment, and their locations were alternated between the treatment times by first inserting the spatula along the front-side tube wall, then the back-side wall, then the right-side wall, and finally, the left-side wall (Figure 1b). Both the needles and spatulas were allowed to reach the bottoms of the microcentrifuge tubes.

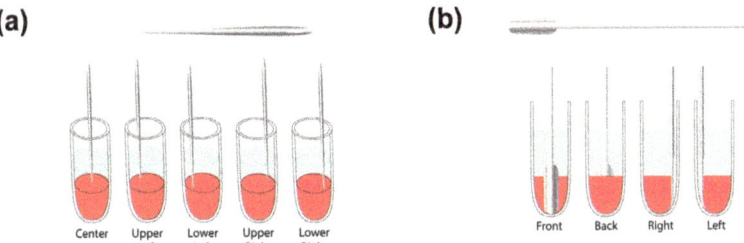

Figure 1. Mechanical stimulation schemes showing the approximate locations of the (**a**) perforations by the 5 needle jabs and (**b**) compressive spatula insertions. The perforated samples were jabbed multiple times during each treatment, while those subjected to compression were compressed once per treatment, either on the front, back, left, or right side of the sample.

To also examine the effects of mechanical stimulation frequency, another release experiment was conducted, where the stimulation was performed daily (with the first treatment being performed 1 d into the release process), and deionized water again served as the release medium. Since increasing the mechanical stimulation frequency without varying the intensity of each treatment increases the total amount of stimulation performed, we decreased the daily perforation in this experiment to 2 needle jabs (where the insertion locations resembled the Number 2 face on a game die and were rotated by ~90° each day). Lastly, to test the release medium's effect on the mechanical stimulation efficacy, the release experiment was repeated using 1 × PBS (pH 7.4) as the release medium. Here, the coacervates were stimulated daily with either needle jabs or through compression but using PBS as the swelling/release medium instead of deionized water. In each of the above experiments, release from the mechanically stimulated coacervates was compared with that from stimulation-free controls, and all release conditions were analyzed using at least three replicate samples.

2.4. Gravimetric Analysis

To gain insights into the mechanical stimulation effects on the coacervate stability, and possibly correlate the release performance to variations in coacervate swelling, changes in the wet coacervate weight were analyzed by gravimetry each time that the release media was replaced. During each such analysis, the supernatant release media was removed from the microcentrifuge tubes, whereupon each coacervate surface was rinsed with fresh release medium (either deionized water or 1 × PBS). To minimize the risk of unintended mechanical stimulation, the rinse stream was aimed at the microcentrifuge tube walls (rather than the coacervate surface) and allowed to gently flow down the tube sides. Any solvent remaining after the rinse step was then carefully removed with a Kimwipe^{TM} (by aspiring it into the corners of the wipes through capillary action while avoiding direct contact with the coacervate surface) before weighing the coacervate phase. The coacervate weights at each time point were normalized to their initial weight before being contacted with the release medium. These gravimetric analyses were supplemented with digital photography, which supplied further evidence of any changes in coacervate size. Like with the spectroscopic release measurements, all gravimetric measurements were performed using three replicate coacervate samples.

3. Results and Discussion

3.1. Effect of Mechanical Stimulation

The release into deionized water was highly sensitive to mechanical stimulation (Figure 1a). After rapidly releasing an average of ~2 µg RhB within the first day, the control (unstimulated) coacervates rapidly diminished their release rate to significantly less than 1 µg/d and released only a few µg total RhB after 1 month (blue diamonds in Figure 2a,b). Coacervates that were mechanically stimulated every 3 d, on the other hand (regardless of whether this was done through perforation or compression), produced pulsatile release profiles where, on the day following mechanical stimulation, there was an order of magnitude increase in the measured RhB released (such that ~2–4 µg of RhB were released in the day following the stimulation; see red circles and grey squares in Figure 2a,b). After this initial increase, however, the release rates returned to their near-baseline level. Further evidence of this pulsatility of the mechanically stimulated release process came from the plumes of pink RhB dye that were visually seen rising from the coacervate pellets upon their perforation (especially early in the release process and when the perforation occurred at the center rather than near the edges of the pellet). Though this reactivation effect became weaker with repeated application, it produced a discernable effect over ~1 month (as evident in Figure 2b), and the cumulative effect of the mechanical stimulation steps was a severalfold increase in the cumulative release over the ~ 1-month experiment (Figure 2a).

Besides the release profiles, the mechanical stimulation had a marked impact on the coacervate swelling. The control PAH/TPP coacervate pellets swelled significantly when placed in contact with the deionized water release media (blue diamonds in Figure 3a). This swelling caused the coacervate weight to monotonically increase with time and the coacervate pellets to increase in size and become wispy (see Figure 3bi). Such a response to deionized water was qualitatively consistent with that seen for polyanion/polycation coacervates (complexes of oppositely charged polyelectrolytes) in deionized water [32,33], and bactericide-loaded PAH/TPP coacervates in tap water [21,27], and—together with the continued slow release from these coacervates in Figure 2—suggested that the swollen coacervates were composed of water-rich pores dispersed in a continuous coacervate phase. These pores were likely gradually generated due to (1) the high osmotic pressure created by the encapsulated RhB and (possibly) unassociated PAH and TPP, and (2) the greater permeability of the (much smaller) water molecules through the coacervate phase, which allowed the water from the release medium to fill these pores. The view of the PAH/TPP coacervate being more permeable to water than other solutes (e.g., RhB or any uncomplexed PAH) was supported by the observation that these coacervates do not take very long to dry when exposed to air and was consistent with prior studies on the swelling of hydrogels, which showed the transport of water into swelling polymer networks to be faster than the release of osmotic pressure-causing organic solutes [34–37].

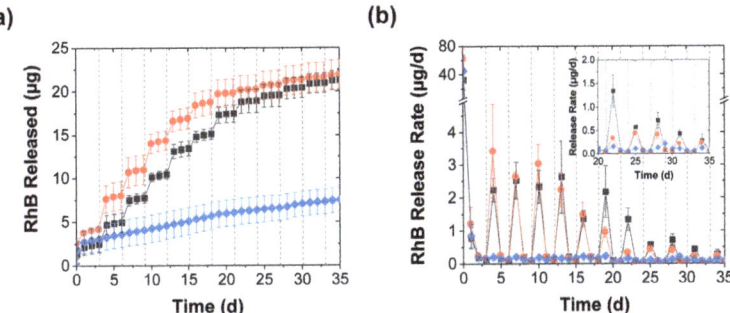

Figure 2. RhB release from coacervates into deionized water achieved with (■) periodic perforation by five needle jabs, (●) periodic compression, and (♦) without mechanical stimulation and shown in terms of both (**a**) the total RhB mass released and (**b**) the release rate (mean ± SD). The inset provides a closeup of the slower release rates near the end of the experiment. The coacervates in this experiment were stimulated every 3 d using either five needle jabs or a spatula. The solid lines are guides to the eye, while the dashed vertical lines mark the mechanical stimulation times.

More importantly, the "porous coacervate phase" interpretation was supported by the effect of mechanical stimulation on the normalized coacervate weight. Each mechanical stimulation step (regardless of whether performed by perforation or compression; grey squares and red circles in Figure 3a and images in Figure 3bii,iii), produced a sudden reduction in swelling, evidently due to the rupture/collapse of the water-rich pores (see Figure 3c,d). Upon the first mechanical stimulation of the coacervates, which occurred 3 d into the release profile (i.e., when the swelling was the fastest), the compression treatment reduced the slope of the coacervate weight versus time curves (cf. red circles and blue diamonds in Figure 3a), while the perforation (grey squares in Figure 3a) did not produce a significant effect. At longer times, however—starting with the second stimulation (performed 6 d into the release experiment), the coacervate weights decreased after each stimulation step. These decreases remained sharp over ~2 weeks, but then gradually became more subtle (Figure 3a). Moreover, their timing and intensity coincided with the spikes in the release rate (Figure 2b), which—along with the aforementioned plumes of RhB dye that were evident during early-stage perforation—suggested that the accelerated release was stimulated by the rupture of the RhB-loaded pores.

3.2. Effect of Mechanical Stimulation Frequency

With more frequent (daily) stimulation, the release became more uniform with, at least when the eluted RhB was measured daily, smooth cumulative release profiles (Figure 4a). Since the first mechanical stimulation was applied after 1 d, the onset of accelerated release became evident after the second day. With the compression (whose application was identical to that in Figure 2), the daily release rate first increased more than tenfold, and then decreased, as the effect of further mechanical stimulation evidently became less pronounced. In contrast, the perforation effect—which was reduced to two needle jabs per treatment (from the five jabs used when stimulated every 3 d) to maintain a similar average rate of needle jabs to that in Figure 2—generated a sigmoidal release profile where the release rate increased over the first five stimulation steps and then gradually decreased (Figure 4). Though the release rate with this perforation procedure did not rise as sharply (and did not reach as high of a crest) as that achieved through compression, this milder mechanical stimulation kept the average release rate above 1 µg/d longer than the (more intense) daily compression. Though after the first two weeks the mechanical stimulation effect became less pronounced for both activation procedures, this effect remained measurable (with a multifold increase in the release rate) throughout the 20-d experiment (see inset in Figure 4b).

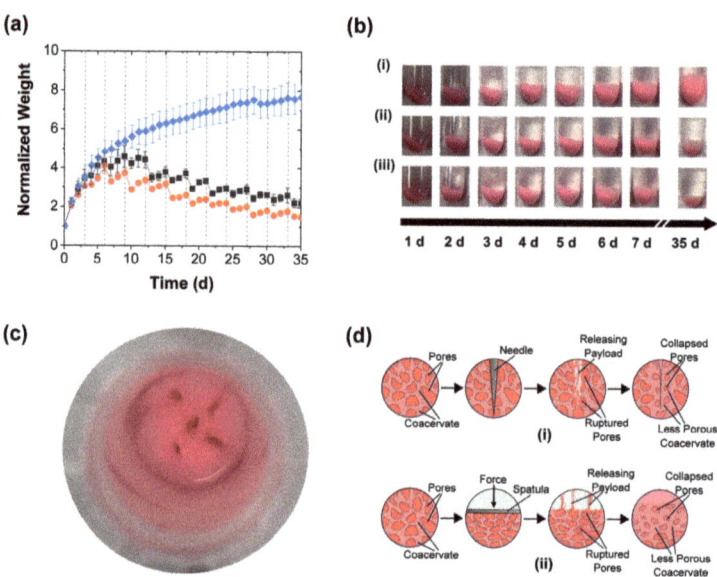

Figure 3. Coacervate swelling during the RhB release into deionized water achieved with (■) periodic perforation, (•) periodic compression, and (♦) without mechanical stimulation and characterized by (**a**) gravimetric analysis of the evolutions in normalized weights (mean ± SD) and (**b**) digital photography (i) without mechanical stimulation, (ii) with periodic perforation, and (iii) periodic compressions. Also shown are (**c**) a top view of a coacervate sample after a perforation treatment and (**d**) schemes of the solvent-filled pores collapsing after each mechanical treatment. The coacervates in this experiment were stimulated every 3 d using either five needle jabs or a spatula. All coacervate weights are normalized to their initial values at the start of the release experiment. The solid lines are guides to the eye, while the dashed vertical lines mark the mechanical stimulation times.

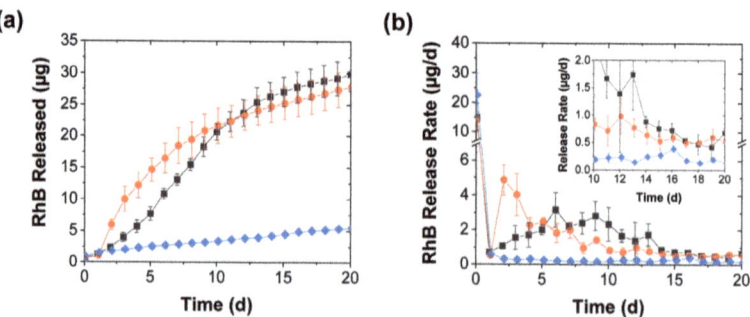

Figure 4. RhB release from coacervates into deionized water achieved with (■) daily perforation with two needle jabs, (•) daily compression, and (♦) without mechanical stimulation and shown in terms of both (**a**) the total RhB mass released and (**b**) the release rate (mean ± SD). The inset provides a closeup of the slower release rates near the end of the experiment, while the lines are guides to the eye.

Like in the case with the less frequent stimulation (in Figures 2 and 3), the accelerated release was correlated with the reductions in swelling produced by the daily stimulation, and the reduction in swelling became apparent earlier with the compression than with the perforation (Figure 5). This correlation again supported the view that the accelerated release reflected the rupture and collapse of the RhB-loaded pores. Moreover, it suggested that, besides accelerating the slow payload release, the mechanical stimulation (as shown by the images in Figures 3b and 5b) provides a potential approach to preventing excessive coacervate swelling, which can cause practical complications such as the blocking of fluid flow over the coacervate. Indeed, this swelling reduction persisted even after the coacervate stimulation stopped. Though after the experiment the coacervates—regardless of the stimulation type or frequency—swelled significantly when left in the deionized water for several weeks without further perforation or compression, the swelling of the stimulated samples remained much lower than that of the control samples. This continued reduction in coacervate swelling may have stemmed from the stimulation-driven collapse/elimination of many of their pores. While a detailed analysis of these post-stimulation swelling effects was beyond the scope of this study, their presence was clear from the continued visual observation of the water-immersed samples (as seen in the photographs in Supplementary Material, Figure S1).

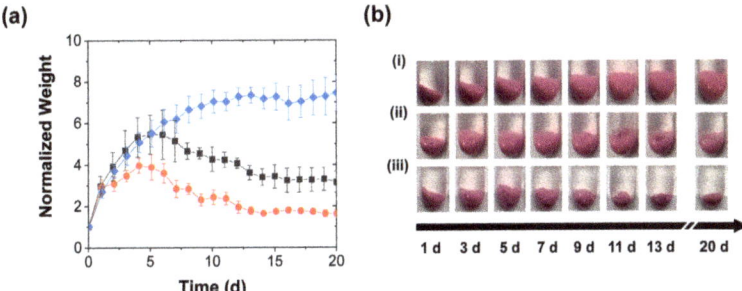

Figure 5. Swelling of coacervates during RhB release into deionized water achieved with (■) daily perforation by two needle jabs, (●) daily compression, and (♦) without mechanical stimulation and characterized by (**a**) gravimetric analysis of the evolutions in normalized weights (mean ± SD) and (**b**) digital photography (i) without mechanical stimulation, (ii) with periodic perforation, and (iii) periodic compression. All coacervate weights are normalized to their initial values at the start of the release experiment. The lines are guides to the eye.

3.3. Effect of Release Media

Collectively, the findings in Figures 2–5 suggest that the mechanical stimulation effect on the PAH/TPP coacervate release performance is (at least partially) caused by the mechanical stimulation impact on the coacervate swelling. This observation raises the question of whether the accelerated release effect requires coacervate swelling. Since PAH/TPP coacervate swelling is much lower in PBS (which also models physiological conditions) [27], some of the above release experiments were repeated in PBS. Consistent with previous work, the PBS greatly diminished the PAH/TPP coacervate swelling, such that—possibly due to its PAH-complexing/PAH coacervation-promoting phosphate ions [38,39]—the average coacervate weight increased by no more than 50% over 14 d (Figure 6a). Thus, while (like in the cases where the RhB was being released into deionized water) the mechanical activation produced a reduction in the swelling, the changes in swelling were significantly less pronounced than those in Figures 3 and 5, regardless of the mechanical stimulation. Indeed, as illustrated in Figure 6b, the mechanical activation (despite some variability in the angle from which the coacervate-bearing tubes were photographed) had no impact on the visual appearance of the coacervates.

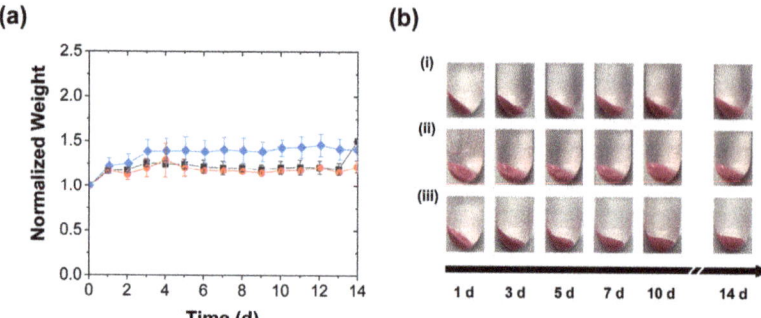

Figure 6. Coacervate swelling in 1 × PBS with (■) daily perforation with five needle jabs, (●) daily compression, and (♦) without mechanical stimulation and characterized by (**a**) gravimetric analysis of the evolutions in normalized weights (mean ± SD) and (**b**) digital photography (**i**) without mechanical stimulation, (**ii**) with periodic perforation, and (**iii**) with periodic compression. All coacervate weights are normalized to their initial values at the start of the release experiment. The lines are guides to the eye.

As expected, the diminished effect of the mechanical stimulation on the swelling was also reflected in its effect on the release profiles (Figure 7a,b). After the burst release in the first hour, the average release rates rapidly dropped to approximately 1 μg/d within 1 d of contact time and continued to fall thereafter. Though this further drop in release rate was initially inhibited upon the first mechanical stimulation (after 1 d of contact time), the magnitude and duration of this release-promoting effect were much smaller than that seen in deionized water (cf. Figures 4 and 7). Unlike the sharp acceleration of the RhB release seen in deionized water (which increased the early day release rates by more than tenfold and remained substantial over weeks), both types of mechanical stimulation in PBS produced only a roughly 2–3× initial increase, which became less pronounced with time, and after a week these release rates from the mechanically stimulated coacervates became either only slightly (less than 2×) higher than or indistinguishable from the controls (Figure 7b). This reduction in the stimulation effect indicates that the mechanical stimulation is most effective in accelerating release from PAH/TPP coacervates under conditions that promote pronounced swelling.

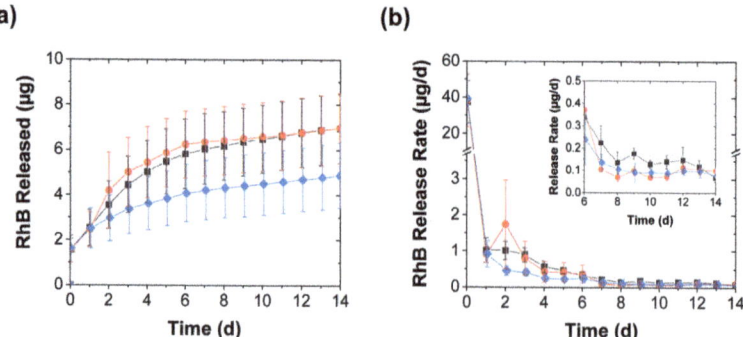

Figure 7. RhB release from coacervates into 1 × PBS achieved with (■) daily perforation with five needle jabs, (●) daily compression, and (♦) without mechanical stimulation and shown in terms of both (**a**) the total RhB mass released and (**b**) the release rate (mean ± SD). The inset provides a closeup of the slower release rates near the end of the experiment, while the lines are guides to the eye.

3.4. Further Discussion

Overall, the above experiments suggest that mechanical stimulation of release from complex coacervate matrices works well in media where significant swelling occurs (exemplified by the deionized water in this work) but has limited efficacy under conditions that inhibit coacervate swelling (e.g., in PBS, when used with PAH/TPP coacervates). Given this inhibitory PBS effect, the mechanical stimulation will (when used with PAH/TPP coacervates) likely work best in applications without high phosphate ion concentrations (i.e., outside of physiological media). It may therefore be better-suited to household and industrial applications, where the coacervates might either be in contact with low-ionic-strength solutions or with salts that promote PAH/TPP coacervate swelling [27]. Under such conditions, the mechanical stimulation strategies explored herein can produce a multifold increase in release rates and—in cases where the duration of efficacious sustained release is limited by the release rate ultimately becoming too slow (e.g., in sustained disinfections [21])—could extend the duration of the desired effect.

More broadly, to optimize this mechanical stimulation effect, the coacervates should likely be prepared under conditions that promote their swelling (e.g., by mixing the complex coacervate-forming species in nonstoichiometric charge ratios [33]). Moreover, to prevent the loss of mechanical stimulation efficacy under physiological conditions (with phosphate contents akin to PBS), coacervates can also be prepared from polycations that (unlike PAH [38,39]) do not complex and undergo complex coacervation with phosphate ions. Provided that conditions for coacervate swelling and mechanically activated pore collapse can be achieved, the mechanical stimulation procedures reported herein provide an attractive potential strategy for accelerating release from complex coacervates when it (over time) becomes too slow. Besides being able to reactivate the sustained release once it slows down, this coacervate stimulation might be useful in situations where the active payload release from the coacervates is too slow to be efficacious from the start, or when a periodic/pulsatile dosing might be desired (e.g., for pulsatile drug delivery [40–42] or to periodically disinfect devices such as water shower heads/hoses [43,44] or dental unit waterlines [45,46] without continuously exposing their users to chemical biocides). Likewise, it may have applications outside of traditional sustained release applications, such as sensing (e.g., where a dye such as the RhB used in this work is released each time that a device is touched to generate a colorimetric signal).

4. Conclusions

We have shown that, when the sustained release of small molecules from complex coacervate matrices is slower than desired, it can be accelerated through mechanical stimulation, akin to flavor release from a piece of chewing gum. When this strategy is used with PAH/TPP coacervates, the release of small, water-soluble molecules into low-ionic-strength water can be greatly accelerated when the coacervate is periodically perforated or compressed. The release profile achieved through this strategy can be varied by tailoring the frequency and type (e.g., perforation versus compression) of the mechanical stimulation.

The accelerated release evidently stems from the rupture and subsequent collapse of payload solution-filled pores within the coacervate phase and is—at least under the conditions examined herein—most pronounced under conditions where there is significant (pore volume-enhancing) swelling. Besides increasing the overall dosing of the slowly releasing payload, the mechanical stimulation enables pulsatile release, where the rate increases sharply immediately upon stimulation and then returns to baseline levels, and (since the coacervate volume is reduced by the mechanical rupture/collapse of the pores) provides a pathway to moderate coacervate swelling in cases where it becomes excessive. Collectively, these findings show that mechanical stimulation of coacervate matrices provides a potential approach to overcoming insufficient release rate problems in their applications, such as sustained disinfection or drug release, and could open doors to new coacervate-based technologies.

Supplementary Materials: The following supporting information can be downloaded at: https://www.mdpi.com/article/10.3390/polym15030586/s1, Figure S1: Representative digital photographs of coacervate swelling after a mechanically stimulated release experiment in deionized water.

Author Contributions: Conceptualization: W.A.H. and Y.L.; Experimental work: W.A.H.; Data analysis: W.A.H. and Y.L.; Paper writing and editing: W.A.H. and Y.L. All authors have read and agreed to the published version of the manuscript.

Funding: The authors gratefully acknowledge the National Science Foundation (IIP-1701104) for supporting this work. W.A.H. was partially supported on a Research Experience for Undergraduates (REU) supplement to the same grant.

Data Availability Statement: The data presented in this study are available upon reasonable request from the corresponding author.

Conflicts of Interest: Y.L. declares financial interest in a patent on PAH/TPP coacervate use in underwater adhesion and sustained release applications. W.A.H. has no conflicts of interest to declare.

References

1. De Jong, H.G.B. Complex colloid systems. In *Colloid Science*; Kruyt, H.R., Ed.; Elsevier: Amsterdam, The Netherlands, 1949; Volume II, pp. 335–432.
2. Menger, F.M.; Sykes, B.M. Anatomy of a coacervate. *Langmuir* **1998**, *14*, 4131–4137. [CrossRef]
3. Kizilay, E.; Kayitmazer, A.B.; Dubin, P.L. Complexation and coacervation of polyelectrolytes with oppositely charged colloids. *Adv. Colloid Interface Sci.* **2011**, *167*, 24–37. [CrossRef] [PubMed]
4. Momeni, A.; Filiaggi, M.J. Rheology of polyphosphate coacervates. *J. Rheol.* **2016**, *60*, 25–34. [CrossRef]
5. Wang, X.; Lee, J.; Wang, Y.-W.; Huang, Q. Composition and rheological properties of β-lactoglobulin/pectin coacervates: Effects of salt concentration and initial protein/polysaccharide ratio. *Biomacromolecules* **2007**, *8*, 992–997. [CrossRef] [PubMed]
6. Liu, Y.; Winter, H.H.; Perry, S.L. Linear viscoelasticity of complex coacervates. *Adv. Colloid Interface Sci.* **2017**, *239*, 46–60. [CrossRef] [PubMed]
7. Blocher, W.C.; Perry, S.L. Complex coacervate-based materials for biomedicine. *WIREs Nanomed. Nanobiotechnol.* **2017**, *9*, e1442. [CrossRef]
8. Johnson, N.R.; Wang, Y. Coacervate delivery systems for proteins and small molecule drugs. *Expert Opin. Drug Deliv.* **2014**, *11*, 1829–1832. [CrossRef]
9. Lawrence, P.G.; Patil, P.S.; Leipzig, N.D.; Lapitsky, Y. Ionically cross-linked polymer networks for the multiple-month release of small molecules. *ACS Appl. Mater. Interfaces* **2016**, *8*, 4323–4335. [CrossRef]
10. Black, K.A.; Priftis, D.; Perry, S.L.; Yip, J.; Byun, W.Y.; Tirrell, M. Protein encapsulation via polypeptide complex coacervation. *ACS Macro Lett.* **2014**, *3*, 1088–1091. [CrossRef] [PubMed]
11. Lawrence, P.G.; Lapitsky, Y. Ionically crosslinked poly(allylamine) as a stimulus-responsive underwater adhesive: Ionic strength and pH effects. *Langmuir* **2015**, *31*, 1564–1574. [CrossRef]
12. Scott, W.A.; Gharakhanian, E.G.; Bell, A.G.; Evans, D.; Barun, E.; Houk, K.; Deming, T.J. Active Controlled and Tunable Coacervation Using Side-Chain Functional α-Helical Homopolypeptides. *J. Am. Chem. Soc.* **2021**, *143*, 18196–18203. [CrossRef] [PubMed]
13. Wang, Q.; Schlenoff, J.B. The polyelectrolyte complex/coacervate continuum. *Macromolecules* **2014**, *47*, 3108–3116. [CrossRef]
14. Bagaria, H.G.; Wong, M.S. Polyamine-salt aggregate assembly of capsules as responsive drug delivery vehicles. *J. Mater. Chem.* **2011**, *21*, 9454–9466. [CrossRef]
15. De Silva, U.K.; Brown, J.L.; Lapitsky, Y. Poly(allylamine)/tripolyphosphate coacervates enable high loading and multiple-month release of weakly amphiphilic anionic drugs: A case study with ibuprofen. *RSC Adv.* **2018**, *8*, 19409–19419. [CrossRef]
16. Zhao, M.; Zacharia, N.S. Sequestration of Methylene Blue into polyelectrolyte complex coacervates. *Macromol. Rapid Commun.* **2016**, *37*, 1249–1255. [CrossRef]
17. Wang, Y.f.; Gao, J.Y.; Dubin, P.L. Protein separation via polyelectrolyte coacervation: Selectivity and efficiency. *Biotechnol. Prog.* **1996**, *12*, 356–362. [CrossRef]
18. Schmitt, C.; Turgeon, S.L. Protein/polysaccharide complexes and coacervates in food systems. *Adv. Colloid Interface Sci.* **2011**, *167*, 63–70. [CrossRef] [PubMed]
19. Shao, H.; Stewart, R.J. Biomimetic underwater adhesives with environmentally triggered setting mechanisms. *Adv. Mater.* **2010**, *22*, 729–733. [CrossRef]
20. Stewart, R.J.; Wang, C.S.; Shao, H. Complex coacervates as a foundation for synthetic underwater adhesives. *Adv. Colloid Interface Sci.* **2011**, *167*, 85–93. [CrossRef]
21. Alam, S.S.; Seo, Y.; Lapitsky, Y. Highly sustained release of bactericides from complex coacervates. *ACS Appl. Bio Mater.* **2020**, *3*, 8427–8437. [CrossRef]
22. Yeo, Y.; Bellas, E.; Firestone, W.; Langer, R. Complex coacervates for thermally sensitive controlled release of flavor compounds. *J. Agric. Food Chem.* **2005**, *53*, 7518–7525. [CrossRef] [PubMed]

23. Martins, I.M.; Barreiro, M.F.; Coelho, M.; Rodrigues, A.E. Microencapsulation of essential oils with biodegradable polymeric carriers for cosmetic applications. *Chem. Eng. J.* **2014**, *245*, 191–200. [CrossRef]
24. Hoare, T.R.; Kohane, D.S. Hydrogels in drug delivery: Progress and challenges. *Polymer* **2008**, *49*, 1993–2007. [CrossRef]
25. Tavera, E.M.; Kadali, S.B.; Bagaria, H.G.; Liu, A.W.; Wong, M.S. Experimental and modeling analysis of diffusive release from single-shell microcapsules. *AIChE J.* **2009**, *55*, 2950–2965. [CrossRef]
26. Jacobs, M.I.; Jira, E.R.; Schroeder, C.M. Understanding How Coacervates Drive Reversible Small Molecule Reactions to Promote Molecular Complexity. *Langmuir* **2021**, *37*, 14323–14335. [CrossRef]
27. Alam, S.S.; Mather, C.B.; Seo, Y.; Lapitsky, Y. Poly(allylamine)/tripolyphosphate coacervates for encapsulation and long-term release of cetylpyridinium chloride. *Colloids Surf. A.* **2021**, *629*, 127490. [CrossRef]
28. Green, B.K.; Schleicher, L. Oil-containing microscopic capsules and method of making them. U.S. Patent 2,800,457, 1957.
29. Uhlemann, J.; Schleifenbaum, B.; Bertram, H.-J. Flavor encapsulation technologies: An overview including recent developments. *Perfum. Flavor.* **2002**, *27*, 52–61.
30. Rungwasantisuk, A.; Raibhu, S. Application of encapsulating lavender essential oil in gelatin/gum-arabic complex coacervate and varnish screen-printing in making fragrant gift-wrapping paper. *Prog. Org. Coat.* **2020**, *149*, 105924. [CrossRef]
31. Weinbreck, F.; Minor, M.; De Kruif, C. Microencapsulation of oils using whey protein/gum arabic coacervates. *J. Microencapsul.* **2004**, *21*, 667–679. [CrossRef]
32. Fares, H.M.; Wang, Q.; Yang, M.; Schlenoff, J.B. Swelling and inflation in polyelectrolyte complexes. *Macromolecules* **2018**, *52*, 610–619. [CrossRef]
33. Hamad, F.G.; Chen, Q.; Colby, R.H. Linear viscoelasticity and swelling of polyelectrolyte complex coacervates. *Macromolecules* **2018**, *51*, 5547–5555. [CrossRef]
34. Scranton, A.B.; Klier, J.; Peppas, N.A. Soluble chain fractions in hydrophilic polymer networks: Origin and effect on dynamic uptake overshoots. *Polymer* **1990**, *31*, 1288–1293. [CrossRef]
35. Lapitsky, Y.; Eskuchen, W.J.; Kaler, E.W. Surfactant and polyelectrolyte gel particles that swell reversibly. *Langmuir* **2006**, *22*, 6375–6379. [CrossRef] [PubMed]
36. Lee, P.I. Kinetics of drug release from hydrogel matrices. *J. Controlled Release* **1985**, *2*, 277–288. [CrossRef]
37. Wood, J.M.; Attwood, D.; Collett, J.H. The swelling properties of poly(2-hydroxyethyl methacrylate) hydrogels polymerized by gamma-irradiation and chemical initiation. *Int. J. Pharm.* **1981**, *7*, 189. [CrossRef]
38. Marmisollé, W.A.; Irigoyen, J.; Gregurec, D.; Moya, S.; Azzaroni, O. Supramolecular surface chemistry: Substrate-independent, phosphate-driven growth of polyamine-based multifunctional thin films. *Adv. Funct. Mater.* **2015**, *25*, 4144–4152. [CrossRef]
39. Yu, J.; Murthy, V.S.; Rana, R.K.; Wong, M.S. Synthesis of nanoparticle-assembled tin oxide/polymer microcapsules. *Chem. Commun.* **2006**, *10*, 1097–1099. [CrossRef]
40. Kikuchi, A.; Okano, T. Pulsatile drug release control using hydrogels. *Adv. Drug Deliv. Rev.* **2002**, *54*, 53–77. [CrossRef]
41. Bansal, M.; Dravid, A.; Aqrawe, Z.; Montgomery, J.; Wu, Z.; Svirskis, D. Conducting polymer hydrogels for electrically responsive drug delivery. *J. Control. Release* **2020**, *328*, 192–209. [CrossRef]
42. Mertz, D.; Harlepp, S.; Goetz, J.; Bégin, D.; Schlatter, G.; Bégin-Colin, S.; Hébraud, A. Nanocomposite polymer scaffolds responding under external stimuli for drug delivery and tissue engineering applications. *Adv. Therap.* **2020**, *3*, 1900143. [CrossRef]
43. Feazel, L.M.; Baumgartner, L.K.; Peterson, K.L.; Frank, D.N.; Harris, J.K.; Pace, N.R. Opportunistic pathogens enriched in showerhead biofilms. *Proc. Natl. Acad. Sci. USA* **2009**, *106*, 16393–16399. [CrossRef]
44. Moat, J.; Rizoulis, A.; Fox, G.; Upton, M. Domestic shower hose biofilms contain fungal species capable of causing opportunistic infection. *J. Water Health* **2016**, *14*, 727–737. [CrossRef] [PubMed]
45. Ji, X.-Y.; Fei, C.-N.; Zhang, Y.; Liu, J.; Liu, H.; Song, J. Three key factors influencing the bacterial contamination of dental unit waterlines: A 6-year survey from 2012 to 2017. *Int. Dent. J.* **2019**, *69*, 192–199. [CrossRef] [PubMed]
46. Ricci, M.L.; Fontana, S.; Pinci, F.; Fiumana, E.; Pedna, M.F.; Farolfi, P.; Sabattini, M.A.B.; Scaturro, M. Pneumonia associated with a dental unit waterline. *Lancet* **2012**, *379*, 684. [CrossRef] [PubMed]

Disclaimer/Publisher's Note: The statements, opinions and data contained in all publications are solely those of the individual author(s) and contributor(s) and not of MDPI and/or the editor(s). MDPI and/or the editor(s) disclaim responsibility for any injury to people or property resulting from any ideas, methods, instructions or products referred to in the content.

MDPI
St. Alban-Anlage 66
4052 Basel
Switzerland
Tel. +41 61 683 77 34
Fax +41 61 302 89 18
www.mdpi.com

Polymers Editorial Office
E-mail: polymers@mdpi.com
www.mdpi.com/journal/polymers

www.ingramcontent.com/pod-product-compliance
Lightning Source LLC
LaVergne TN
LVHW070743100526
838202LV00013B/1290